SECOND EDITION

BEHAVIORAL GENETICS

A PRIMER

Robert Plomin
Pennsylvania State University

J. C. DeFries
University of Colorado

G. E. McClearn
Pennsylvania State University

W. H. FREEMAN AND COMPANY
New York

Cover image courtesy of Biophoto Associates/Photo Researchers.

Library of Congress Cataloging-in-Publication Data

Plomin, Robert, 1948–
  Behavioral genetics.

  (A Series of books in psychology)
  Bibliography: p.
  Includes indexes.
  1. Behavior genetics. I. DeFries, J. C., 1934–   .
II. McClearn, G. E., 1927–   . III. Title. IV. Series.
QH457.P56 1990   591.1′5   89-11932
ISBN 0-7167-2056-6

Printed in the United States of America

1 2 3 4 5 6 7 8 9 0   VB   8 9

Contents

Preface

This textbook follows an earlier text by G. E. McClearn and J. C. DeFries (*Introduction to Behavioral Genetics*, W. H. Freeman and Company, 1973). Rather than a revision of the earlier text, it is more of a sequel. The earlier text was written primarily for advanced undergraduate and graduate students taking their first course in behavioral genetics. However, it quickly became apparent that the "little green book" was not truly an introductory text. Therefore, as more undergraduates became attracted to the course, we felt a need for a text that assumed less previous exposure to genetics and statistics. As a new member of the faculty of the Institute for Behavioral Genetics, Robert Plomin undertook this task, in collaboration with McClearn and DeFries.

The result was a text published in 1980 that included some sections from the earlier book but was largely new. The present text is a revision of the 1980 book. In addition to generally updating the text, two major additions include a chapter on the "new genetics" of recombinant DNA and a chapter that provides an overview of the results of behavioral genetics research for the major domains of behavior, especially cognitive abilities, psychopathology, and personality.

Although behavioral genetics is a complex field, we have written a book that is as simple as possible without sacrificing honesty of presentation. Most importantly, we have written a book that is fundamentally for students, not for our colleagues. Although our coverage is representative, it is by no means exhaustive or encyclopedic; for those who seek additional details, there are other books (for example, Ehrman and Parsons, 1981; Fuller and Thompson, 1978; Hay, 1985). On the basis of seven years' experience in using the 1980 text in undergraduate classes, we expect that you will find this text challenging but readable. We hope that it will stimulate you to learn more about behavioral genetics and to begin thinking about behavior from the behavioral genetics perspective.

August 1989   **Robert Plomin**
                                                           J. C. DeFries
                                                           G. E. McClearn

BEHAVIORAL GENETICS

A PRIMER

CHAPTER · 1

Overview

Recognition of the importance of hereditary influence on behavior represents one of the most dramatic changes in the social and behavioral sciences during the past two decades. As described later in this chapter, the legacy of John Watson's behaviorism from the 1920s was the detaching of the study of behavior from the budding interest in heredity. A preoccupation with the environmental determinants of behavior continued until the 1970s, when a shift began toward the more balanced contemporary view that recognizes genetic as well as environmental influences on behavior.

The beauty of behavioral genetics lies in its theory and methods, which consider both genetic and environmental influences as sources of behavioral differences among individuals. Our major objective is to explain the theory and methods of behavioral genetics and the results of decades of research, in order to communicate the principle that consideration of both heredity and environment is necessary for understanding the complexity of behavior.

This unique perspective, combining genetics and the behavioral sciences, makes behavioral genetics an exciting interdisciplinary area. (See Figure 1.1.) Although research in behavioral genetics has been conducted for many years, the field has only recently emerged as a distinct discipline. The field-defining monograph *Behavior Genetics*, by John L. Fuller and W. Robert Thompson, was published in 1960. Since that date, research in behavioral genetics has undergone exponential growth.

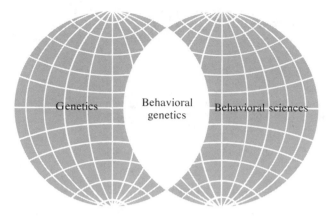

Figure 1.1
Behavioral genetics as the intersection between genetics and the behavioral sciences.
(From *Introduction to Behavioral Genetics* by G. E. McClearn and J. C. DeFries,
W. H. Freeman and Company. Copyright © 1973.)

PLAN OF THE BOOK

We begin with a historical perspective on evolution and genetics, followed by
three chapters on basic principles. Behavior is a *phenotype*—that is, an
observable characteristic we can measure. The basic principles of heredity are
the same, regardless of the phenotypes we choose to study. Thus, much of this
book is devoted to genetic concepts that are not unique to behavioral characters. On the other hand, behavior is not just another phenotype. Because
behavior involves the action of the whole organism rather than the action of a
single molecule, a single cell, or a single organ, behavior is the most complex
phenomenon that can be studied genetically. Unlike some physical characteristics, behavior is dynamic, changing in response to the environment—
indeed, behavior is at the cutting edge of evolution for this very reason. Thus,
this book is not simply another genetics text, because its focus is on the
complexity of behavioral phenotypes. Although behavioral genetics is relevant to the perspectives of anthropologists, economists, educators, political
scientists, sociologists, and others, the behavioral phenotypes we discuss necessarily reflect the fact that most behavioral genetics research currently
centers on topics traditionally defined within the domain of psychology.

The historical perspective of Chapter 2 describes the work of Charles
Darwin and his cousin Francis Galton. The next two chapters provide the
basic principles of genetics. Chapter 3 focuses on Gregor Mendel's classic
experiments and on single-gene influences on behavior. In Chapter 4 we
present a basic description of genes and how they work, with particular
reference to the heredity of behavior. Chromosomes and their influence on
behavior are dealt with in Chapters 5 and 6. Chapter 7 presents an overview of
the "new genetics" of recombinant DNA, which indicates its usefulness for

behavioral genetics research. The next two chapters focus on population genetics and quantitative genetics, two topics that tend to be difficult for students who are not statistically inclined. For this reason, we have attempted to discuss these concepts rather than to present them algebraically. Population genetics (Chapter 8) considers the transmission of genes from the population perspective rather than from the individual perspective. Quantitative genetics (Chapter 9) generalizes the single-gene, Mendelian model to a *multifactorial model* that considers the effects of many genes and many environmental factors. The multifactorial model is most relevant to the study of behaviors that interest behavioral scientists.

In Chapters 10 to 13 we look at the methods and findings of behavioral genetics in greater detail. In Chapter 10 we consider methods specific to the analysis of nonhuman behavior and present results for various behaviors as examples of these methods. Chapters 11, 12, and 13 are devoted, respectively, to the three major human behavioral genetics methods—family studies, twin studies, and adoption studies. In these chapters the two most widely studied domains of behavior, IQ and schizophrenia, are used as examples of the application of these methods. The final chapter, Chapter 14, presents an overview of human behavioral genetics research and addresses some of the controversial issues in behavioral genetics. These include the origin of group differences (ethnic and class differences) and the societal implications of behavioral genetics studies. It is likely that you are familiar with some of the controversies surrounding behavioral genetics, and you may have some preconceptions about them. So a few words are given about the origin of some of the controversial issues.

NATURE AND NURTURE

Some controversy can be expected surrounding any new approach that questions prevailing views, and behavioral genetics is certainly no exception. Much of the controversy concerning behavioral genetics stems from a misunderstanding of what it means to say that genes influence behavior. For example, behavioral genetics is often mistakenly viewed as pitting nature (genes) against nurture (environment), as if behavior were influenced solely by one or the other. The story behind this view is an interesting bit of history, and it helps to explain the frequent appearance of this view even today.

Historically, the controversy began at the turn of the century with the development of the *behavioristic* point of view, which assumed a dominant role in the developing discipline of psychology, particularly in the United States. Behaviorism arose as a protest against all forms of "introspective psychology," which was concerned with mental states such as consciousness and will. The term *behaviorism* refers to a strict focus on observable behavioral responses. Because its emphasis on observable responses led to an emphasis on observable environmental stimuli, behaviorism came to imply *environmentalism.* Stimulus–response chains eventually became the only acceptable explanation of behavior. Behaviorism also moved in the direction of environmentalism with its rejection of the instinct doctrine of W. McDougall (1908). At that time, instincts were thought of as inherited patterns of behavior, and the behaviorists attacked this position as redundant and circular. However, in rejecting this naive view of instincts, the behaviorists also discarded the notion that heredity can influence behavior. Thus, the behaviorists explained individual differences completely by environmental factors.

J. B. Watson, the founder of behaviorism, said:

> So let us hasten to admit—yes, there are heritable differences in form, in structure. . . . These differences are in the germ plasm and are handed down from parent to child. . . . But do not let these undoubted facts of inheritance lead us astray as they have some of the biologists. The mere presence of these structures tells us not a thing about function. . . . Our hereditary structure lies ready to be shaped in a thousand different ways—the same structure— depending on the way in which this child is brought up. . . .
>
> Objectors will probably say that the behaviorist is flying in the face of the known facts of eugenics and experimental evolution—that the geneticists have proven that many of the behavior characteristics of the parents are handed down to the offspring. . . . Our reply is that the geneticists are working under the banner of the old "faculty" psychology. One need not give very much weight to any of their present conclusions. We no longer believe in faculties nor in any stereotyped patterns of behavior which go under the names of "talent" and inherited capacities. . . .
>
> Our conclusion, then, is that we have no real evidence of the inheritance of traits. I would feel perfectly confident in the ultimately favorable outcome of careful upbringing of a *healthy, well-formed* baby born of a long line of crooks, murderers and thieves, and prostitutes. Who has any evidence to the contrary? (Watson, 1930, pp. 97–103)

Watson followed this with the familiar and frequently quoted challenge:

> I should like to go one step further now and say, "Give me a dozen healthy infants, well-formed, and my own specified world to bring them up in and I'll guarantee to take any one at random and train him to become any type of specialist I might select—doctor, lawyer, artist, merchant-chief and, yes, even beggar-man and thief, regardless of his talents, penchants, tendencies, abilities, vocations, and race of his ancestors." I am going beyond my facts and I admit it, but so have the advocates of the contrary and they have been doing it for many thousands of years. (Watson, 1930, p. 104)

R. S. Woodworth (1948) pointed out that this extreme environmentalism was not a necessary consequence of the behavioristic position. He suggested that Watson's stand was taken, in part, "to shake people out of their complacent acceptance of traditional views" (p. 92). For whatever reason, Watson sought to exorcise genetics from psychology, and he succeeded to a remarkable degree. His position in his book *Behaviorism* soon became the traditional view that was complacently accepted by the majority of behavioral scientists.

But this majority view was not without opposition. In fact, since Watson's pronouncement, not a single year has passed without publication of some evidence showing it to be wrong. Collectively, this research has demonstrated the important role of heredity in many varieties of behavior and in many kinds of organisms.

INTERACTIONISM AND INDIVIDUAL DIFFERENCES

During the past two decades, there have been clear signs that the behavioral sciences are beginning to accept the theory and methodology of genetics. Rarely does anyone concur with Watson's conclusion that genes are unimportant in behavior. Instead, it is now generally agreed that both nurture and nature play a role in determining behavior. However, the mistaken notions of the nature–nurture argument have too often been replaced with the equally mistaken view that the effects of heredity and environment cannot be analyzed separately, a view called interactionism (Plomin, DeFries, and Loehlin, 1977). This topic will be discussed in Chapter 9, but we shall say a few words about it here.

Obviously, there can be no behavior without both an organism and an environment. The scientifically useful question is: For a particular behavior, what causes differences among individuals? For example, what causes individual differences in cognitive ability? Various environmental hypotheses leap to mind: Families differ in the stimulation they offer for cognitive growth; environments differ in motivating people toward intellectual goals; educational experiences differ. However, genetic hypotheses should also be considered.

Research in behavioral genetics is directed toward understanding differences in behavior. Methods are employed that consider both genetic and environmental influences, rather than assuming that one or the other is solely important. As a first step, behavioral genetics research studies whether individual behavioral differences are influenced by hereditary differences and estimates the relative influences of genetic and environmental factors.

It is critically important to keep in mind that behavioral genetics aims to explain differences among individuals. For example, differences among us in height are for the most part due to genetic differences, not to environmental differences. This means that if you are taller than average, most of that difference between you and other people is genetic in origin. This statement does not apply to you in isolation; it does not mean that if you are six feet tall that heredity explains all but an inch or so of your height. Nor does the statement of heritable influence on individual differences apply to differences between groups. That is, even though individual differences are highly heritable, average differences between groups are not necessarily hereditary. For example, we shall see that vocabulary is one of the most highly heritable facets of cognitive ability. This does not mean that vocabulary words are transmitted by heredity; nor does it imply that the average superiority of females in vocabulary is genetic in origin. To say that vocabulary is heritable means that individuals in a particular population differ on scores on vocabulary tests and that these vocabulary differences are due in part to genetic differences among individuals in that population. Even though vocabulary differences among individuals are heritable, the average vocabulary difference between girls and boys could be entirely environmental in origin. For example, it is possible that girls are encouraged and rewarded for language skills to a greater extent than boys.

Interest in behavioral genetics depends on wanting to know why people differ. There are three reasons to be interested in this topic. First, individual differences are substantial. For example, even though most members of our species are able to speak, people differ considerably in scores on vocabulary tests. On a moderately difficult test in a normal sample, the top scorers will get five times as many words correct as will low scorers. Moreover, differences among individuals are far greater than average differences between groups. Although girls on average perform better than boys on vocabulary tests, this statistically significant difference is small. That is, if all you know is whether a child is a boy or a girl, you know very little about that child's verbal ability, because the individual differences among boys and among girls is great whereas the average difference between boys and girls is small. The second reason is that societally relevant issues in the behavioral sciences are usually issues of individual differences. For example, although it is interesting to ask why the human species uses language, more relevant to society are questions such as why some children are delayed in the use of language, why some are reading disabled, and why some people are more verbally fluent than others. Third, the causes of individual differences are not necessarily related to the causes of average differences between groups. For instance, the causes of individual differences in tests of verbal ability could be substantially influenced by genetic factors, and yet the average difference between girls and boys on these tests could be environmental in origin.

GENES AND BEHAVIOR

Some people are disturbed by the idea that genes can influence behavior. They don't understand the workings of genes and probably picture them as master puppeteers within us, pulling our strings. To the contrary, genes are merely chemical structures. However, encoded in these structures are the messages that enable genes to do their marvelous job of reliably replicating themselves and controlling development. In Chapter 4 we shall see that there is no such thing as a gene for behavior; nor is there a gene for the length of one's nose. Genes are blueprints for the assembly and regulation of proteins, which are the building blocks of our bodies, including the nervous system. Each gene codes for a specific sequence of amino acids that the body assembles to form a protein. If even the smallest part of this chain is altered, the entire protein can malfunction. We shall discuss this in much greater detail in Chapter 4.

Here, we want to emphasize that genes are not mystical entities. Genes do not magically blossom into behavior or anything else. They are stretches of chemicals that code for protein production or regulate the activity of other genes. In this sense, all aspects of human beings—behavior as well as our bones—are part of this process. As illustrated in Figure 1.2, proteins do not directly cause behavior. For example, one gene (G_2) codes for a particular protein (P_2). However, that protein does not cause a particular type of behavior. There is no gene or protein, for example, that repeatedly causes a person to lift shot glasses and perhaps become an alcoholic. Proteins interact with other physiological intermediaries (such as I_2), which may be other proteins, such as hormones or neurotransmitters, or may be structural properties of the nervous system. Environmental factors (such as E_2, which might represent nutrition) may also be involved. These influences can ultimately and indirectly influence behavior in a certain direction. For example, differences in neural sensitivity to ethanol may tip the scale in the direction of alcoholism for an individual who imbibes frequently. Various genes, chemical and structural brain differences, and environmental factors may be at the root of such differences in neural sensitivity.

So when we talk about genetic influences on behavior, we do not mean robotlike, hard-wired circuits. We are referring to indirect and complex paths

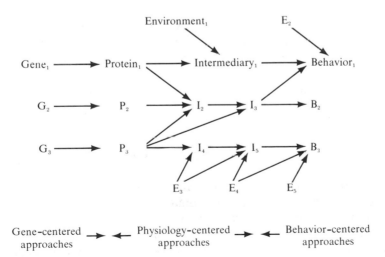

Figure 1.2
Genes do not directly cause behavior; they work indirectly via physiological systems.

between genes and behavior via proteins and physiological systems. Ernst Mayr (1974) distinguished *closed behavioral programs* from *open behavioral systems* to replace the hopelessly entangled arguments concerning instinct and learning. Closed behavioral programs are relatively impervious to individual experiences. Even so, there is no implication that a gene directly produces a particular behavior; genes always work by controlling the production of proteins. Open behavioral systems are more susceptible to individual experience. Closed and open systems differ only in the degree of flexibility. Behavioral systems of higher organisms tend toward the "open" end of this continuum.

Figure 1.2 also points out different approaches to the study of behavioral genetics. Researchers using these different approaches tend to have different goals. The *gene-centered approach* starts with a single gene and studies its effect on behavior. For example, we can study a single-gene mutation and observe its behavioral effect. Another approach, which characterizes the field of biopsychology, can be called *physiology centered.* This focuses on the physiological intermediaries between genes and behavior. The physiology-centered approach ideally works back to genes and forward to behavior. (See Figure 1.2.) In both the gene-centered and physiology-centered approaches, behavior is really only a tool for understanding the workings of both genes and various physiological systems. The third approach begins with behavior. The behaviors are not selected for their genetic or physiological simplicity, but rather because of their intrinsic interest or social relevance. A first step in understanding the etiology of such behavior is to ask the extent to which genetic and environmental influences make a difference for a particular behavior. Once in a great while, we find relatively simple genetic and physiological systems. This is what happened for one type of mental retardation, phenylketonuria (PKU), which is caused by a single-gene defect. More often, particularly for the complex behaviors that interest behavioral scientists, we find that both genetic and environmental differences are important. Usually, many genes are involved, and we cannot identify the physiological systems involved. However, it is an important first step to know that genetic factors influence a particular behavior, just as it is important to narrow the search for salient environmental influences.

DO GENES DETERMINE ONE'S DESTINY?

This text will demonstrate that genetic differences can account for a substantial portion of individual variation in many important behaviors. Some people do not look favorably on such findings. These people believe that if genes are shown to influence behavior, there is nothing that can be done to alter that behavior short of eugenic (breeding) intervention or genetic engineering. Thus, if a certain form of psychopathology was shown to be caused primarily by genes, it might be mistakenly assumed that psychotherapy and other environmental intervention would be useless.

   This pessimistic view is simply wrong. Genes do not determine one's destiny. A genetically determined behavioral problem may be bypassed, ameliorated, or remediated by environmental interventions. The best example is PKU, a single-gene defect that formerly resulted in severe retardation and was responsible for about 1 percent of institutionalized retarded individuals. Biochemical studies of the gene-behavior pathways indicated that the ultimate cause of the retardation was the inability to break down a particular chemical, phenylalanine, which led to its accumulation at high levels in the blood. This caused severe damage to the developing brain. As we will discuss in Chapter 4, PKU individuals do not suffer retardation if a diet low in phenylalanine is provided during the developing years. Thus, an environmental intervention was successful in bypassing a genetic problem. This important discovery was made possible by recognition of the genetic basis for this particular type of retardation.

"WHAT IS" VERSUS "WHAT COULD BE"

Most research in human behavioral genetics has involved analyzing behavioral differences among individuals and estimating the relative extent to which these differences are due to heredity and environment, given the mix of genetic and environmental influences at a particular time. If either the genetic or the environmental influences change, the relative impact of genes and environment will change. Thus, if a new environmental treatment were introduced, a behavior strongly influenced by heredity could nonetheless be altered. In other words, behavioral genetics research tells us about *what is*—the genetic and environmental origins of individual differences in a population—not *what could be*—whether, for example, a particular intervention will work. The two should be viewed as complementary. Knowledge about *what is* can help to guide research concerning *what could be*.

   Behavioral genetic research most certainly does not tell us *what should be*. For example, evidence for substantial genetic influence for a behavior is compatible with a wide range of social and political views, most of which depend on values, not facts. This issue is discussed in Chapter 14.

INDIVIDUAL DIFFERENCES AND EQUALITY

Another reason some people have difficulty in accepting the role of genes in influencing behavior is that they do not recognize the wide range of individual differences for most behaviors. One mouse is not like every other mouse; nor is every human like every other. In fact, as we shall see, the principles of genetics demonstrate that there has never been nor will ever be another human being who is genetically exactly like you. Variability is the key to understanding evolution and genetics, and it provides a needed perspective for behavioral scientists.

J. Hirsch (1963) outlined three different approaches that have been used to study behavior. The three approaches differ in the extent to which they recognize variability. The first view recognizes no important differences between or within species. This is a position suggested by much of the older research in learning, which assumed that the laws of learning for one species would generalize to all other species. This approach is typified by titles such as *The Behavior of Organisms* (Skinner, 1938). This view has largely given way to a form of *typological thinking* (Mayr, 1982) that recognizes differences *between*, but not *within*, species. It characterizes comparative psychology and, indirectly, much current research on human behavior. Researchers interested in human behavior may study humans rather than other animals because they assume that important species differences exist, but they do not study differences within the human species. The third approach, which considers variability between *and* within species, is characteristic of behavioral genetics.

To some extent, these three views are like the different powers of a telescope. To a visitor from another planet, peering at us from an orbiting space capsule, we humans might not appear all that different from squirrels. Humans are bigger and have somewhat less fur, but humans and squirrels alike have two eyes, a nose, and four limbs. If the two species were examined with a more powerful lens, the differences between humans and squirrels would readily emerge. At this power setting, however, all humans would seem essentially identical. Some scientists choose this power setting to study behavior — clearly, there are some common features among all humans in the way we perceive stimuli and learn. However, it is our belief that a powerful science of behavior will not emerge until we switch to a higher resolving power that reveals differences within, as well as between, species. The questions that most often confront scientists studying human behavior are those dealing with differences among people. And genetics, the study of the variation of organisms, is uniquely qualified to aid us in analyzing these individual differences.

This focus on individual differences may be difficult for you to accept philosophically. Are not all men created equal? This was a self-evident truth to the signers of the Declaration of Independence. Arguments that some men are inherently more able to learn that others may appear to be at odds with the democratic ideal and to imply the principle of rule by the elite (even though it is obvious that there are inherent differences in height, strength, and other characteristics). A more thorough understanding of behavioral genetics suggests a very different philosophical conclusion.

If there is a central message from behavioral genetics, it involves individual differences. With the trivial exception of members of identical multiple births, each one of us is a unique genetic experiment, never to be repeated again. Here is the conceptualization on which to build a philosophy of the dignity of the individual! Human variability is not simply imprecision in a process that, if perfect, would generate unvarying representatives of a Platonic ideal. Individuality is the quintessence of life; it is the product and the agency of the grand sweep of evolution.

It is important not to confuse biological identity with the political concept of equality. The two concepts are very different, though the problem is to an extent a semantic one. The Greeks had not one, but many words for equality, distinguishing, for example (Hutchins and Adler, 1968, p. 305):

isonomia: equality before the law
isotimia: equality of honor
isopoliteia: equality of political rights
isokratia: equality of political power
isopsephia: equality of votes or suffrage
isegoria: equality in right to speak
isoteleia: equality of tax or tribute
isomoiria: equality of shares or partnership
isokleria: equality of property
isodaimonia: equality of fortune

Perhaps we confuse ourselves by using only the word *equality.*

SUMMARY

Behavioral genetics lies at the interface between genetics and the behavioral sciences. In this book we present basic genetic concepts, including transmission genetics, chromosomes, population genetics, and quantitative genetic theory and methods, but always with an eye toward behavior.

Several issues are introduced in this chapter:

1. Controversy arises when nature (genes) is pitted against nurture (environment). One of the oldest schools of thought in psychology is behaviorism. It led to an environmentalism that rejected the possibility of genetic influences on behavior.
2. Although genetic and environmental influences interact within an individual, this does not imply that the separate effects of genes and environment cannot be untangled when we focus on differences among individuals.
3. Genes do not act as master puppeteers within us. They are chemical structures that control the production of proteins, thereby indirectly affecting behavior.

4. Genes do not determine one's destiny. A genetically determined behavioral problem can sometimes be alleviated environmentally.

5. Behavioral genetics research usually describes *what is*—the relative impact of genes and environment at a particular time for a particular population—although it may help us understand *what could be.*

6. Finally, recognition of individual differences—regardless of their environmental or genetic etiology—is independent of acknowledgment of the values of political equality.

CHAPTER · 2

Historical Perspective

To illustrate a point concerning the inheritance of gestures, Darwin quoted an interesting case that had been brought to his attention by Galton:

> A gentleman of considerable position was found by his wife to have the curious trick, when he lay fast asleep on his back in bed, of raising his right arm slowly in front of his face, up to his forehead, and then dropping it with a jerk so that the wrist fell heavily on the bridge of his nose. The trick did not occur every night, but occasionally.

Nevertheless, the gentleman's nose suffered considerable damage, and it was necessary to remove the buttons from his nightgown cuff in order to minimize the hazard.

> Many years after his death, his son married a lady who had never heard of the family incident. She, however, observed precisely the same peculiarity in her husband; but his nose, from not being particularly prominent, has never as yet suffered from the blows. . . . One of his children, a girl, has inherited the same trick. (Darwin, 1872, p. 34)

Probably everyone could cite some examples, perhaps less quaint than Galton's, in which some peculiarity of gait, quality of temper, degree of talent, or other trait is characteristic of a family. Phrases like "a chip off the old block," "like father, like son," and "it runs in the family" indicate the notion that behavioral traits, like physiological ones, can be inherited. The aim of this chapter is to consider the history of scientific inquiry into these matters, with major emphasis on developments during the latter half of the nineteenth and the early twentieth centuries.

It is difficult to pinpoint the earliest expression of a view concerning any subject. While the present topic is no exception, we should note that its origins must be very remote indeed. The concept that "like begets like" has had great practical importance in the development of domesticated animals, which have been bred for behavioral, as well as morphological, characteristics. We can postulate that the notion of inheritance, including inheritance of behavioral traits, may have appeared in human thought as early as 8000 B.C., when the domestication of the dog began.

Biological thought during recorded history was dominated by Aristotle's pronouncements on natural history, and by the teachings of Galen, a Roman, concerning anatomy. Progress in understanding biological phenomena was virtually halted during the general stagnation of secular pursuits that typified the Middle Ages. Then came the Renaissance. Leonardo da Vinci's study of anatomy in connection with art characterized the far-ranging inquisitiveness of the Renaissance scholars. Less well known is a family incident that reveals a deep conviction about the workings of heredity. Leonardo was an illegitimate child resulting from a liaison between Piero, a notary from the village of Vinci, and a peasant girl named Caterina. As a modern biographer of Leonardo puts it:

> There was an interesting and deliberate attempt to repeat the experiment. Leonardo had a step-brother, Bartolommeo, by his father's third wife. The step-brother was forty-five years younger than Leonardo, who was already a legend when the boy was growing up and was dead when the following experiment took place. Bartolommeo examined every detail of his father's association with Caterina and he, a notary in the family tradition, went back to Vinci. He sought out another peasant wench who corresponded to what he knew of Caterina and, in this case, married her. She bore him a son but so great was his veneration for his brother that he regarded it as profanity to use his name. He called the child Piero. Bartolommeo had scarcely known his brother whose spiritual heir he had wanted thus to produce and, by all accounts, he almost did. The boy looked like Leonardo, and was brought up with all the encouragement to follow his footsteps. Pierino da Vinci, this experiment in heredity, became an artist and, especially, a sculptor of some talent. He died young. (Ritchie-Calder, 1970, pp. 39–40)

THE ERA OF DARWIN

The applied husbandry of the da Vinci family cannot be said to have had a pivotal effect on the subsequent development of biological thought. However, there were other concurrent developments that did. Andreas Vesalius's exhaustive work on human anatomy, published in 1543, was based on detailed and painstaking dissection of human bodies. In 1628 William Harvey made his momentous discovery of the circulation of the blood. These findings were of far-reaching importance, for they opened the way to experimentation on the phenomena of life.

After Harvey's discovery, the pace of biological research quickened, and many fundamental developments in technique and theory ensued in the following century. One of the cornerstones of biology was laid by the Swede Karl von Linné (better known as Linnaeus), who, in 1735, published *Systema Naturae,* in which he established a system of taxonomic classification of all known living things. In so doing, Linnaeus emphasized the separateness and distinctness of species. As a result, the view that species were fixed and unchanging became the prevailing one. This notion, of course, fit the biblical account of creation. However, not everyone was persuaded that species are unchangeable. For example, the Englishman Erasmus Darwin suggested, in the latter part of the eighteenth century, that plant and animal species appear capable of improvement, although he believed that God had so designed life. Another view on this subject was promoted by the Frenchman Jean Baptiste Lamarck, who argued that the deliberate efforts of an animal could result in modifications of the body parts involved, and that the modification so acquired could be transmitted to the animal's offspring. For example:

> We perceive that the shore bird, which does not care to swim, but which, however, is obliged . . . to approach the water to obtain its prey, will be continually in danger of sinking in the mud, but wishing to act so that its body shall not fall into the liquid, it will contract the habit of extending and lengthening its feet. Hence it will result in the generations of these birds which continue to live in this manner, that the individuals will find themselves raised as if on stilts, on long naked feet: namely, denuded of feathers up to and often above the thighs. (In Packard, 1901, p. 234)

Changes of this sort were presumed to accumulate, so that eventually the characteristics of the species would change, although Lamarck did not believe that species became extinct. Lamarck was not the first to assume that changes acquired in this manner could be transmitted to the next generation, but he crystallized the notion. Thus, it has come to be called Lamarckism, or the law of use and disuse. As we shall see, this is an incorrect view of evolution. But it was significant in that it questioned the prevailing view that species do not change.

The strict and literal interpretation of the account in Genesis of the creation of the earth and its inhabitants was being challenged most seriously on the basis of geological evidence. The discovery of fossilized animal bones deep in strata beneath the earth's surface proved difficult to accommodate to Bishop Ussher's calculations that the earth had been created in 4004 B.C. A theory of "catastrophism" was put forward to account for these fossils. The Deity was regarded as having created and extinguished life on many successive occasions, with catastrophes such as floods and violent upheavals, which

Figure 2.1
Charles Darwin as a young man.
(Courtesy of Trustees of the
British Museum [Natural
History].)

caused the bones to be buried at various depths. Many geologists questioned, however, whether catastrophic events were responsible for the geological record. A school of "uniformitarians" argued that the processes at work in the past were the same as those of the present, and thus that the accumulation of strata required millions of years rather than the six thousand–odd years derived from Bishop Ussher's postulated date of creation. A leader of this uniformitarian school of thought, and one of the dominant intellects of the time, was Charles Lyell (see Eiseley, 1959). Lyell published the first volume of his *Principles of Geology* in 1830, and one of the early copies found its way into the baggage of a young man about to embark on what was probably the most important voyage in the history of biological thought.

Erasmus Darwin's grandson Charles (Figure 2.1) had been a student of medicine at Edinburgh but was so unnerved by the sight of blood during surgery that he gave up further medical study. He then went to Cambridge, where, although a student of mediocre accomplishment, he received a degree in 1831. Darwin appeared to be destined for a career as a clergyman, when suddenly and unexpectedly, through the recommendation of one of his old professors, he was nominated for the unpaid post of naturalist aboard H.M.S. *Beagle*, a survey ship of the Royal Navy about to embark on a long voyage. It was not uncommon for a naturalist to be taken on trips of this kind, and the young and devout captain of the *Beagle*, Captain Robert Fitz-Roy, was pleased at the prospect, for he expected a naturalist to be able to produce yet more data in support of "natural theology." A central theme of natural theology was the so-called argument from design, which viewed the adaptation of animals and plants to the circumstances of their lives as evidence of the Creator's wisdom. Such exquisite design, so the argument went, implied a "Designer." As exploration opened up hitherto unexplored parts of the world, new evidence of the Designer's works was uncovered. It was with this end in mind that Captain Fitz-Roy welcomed the young Darwin.

During the next five years, Darwin experienced chronic seasickness, tropical fever, volcanic eruptions, earthquakes, tidal waves, and the high adventure of encounters with rebels and life with Argentine gauchos. He filled many notebooks with observations on fossils, primitive men, and various species of animals—and their remarkable and specific adaptation to their environments. He made particularly compelling observations about the 14 species of finch found in a small area on the Galápagos Islands. The principal differences among these finches were in their beaks, and each was exactly appropriate for the particular eating habits of the species (see Figure 2.2). Somehow, thought Darwin, these birds derived from a common ancestral group. "Seeing this gradation and diversity of structure in one small, intimately related group of birds, one might really fancy that from an original paucity of birds in this archipelago, one species had been taken and modified for different ends" (Darwin, 1896, p. 380). However, influenced by Lyell's book, and his own observations on geology and biology, Darwin was not inclined to make the argument in favor of the "argument from design." On his return to England, Darwin began work on several reports summarizing his observations on coral reefs, barnacles, and other matters. Meanwhile, he gradually and systematically marshaled evidence that species evolve one from another and pondered the possible mechanisms through which this evolution could occur. He shared his developing theory with a few friends, including Lyell himself, an eminent botanist named J. D. Hooker, and T. H. Huxley. He gradually convinced some but not all of them of the merit of his theory.

Realizing the kind of opposition that a theory contradicting the biblical account of creation would encounter, Darwin hesitated. He planned a monumental work in which he would present an overwhelming mass of evidence. Though his friends warned him that he should publish a brief version immediately lest someone anticipate him, he continued to work slowly and carefully on amassing the evidence and anticipating the objections. He did, however, take time to write out short sketches in correspondence with his friends and confidants. Finally, in 1858, a blow fell. A young man named Alfred Wallace sent Darwin a manuscript for his comments. In it, with much less evidence in hand than Darwin had, Wallace arrived at essentially the same theory that Darwin had been developing for more than two decades. Darwin was greatly concerned over the course of action he should take. As he said that year in a letter to Lyell:

> I should be extremely glad now to publish a sketch of my general views in about a dozen pages or so; but I cannot persuade myself that I can do so honourably. Wallace says nothing about publication, and I enclose his letter. But as I had not

Figure 2.2
The 14 species of Galápagos and Cocos Island finches. (a) A woodpeckerlike finch
that uses a twig or cactus spine instead of its tongue to dislodge insects from
tree-bark crevices. (c, d, e) Insect-eaters. (f, g) Vegetarians. (h) The Cocos Island
finch. The birds on the ground eat mostly seeds. Note the powerful beak of (i),
which lives on hard seeds. (From "Darwin's finches" by D. Lack. Copyright © 1953
by Scientific American, Inc. All rights reserved.)

intended to publish any sketch, can I do so honourably, because Wallace has sent me an outline of his doctrine? I would far rather burn my whole book, than that he or any other man should think that I had behaved in a paltry spirit. (Darwin, 1888, p. 117)

Lyell and Hooker took the initiative and resolved the issue by arranging for the simultaneous presentation, at a meeting of the Linnean Society in 1858, of a sketch Darwin had prepared in 1844 and Wallace's paper.

With the theory now out in the open, Darwin began work on what he called an abstract. This "abstract," published in 1859 under the title *On the Origin of Species by Means of Natural Selection, or the Preservation of Favoured Races in the Struggle for Life*, proved to be one of the most influential books ever written. A contemporaneous *London Times* review of his book, which sold out on the first day, is excerpted in Box 2.1.

Darwin was honored by being buried near Sir Isaac Newton in Westminster Abbey, primarily because he convinced the world of the reasonableness of evolution. However, his theory of the evolution of species had some serious gaps, mainly because the mechanism for heredity, the gene, was not yet understood.

The elements of Darwin's theory can be stated as follows. Within any species, many more individuals are born each generation than survive to maturity. Great variation exists among the individuals of a population. These individual differences are due, at least in part, to heredity. If the likelihood of surviving to maturity and reproducing is influenced, even to a slight extent, by a particular trait, offspring of the survivors and reproducers should manifest slightly more of the trait than their parents' generation. Thus, bit by bit, the characteristics of a population can change. Over a sufficiently long period, the cumulative changes are so great that in retrospect the latter and the earlier populations are, in effect, different.

Darwin's theory of evolution thus has three components, as indicated in Table 2.1. Some mechanism causes and maintains variation (as discussed later in this chapter), and natural selection uses this variability to shape a species. We shall see that Darwin and his contemporaries knew little about the source of variability or the process by which it is maintained and transmitted from one generation to another.

Darwin's most notable contribution to the theory of evolution was his principle of natural selection. He used the phrase "survival of the fittest" to characterize the principle, but it could more appropriately be called "differential reproduction of the fittest" — or more simply, "reproductive fitness." Mere survival is necessary, but it is not sufficient. As Darwin himself put it:

Owing to this struggle [for life], variations, however slight and from whatever cause proceeding, if they be in any degree profitable to the individuals of a species, in their infinitely complex relations to other organic beings and to their physical conditions of life, will tend to the preservation of such individuals, and will generally be inherited by the offspring. The offspring, also, will thus have a

Box 2.1

1859 Review of Darwin's *The Origin of Species*

The Origin of Species made an immediate impact on the scientific world. The following is a short excerpt from a long review (over 5,000 words) of Darwin's book that appeared on December 26, 1859, in the *Times* (London).

There is a growing immensity in the speculations of science to which no human thing or thought at this day is comparable. . . . Hence it is that from time to time we are startled and perplexed by theories which have no parallel in the contracted moral world. . . . The hypothesis to which we point, and of which the present work of Mr. Darwin is but the preliminary outline, may be stated in his own language as follows: "Species originated by means of natural selection, or through the preservation of the favored races in the struggle for life." . . . When we know that living things are formed of the same elements as the inorganic world, that they act and react upon it, bound by a thousand ties of natural piety, is it probable, nay is it possible, that they, and they alone, should have no order in their seeming disorder, no unity in their seeming multiplicity, should suffer no explanation by the discovery of some central and sublime law of mutual connexion?

Questions of this kind have assuredly often arisen, but it might have been long before they received such expression as would have commanded the respect and attention of the scientific world, had it not been for the publication of the work which prompted this article. Its author, Mr. Darwin, inheritor of a once celebrated name, won his spurs in science when most of those now distinguished were young men, and has for the last 20 years held a place in the front ranks of British philosophers. After a circumnavigatory voyage, undertaken solely for the love of his science, Mr. Darwin published a series of researches which at once arrested the attention of naturalists and geologists; his generalizations have since received ample confirmation, and now command universal assent, nor is it questionable that they have had the most important influence on the progress of science. More recently Mr. Darwin, with a versatility which is among the rarest of gifts, turned his attention to a most difficult question of zoology and minute anatomy; and no living naturalist and anatomist has published a better monograph than that which resulted from his labours. Such a man, at all events, has not entered the sanctuary with unwashed hands, and when he lays before us the results of 20 years' investigation and reflection we must listen even though we be disposed to strike. But, in reading his work it must be confessed that the attention which might at first be dutifully, soon becomes willingly given, so clear is the author's thought, so outspoken his conviction, so honest and fair the candid expression of his doubts.

Table 2.1
Model of evolution: what Darwin knew and what we know now

	Darwin	Now
Induction of variation	Environmental modification and "use and disuse"	Mutation of DNA
Maintenance of variation	Pangenesis	Segregation and genetic transmission
Selection of variation	Natural selection — "survival of the fittest" and "reproductive fitness"	Natural selection — "reproductive fitness" (see Chapter 8)

better chance of surviving, for, of the many individuals of any species which are periodically born, but a small number can survive. (Darwin, 1859, pp. 51–52)

It is clear that Darwin considered behavioral characteristics to be just as subject to natural selection as physical traits. In *The Origin of Species* an entire chapter is devoted to the discussion of instinctive behavior patterns. A later book, *The Descent of Man and Selection in Relation to Sex*, gave detailed consideration to comparisons of mental powers and moral senses of animals and humans, as well as to the development of intellectual and moral faculties in humans. In these discussions Darwin was satisfied that he had demonstrated that the difference between the mind of a human being and the mind of an animal "is certainly one of degree and not of kind" (1871, p. 101). This is an essential point, since one of the strongest objections to the theory of evolution was the qualitative gulf that was supposed to exist between the mental capacities of humans and of lower animals.

In an explicit summary statement, based largely on observations of "family resemblance," Darwin said:

So in regard to mental qualities, their transmission is manifest in our dogs, horses, and other domestic animals. Besides special tastes and habits, general intelligence, courage, bad and good temper, etc., are certainly transmitted. With man we see similar facts in almost every family; and we now know, through the admirable labours of Mr. Galton, that genius which implies a wonderfully complex combination of high faculties, tends to be inherited; and, on the other hand, it is too certain that insanity and deteriorated mental powers likewise run in families. (Darwin, 1871, p. 414)

Darwin's writings focused on selection of individuals and their characteristics. A different perspective has been emphasized by researchers in an area called sociobiology, which applies evolutionary theory to the study of social behavior. Sociobiologists look at evolution from the standpoint of the gene rather than the individual. Just as the chicken may only be the egg's way of producing other eggs, sociobiology suggests that the individual may only be the gene's way of producing more genes. This way of thinking suggests an interesting analysis of certain social behaviors, such as altruism, as well as the importance of selection for genetically similar individuals, a theory called *kinship selection.* This view of evolution is discussed in Chapter 8.

GALTON'S CONTRIBUTIONS

Among the supporters and admirers of Darwin at this time was another one of Erasmus Darwin's grandsons, Francis Galton (Figure 2.3). Galton had already established something of a reputation as a geographer, explorer, and inventor. By the time *The Origin of Species* was published, he had invented a printing electric telegraph, a type of periscope, and a nautical signaling device. The effect on him of Darwin's work is revealed in a letter he later wrote to Darwin:

> I always think of you in the same way as converts from barbarism think of the teacher who first relieved them from the intolerable burden of their superstition. I used to be wretched under the weight of the old-fashioned arguments from design; of which I felt though I was unable to prove to myself, the worthlessness. Consequently the appearance of your *Origin of Species* formed a real crisis in my life; your book drove away the constraint of my old superstition as if it had been a nightmare and was the first to give me freedom of thought. (In Pearson, 1924, vol. I, plate II)

Figure 2.3
Francis Galton, in 1840, from a portrait by O. Oakley. (Courtesy of the Galton Laboratory.)

The Origin of Species directed Galton's immense curiosity and talents to biological phenomena, and he soon developed what was to be a central and abiding interest for the rest of his life: the inheritance of mental characteristics.

Hereditary Genius

In 1865 two articles by Galton, jointly entitled "Hereditary Talent and Character," were published in *Macmillan's Magazine.* Four years later a greatly expanded discussion was published under the title, *Hereditary Genius: An Inquiry into Its Laws and Consequences.* The general argument presented in this work is that a greater number of extremely able individuals is found among the relatives of persons endowed with high mental ability than would be expected by chance. Furthermore, Galton discovered that the closer the family relationship, the higher the incidence of such superior individuals. Galton applied Quetelet's "law of deviation from an average," which at the time was a recent development, but later became familiar as the normal curve. Galton distinguished 14 levels of human ability, ranging from idiocy through mediocrity to genius.

Since there was no satisfactory way of quantifying natural ability, Galton had to rely on reputation as an index. By "reputation," he did not mean notoriety for a single act, nor mere social or official position, but "the reputation of a leader of opinion, of an originator, of a man to whom the world deliberately acknowledges itself largely indebted" (1869, p. 37). The designation "eminent" was applied to those individuals who constituted the upper 250-millionths of the population (i.e., 1 in 4,000 persons would attain such a rank), and the discussion focused on such men. Indeed, the majority of individuals Galton presented in evidence were, in his estimation, the cream of this elite group, and were termed "illustrious." These were men whose talents ranked them one in a million.

On the basis of biographies, published accounts, and direct inquiry, Galton evaluated the accomplishments of well-known judges, statesmen, peers, military commanders, literary men, scientists, poets, musicians, painters, Protestant religious leaders, and Cambridge scholars. (Oarsmen and wrestlers of note were also examined to extend the range of inquiry from brain to brawn.) The approximately 1,000 men who were designated as "eminent" were found to belong to 300 families. With the overall incidence of eminence only 1 in 4,000, this result clearly illustrated the tendency for eminence to be a familial trait.

Taking the most eminent man in each family as a reference point, the other individuals who attained eminence (in the same or in some other field of endeavor) were tabulated with respect to closeness of family relationship, as indicated in Figure 2.4. Briefly stated, the results showed that eminent status was more likely to appear in close relatives, with the likelihood of eminence decreasing as the degree of relationship became more remote. Eminence was

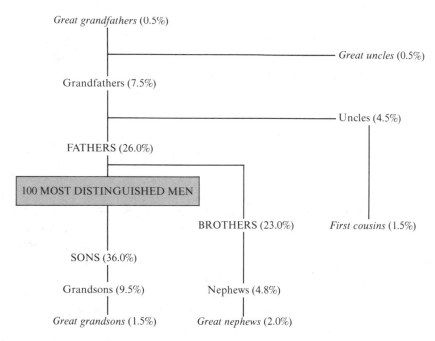

Great grandfathers (0.5%)

Great uncles (0.5%)

Grandfathers (7.5%)

Uncles (4.5%)

FATHERS (26.0%)

100 MOST DISTINGUISHED MEN

BROTHERS (23.0%)   *First cousins* (1.5%)

SONS (36.0%)

Grandsons (9.5%)   Nephews (4.8%)

Great grandsons (1.5%)   *Great nephews* (2.0%)

Figure 2.4
Percentage of eminent men among relatives of the 100 most distinguished men in
Galton's study. (From *Hereditary Genius* by Francis Galton, 1869.)

attained by 26 percent of the fathers of the 100 most distinguished men, 23
percent of their brothers, and 36 percent of their sons. Second-degree relatives
such as grandfathers, uncles, nephews, and grandsons achieved eminence to a
much lower degree (7.5 percent, 4.5 percent, 4.8 percent, and 9.5 percent,
respectively). However, these percentages are still high when compared to the
overall incidence of only 1 in 4,000 (0.025 percent).

Galton was aware of the possible objection that relatives of eminent
men share social, educational, and financial advantages, and he knew that the
results of his investigation might be interpreted as showing the effectiveness of
such environmental factors. Three counterarguments were presented. First,
Galton stressed the fact that many men had risen to high rank from humble
family backgrounds. Second, it was noted that the proportion of eminent
writers, philosophers, and artists in England was not less than that in the
United States, where education of the middle and lower socioeconomic
classes was more advanced. He felt that the educational advantages in America had spread culture more widely without producing more persons of
eminence. Finally, Galton compared the success of adopted kinsmen of
Roman Catholic popes, who were given great social advantages, with the sons
of eminent men, and the latter were judged to be more distinguished. This last
point presages the use of the adoption design to disentangle genetic and

environmental contributions to family resemblance. Nonetheless, these counterarguments do not today justify Galton's assertion that genius is solely a matter of nature (heredity) rather than nurture (environment).

Pioneering Research in Psychology and Statistics

In order to further his researches, it was necessary for Galton to find ways of assessing mental characteristics. In a prodigious program of research, he developed apparatus and procedures for measuring auditory thresholds, visual acuity, color vision, touch, smell, judgment of the vertical, judgment of length, weight discrimination, reaction time, and memory span. In addition, he employed a questionnaire technique to investigate mental imagery.

The problems of properly expressing and evaluating the data obtained from such researches were formidable, and Galton also turned his remarkable energies to statistics. He pioneered the development of the concepts of the median, percentiles, and correlation.

Since it was, of course, desirable to have data from large numbers of individuals, Galton employed various stratagems to this end. For example, he arranged for an "Anthropometric Laboratory for the measurement in various ways of Human Form and Faculty" to be located at an International Health Exhibition during 1884 and 1885. Some 9,337 people paid threepence or fourpence each for the privilege of being measured for various bodily and sensory characteristics. (See Figure 2.5.) After the International Health Exhibition closed, a permanent Anthropometric Laboratory was established. During the next seven years, another 7,500 individuals were tested (at the bargain rate of one penny each), including Galton himself. His data are reproduced in Figure 2.6; data from the Anthropometric Laboratory have been shown to be highly reliable for most variables, and even though a century old, they provide some of the only available information on sibling resemblance for some of these measures (Johnson ct al., 1985). On another occasion a contest was sponsored in which awards of £7 were given to those submitting the most careful and complete "Extracts from their own Family Records." In this way, Galton was able to obtain a large number of pedigrees that he could examine for evidence of human inheritance.

Twins and the Nature–Nurture Problem

Galton introduced (1876) the use of twins to assess the roles of nature (inheritance) and nurture (environment). The essential question in his examination of twins was whether twins who were alike at birth became more dissimilar as a consequence of any dissimilarities in their nurture. Conversely, did twins who were unlike at birth become more similar as a consequence of similar nurture? Galton acknowledged two types of twins—those arising

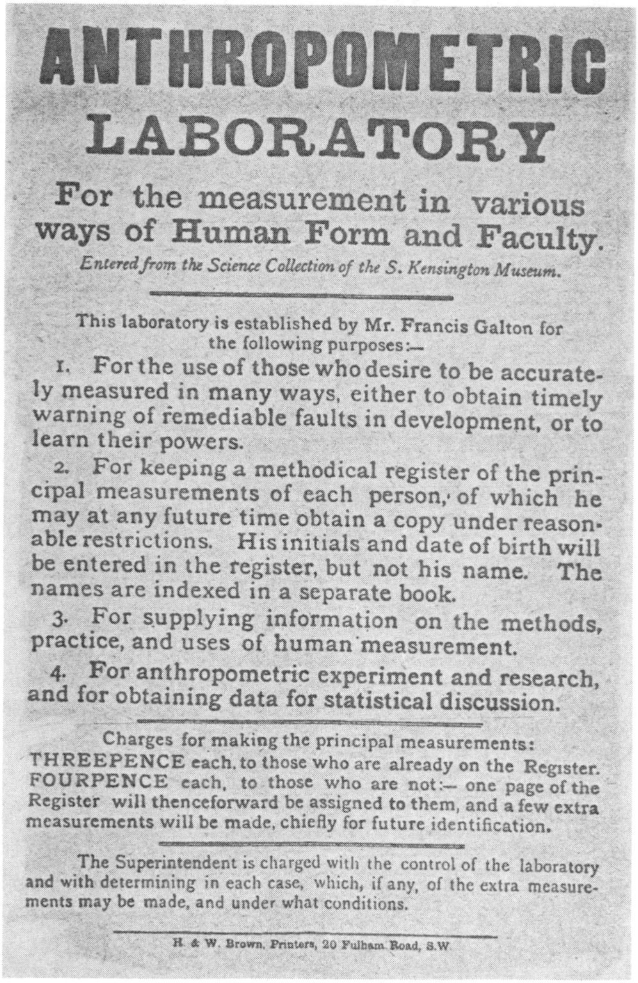

ANTHROPOMETRIC LABORATORY

For the measurement in various ways of Human Form and Faculty.

Entered from the Science Collection of the S. Kensington Museum.

This laboratory is established by Mr. Francis Galton for the following purposes:—

1. For the use of those who desire to be accurately measured in many ways, either to obtain timely warning of remediable faults in development, or to learn their powers.

2. For keeping a methodical register of the principal measurements of each person, of which he may at any future time obtain a copy under reasonable restrictions. His initials and date of birth will be entered in the register, but not his name. The names are indexed in a separate book.

3. For supplying information on the methods, practice, and uses of human measurement.

4. For anthropometric experiment and research, and for obtaining data for statistical discussion.

Charges for making the principal measurements:
THREEPENCE each, to those who are already on the Register.
FOURPENCE each, to those who are not:— one page of the Register will thenceforward be assigned to them, and a few extra measurements will be made, chiefly for future identification.

The Superintendent is charged with the control of the laboratory and with determining in each case, which, if any, of the extra measurements may be made, and under what conditions.

H. & W. Brown, Printers, 20 Fulham Road, S.W.

Figure 2.5
Handbill of 1884 announcing Galton's Anthropometric Laboratory. (Courtesy of Cedric A. B. Smith, the Galton Laboratory, University College, London.)

from separate eggs, and those arising from the same egg. Yet he did not distinguish between the two types in his discussion, except as they fell into his categories, "alike at birth" or "unlike at birth."

Galton gathered his evidence from answers to questionnaires and biographical and autobiographical material. He observed that, among 35 pairs who had been alike at birth, and who had been reared under highly similar conditions, the similarities within the twinship persisted after the members had grown to adulthood and gone more-or-less separate ways. From 20 pairs of originally dissimilar twins, there was no compelling evidence that any had

F Galton   12

Initials of			Father and mother first cousins?	Date of birth	Eye color	Sex	Married Single Widowed	Color sense	Occupation
Self	Father	Mother							
FG	G	D	No	16 2 22	light blue	M	M	Normal	Private Gen.

Date of measurement:			Head length maximum from root of nose		Head breadth maximum		Height standing less heels of shoes		Span of arms from opposite fingertips		Weight in ordinary clothing	Strength of squeeze Right hand	Left hand	Breathing capacity	Keenness of sight distance of reading diamond numerals Right eye	Left eye	Snellen's type read at 90 feet
Day	Month	Year	Inches	Tenths	Inches	Tenths	Inches	Tenths	Inches	Tenths	Pounds	Pounds	Pounds	Cubic inches	Inches	Inches	Number of type
25	2	88	7	1	6	1½	68	7	71	8	171	80	71	159	17	19	D9

Height sitting above seat of chair		Height of top of knee when sitting less heels left arm		Length of elbow to finger of left hand		Length of middle finger of left hand	
Inches	Tenths	Inches	Tenths	Inches	Tenths	Inches	Tenths
36	5	21	6	18	4	4	4

HEARING		REACTION TIME	
Keenness of hearing	Highest audible note	To sight	To sound
	Vibrations per second	Thousandths of second	Thousandths of second

Figure 2.6

Galton's own data from the records of his Anthropometric Laboratory. For some reason, hearing and reaction-time data were not recorded for Galton. (Courtesy of Cedric A. B. Smith, the Galton Laboratory, University College, London.)

become more alike through exposure to similar environments. According to Galton, "There is no escape from the conclusion that nature prevails enormously over nurture when the differences of nurture do not exceed what is commonly to be found among persons of the same rank of society and in the same country. My fear is, that my evidence may seem to prove too much, and be discredited on that account, as it appears contrary to all experience that nurture should go for so little" (1883, p. 241). This claim *is* too much based on the evidence Galton provides. His data would have been more convincing if he had compared identical and fraternal twins, the essence of the twin method, which was discovered 50 years later (Rende, Plomin, and Vandenberg, 1989) and is the topic of Chapter 12.

Galton's Work in Perspective

The ten years between *The Origin of Species* and *Hereditary Genius* had not been sufficient for the idea of man as an animal to be completely accepted. For many of those who accepted Darwin's theory, of course, Galton's work was a natural and logical extension: human beings differ from animals most strikingly in mental powers; humans, like other animals, have evolved; evolution works by inheritance; mental traits are heritable. Galton certainly overstated his case in terms of demonstrating the importance of heredity, and he set up a needless battle by pitting nature against nurture. Nonetheless, his work was pivotal in documenting the range of variation in human behavior and in suggesting that heredity underlies behavioral variation. For those whose faith in the special creation of humankind remained firm, Galton's views were unacceptable, atheistic, and reprehensible.

But there were scholars whose inquiries stemmed from a genuine desire to understand the mind. This philosophical approach was dominated by the British philosophers, whose emphasis was clearly on experience (nurture). They based their views on John Locke's seventeenth-century tabula rasa dictum that ideas are not inborn, but come from experience. The role of experience was also emphasized by experimental psychology, which is usually dated from Wilhelm Wundt's establishment in 1879 of the Psychologisches Institut at Leipzig, Germany. In spite of the fact that Wundt had come to psychology from physiology, his approach was not biological in the same sense as Galton's. The goal of Wundt's institute was the identification, through introspection, of the components of consciousness. Individual differences, which formed the heart of Galton's investigations, were nuisances in this search for principles that could be generally applied to all. One notable exception to this general trend was provided by the American, J. McK. Cattell, who, as a student of Wundt, insisted on studying individual differences. After Cattell left Leipzig, he worked for a while with Galton, with whom he strengthened and confirmed his belief in the importance of individual differences. Cattell had an important influence on the development of

American psychology and inspired some of the earliest experimental work in behavioral genetics.

From the foregoing it is clear that Galton's work was neither completely in step, nor completely out of step, with his times. Galton lived during a period of great intellectual turmoil in biology. His work was both a product and a cause of the advances that were made. Galton was not the first to insist on the importance of heredity in traits of behavior. Nor was Galton the first to place his conclusions in an evolutionary context. But it was Galton who championed the idea of the inheritance of behavior and vigorously consolidated and extended it. In effect, we may regard Galton's efforts as the beginning of behavioral genetics.

PRE-MENDELIAN CONCEPTS OF HEREDITY AND VARIATION

Neither Darwin nor Galton understood the mechanism by which heredity works or how heritable variation is maintained. The answers were being worked out by a contemporary scientist working in what is now Czechoslovakia. But since this research was not known to Darwin or to Galton, they had to work within the prevailing views of heredity and heritable variation.

Heredity

Long before Darwin and Galton, there had been substantial evidence of the importance of heredity, although its laws had proved extremely resistant to analysis. In particular, a vast amount of data had been accumulated from plant and animal breeding. Many offspring bore a closer resemblance to one parent than to the other. It was also common for the appearance of offspring to be intermediate between the two parents. But two offspring from the same parents could be quite unlike. As J. L. Lush described the situation considerably later, the first rule of breeding is that "like produces like," while the second rule is that "like does not always produce like" (1951, p. 496).

In Darwin's time, the theory of heredity that seemed to explain most adequately the confusion of facts at the time was the "provisional hypothesis of pangenesis." In this view, the cells of the body, "besides having the power, as is generally admitted, of growing by self-division, throw off free and minute atoms of their contents, that is gemmules. These multiply and aggregate themselves into buds and the sexual elements" (Darwin, 1868, p. 481). Gemmules were presumably thrown off by each cell throughout its course of development. In embryogenesis and later development, gemmules from the parents, originally thrown off during various developmental periods, would come into play at the proper times, thus directing the development of a new organ like that of the parents. The theory of pangenesis was quite reasonable (although it was wrong). It was particularly compelling because it was compatible with Lamarck's notion of "use and disuse" as the source of variability in evolution (see Table 2.1).

Variation

The source of heritable variation was the most difficult component of the model of evolution for Darwin to explain. Without heritable variation in each generation, evolution could not continue. Because children often exhibit some of the characteristics of each of their parents, it was commonly accepted that characteristics of parents merged or blended in their offspring. The troublesome implication of such a "blending" hypothesis is that variability would be greatly reduced (in fact, roughly halved) each generation. For example, if one parent were tall and the other short, the offspring would be of average height. Thus, the blending hypothesis implies that variability would rapidly diminish to a trivial level if it were not replenished in some manner. Although Darwin worried about the problem, he never resolved it. He suggested two ways in which variability might be induced, but both of them assumed that environmental factors altered the stuff of heredity. The theory of pangenesis suggested that gemmules (miniature replicas of the parents' cells) could reflect changes in environment. Darwin vaguely concluded that changes in the conditions of life in some way altered gemmules in the reproductive systems of animals so that their offspring were more variable than they would have been under stable conditions. Ordinarily, this increased variability would be random. Natural selection would then preserve those deviants that by chance happened to be better adapted as a consequence of their deviation.

Sometimes, however, an environmental condition might induce *systematic* change. Darwin hesitatingly accepted the Lamarckian theory of use and disuse to suggest that acquired characteristics can be inherited. In *The Descent of Man*, Darwin speculated about the alleged longer legs and shorter arms of sailors as compared to soldiers: "Whether the several foregoing modifications would become hereditary, if the same habits of life were followed during many generations, is not known, but it is probable" (1871, p. 418). In some of his writings, Darwin seemed sure that variations in life experiences can increase genetic variability: "there can be no doubt that use in our domestic animals has strengthened and enlarged certain parts, and disuse diminished them; and that such modifications are inherited " (1859, p. 102). Likewise, he stated, with respect to behavioral characteristics, that "some intelligent actions, after being performed during several generations, become converted into instincts and are inherited" (1871, p. 447). However, for the most part, Darwin was unsure of the source of variability: "Our ignorance of the laws of variation is profound. Not in one case out of a hundred can we pretend to assign any reason why this or that part has varied. . . . Habit in producing constitutional peculiarities and use in strengthening and disuse in weakening and diminishing organs, appear in many cases to have been potent in their effects" (1859, p. 122).

THE WORK OF GREGOR MENDEL

While Darwin struggled with these issues, in his files was an unopened manuscript by an Augustinian monk, Gregor Mendel (Allen, 1975). (See Figure 2.7). Through his research on pea plants in the garden of a monastery at Brunn, Moravia, Mendel had provided the answer to the riddle of inheritance and variability.

As we will discuss in detail in Chapter 3, Mendel summarized his many experiments with two laws. The law of segregation states that there are two elements of heredity for a single character. These two elements segregate, or separate, "cleanly" during inheritance so that offspring receive one of the two elements from each parent. The important implication of this law is that the hereditary elements are discrete and do not blend. Because the elements do not blend, the reduction in inherited variation that concerned Darwin does not occur. Mendel's second law, the law of independent assortment, concerns the inheritance of two traits, each with two elements. When we look at the inheritance of two traits at the same time, the elements for each trait assort

Figure 2.7
Gregor Johann Mendel, 1822–1884. A photograph taken at the time of his research. (Courtesy of V. Orel, Mendel Museum, Brno, Czechoslovakia.)

independently of the elements for the other trait. In other words, the inheritance of one trait does not affect the inheritance of the other.

We can translate Mendel's theory into more modern terms. As we will discuss in Chapter 4, a *gene* is a coded stretch of deoxyribonucleic acid (DNA). Genes are located on chromosomes in the nucleus of a cell, and chromosomes occur in pairs—one from the mother, and one from the father. Alternate forms of genes (*alleles*) are at the same place (*locus*) on the matching chromosomes. Most humans have 23 pairs of chromosomes (for example, at the tip of each member of one medium-sized pair (designated chromosome number 9) is an allele (either *A, B,* or *O*) of the gene determining the *ABO* blood group. At this locus on one of the chromosomes is the allele (e.g., *A*) from the mother; on the matching chromosome is the allele (e.g., *O*) from the father. In this example, the offspring would have blood group *AO*. Genes do not blend in inheritance, and genetic variation is maintained. But what initially causes the genetic variation? The major source of genetic variability is *mutation*, or changes in the genetic code of DNA.

Mendel's results and his theory were read to the Brunn Society of Natural Science in 1865, and were later published in the proceedings of the society. The crucial experiments had thus been done prior to Darwin's statement of pangenesis. But Darwin was not alone in overlooking Mendel's ideas. For 34 years, the "Experiments in Plant Hybridization" (Mendel, 1866) remained almost completely unacknowledged.

In 1900, three investigators—C. Correns, Hugo de Vries, and Erich von Tschermak—almost simultaneously "rediscovered" Mendel's work. A period of intensive research was inaugurated, in which the Mendelian theory was confirmed and extended. The vigorously developing area of research was given the name *genetics* by William Bateson in 1905, and in 1909 the name *gene* was proposed for the Mendelian elements by Wilhelm Johannsen. At the same time, Johannsen made a fundamental distinction between *genotype*, which is the genetic composition of the individual, and *phenotype*, which is the apparent, visible, measurable characteristic. The importance of this distinction is that it makes clear that the observable trait is not a perfect index of the individual's genetic properties. Given a number of individuals of the same genotype, we might nonetheless expect differences among them—those caused by environmental agents. Thus, two beans might be from the same "pure line" and have identical genotypes for size, yet one might be larger than the other because of differences in nurture, such as soil conditions. Nevertheless, their genotypes would remain unaffected, and the beans of the plants grown from these two beans would be of the same average size. The inheritance of acquired characteristics obviously has no place in this scheme.

THE LEGACY OF DARWIN AND MENDEL

The most important lesson to be learned from evolution and Mendelian genetics is the pervasiveness of genetic variability. Most evolutionary theorists focus on genetic variability from one species to another. There are now approximately 193 primate species, of which 53 are little animals (such as tree shrews, lemurs, and tarsiers) that do not look much like primates. Although the genetic variability of organisms closest to humans in an evolutionary sense is most interesting to us, it needs to be placed in a larger context. The primate order belongs to the mammalian class, which includes 18 other orders. There are about 4,300 species in the mammalian class alone, but this great variability represents only a small part of the phylum Chordata, which consists primarily of vertebrates. There are about 3,000 amphibian species, 6,000 reptile species, 11,000 species of birds, and 28,000 species of bony fish. All told, there are about 55,000 species of vertebrates. This genetic diversity, although impressive, is minute when compared to the invertebrates. There are well over 1 million invertebrate species.

A microscopic view of single-celled organisms also adds to our perspective on variability. It has been estimated that there are about 100 octillion living cells in the world today (Hockett, 1973). Of these, perhaps as many as 99 octillion (at least 90 octillion) are tied up in single-celled organisms such as bacteria. We, the *metazoan* (multicellular) organisms, are in a decided minority. It should also be pointed out that this variability includes only the survivors of the process of natural selection. Many more species are extinct than extant.

Not only does genetic variability exist among species, but individuals within each species differ genetically. Darwin recognized that hereditary variability among individuals is the key to natural selection. Nonetheless, there is a common, but mistaken, tendency to consider selection as a force that creates an ideal match between a species and its environment and to assume that all members of the species conform to this ideal and show no important variations. Forces of natural selection that maintain genetic variability within species are discussed in Chapter 8. For now, the critical point is that genetic differences among individuals within a species are the building blocks of evolution. As explained in Chapter 7, molecular genetics documents the extensive variability between and within species. Human beings differ from chimpanzees by about one nucleotide base of DNA in 100; unrelated humans differ from one another by one base in 1,000 (DeLisi, 1988). Because each person has about 3.5 billion DNA bases, this means that about 3.5 million DNA bases differ among us. The issue for behavioral genetics is the extent to which this genetic variability causes behavioral differences among individuals.

Consider dogs as a dramatic yet familiar example of genetic variability within species. (See Figure 2.8.) From the three-pound Pekingnese to the Irish Wolfhound weighing more than 100 pounds, dog breeds are members of the same species. Dogs are also an excellent example of the effect of genetic differences within a species on behavior. Although physical differences are most obvious, dogs have been bred for centuries as much for their behavior as for their looks. For example, sheepdogs herd, retrievers retrieve, trackers track, and pointers point with scarcely any training. Breeds also differ strikingly in intelligence and in temperamental characteristics such as emotionality, activity, and aggressiveness. Dogs are not unusual in their genetic diver-

Figure 2.8
Dog breeds illustrate genetic diversity within a species.

sity, although they are unusual in the extent to which different breeds have been intentionally bred to accentuate genetic differences.

The potential variability within the human species is so great that it is next to impossible to imagine that there have ever been two individuals with the same combination of genes. Each of us has the capacity to generate $10^{3,000}$ eggs or sperm with unique sets of genes. If we consider $10^{3,000}$ possible eggs being generated by an individual woman and the same number of sperm being generated by an individual man, the likelihood of anyone else with your set of genes in the past or in the future becomes infinitesimal (Bodmer and Cavalli-Sforza, 1976).

In summary, our legacy from Darwin and Mendel includes an appreciation of genetic diversity among species as well as among individuals within species. This view creates a dynamic picture of genetic variability: Our biological system is not merely tolerant of genetic differences; the system generates differences and depends on them. They are the sine qua non of evolution, the quintessence of life.

SUMMARY

Although ancient concepts of heredity are interesting, the history of behavioral genetics really began with Darwin, Galton, and Mendel. Darwin's theory of natural selection as an explanation for the origin of species made a major impact on scientific thinking. Galton was the first to study the inheritance of mental characteristics and to suggest using twins and adoptees to study nature–nurture problems. Mendel solved the riddle of inheritance with his experiments on garden peas. He demonstrated that heredity involves discrete elements, now called genes, and he formulated two laws of inheritance— segregation and independent assortment. We now know that heritable variability is caused by mutations of DNA, the genetic material, and is maintained and transmitted according to the laws discovered by Mendel.

CHAPTER·3

Single Genes

Although genetics teachers can dream up very complicated problems, the basic principles of Mendelian genetics are elegantly simple. This is not to say, of course, that the field of genetics is without its complexities. The search for detailed understanding of the transmission of hereditary factors and their mode of action has led investigators to look into cytology, embryology, physiology, biochemistry, biophysics, and mathematics. However, the fundamentals of Mendelian genetics were established without knowledge of the physical or chemical nature of the hereditary material. It is still convenient to introduce the principles of genetics by treating the hereditary determinants as hypothetical factors.

MENDEL'S EXPERIMENTS AND LAWS

To understand the logic of Mendel's experiments, it is helpful to remember that in the 1800s no one knew about genes, and that the prevailing theory of inheritance was pangenesis (discussed in Chapter 2). Much of the research on heredity involved crossing plants of different species. A critical drawback to this approach is that the offspring are usually sterile, meaning that succeeding generations cannot be studied. Also, the features of the plants that were investigated were too complex for clear analysis. Mendel's success can be attributed in large part to circumventing these problems. He crossed different varieties of pea plants of the same species and thus obtained fertile offspring that could be crossed to study subsequent generations. In addition, he picked simple qualitative ("either–or") traits that happened to have a number of fortunate characteristics that we shall consider in a moment. Mendel also counted all the progeny rather than being content, as his predecessors had, with a verbal summary of the typical result.

Mendel's research involved two kinds of experiments. The first, which used *monohybrid* crosses, followed the inheritance of one character at a time. These experiments led to Mendel's first law, the *law of segregation.* This law states that there are two "elements" of heredity for each character and that these two elements separate, or segregate, during inheritance. Offspring receive one of the two elements from each parent. These elements do not blend in inheritance, as the theory of pangenesis suggested. The second type of experiment utilized *dihybrid* and *trihybrid* crosses, and traced the inheritance of characters considered two or three at a time. The dihybrid and trihybrid experiments led Mendel to conclude that the hereditary elements for one character assort independently of the elements for other characters. In other words, the inheritance of one trait in no way influences the inheritance of the other. This conclusion is now known as Mendel's second law, the *law of independent assortment.*

Monohybrid Experiments

Mendel looked at seven qualitative traits of the pea plant. Three of these included whether the seed was green or yellow inside (the cotyledon), whether the seed was smooth or wrinkled, and whether the stem was long or short. He obtained 22 varieties of the pea plant that differed in some of these characteristics. For each of the seven traits, he crossed two *truebreeding* varieties. Truebreeding plants are those that always yield the same result when self-pollinated or crossed with the same kind of plant.

In one experiment, Mendel crossed truebreeding plants with round seeds to truebreeding plants with wrinkled seeds. Later in the summer, when he opened the pods containing their offspring (called the F_1, or first filial generation), he found that all of them had round seeds. This result indicated that the traditional view of blending inheritance was not correct. The F_1 did not have seeds that were moderately wrinkled. Because these F_1 offspring were fertile, Mendel was able to take the next step of self-fertilizing plants from the F_1 generation to study their offspring, known as F_2. The results were striking: ¾ of the offspring had round seeds and ¼ had wrinkled seeds. Of the 7,324 seeds from the F_2, 5,474 were round and 1,850 were wrinkled. This result suggests that the factor responsible for wrinkled seeds had not been lost in the F_1 generation, but had been dominated by the factor causing round ones. Excerpts from the paper in which Mendel described these results are presented in Box 3.1.

Given these facts, Mendel deduced a simple explanation involving two hypotheses, summarized in Figure 3.1. First, each individual has two hereditary factors, now called alleles (alternate forms of a gene), that determine

	Observed	Hypothesized
Truebreeding parents	Truebreeding round × Truebreeding wrinkled	$A_1A_1 \times A_2A_2$
F_1	All round	All A_1A_2 (A_1 dominant)
F_2	¾ round, ¼ wrinkled	¼ A_1A_1, ½ A_1A_2 ⎵ round ¼ A_2A_2 ⎵ wrinkled

Figure 3.1
Summary of Mendel's monohybrid experiments.

whether the seed is wrinkled or round. Thus, each parent has two alleles but passes only one of those to its offspring. The second hypothesis was that one allele could dominate the other. These two hypotheses neatly explain the data. The truebreeding parent plant with round seeds has two alleles for round seeds (A_1A_1). The truebreeding parent plant with wrinkled seeds is A_2A_2. Thus, the F_1 offspring will have one allele from each parent (A_1A_2). If A_1 dominates A_2, F_1 will have round seeds. The real test is the F_2 population. Mendel's theory would predict that when F_1's are self-fertilized or crossed with other F_1 individuals (A_1A_2 with A_1A_2), ¼ of the F_2's should be A_1A_1, ½ A_1A_2 and ¼ A_2A_2. However, assuming A_1 dominates A_2, then A_1A_2 should have round seeds like the A_1A_1. Thus, ¾ of the F_2 should have round seeds and ¼ wrinkled, which is exactly what Mendel's data indicated. Mendel completed many experiments with other characteristics and other varieties of pea plants, and they all confirmed this theory.

At this point, we need to introduce several terms in addition to genes and alleles. Individuals with the same alleles for a particular gene are called *homozygotes*; those with different alleles are *heterozygotes*. The truebreeding parental varieties were all homozygotes (A_1A_1 or A_2A_2) and the F_1 were all heterozygotes (A_1A_2). *Genotype* refers to the genetic constitution of an individual. With two alleles for a gene, three possible genotypes exist for that particular gene: A_1A_1, A_1A_2, and A_2A_2. We observe the *phenotype*, not the genotype. For example, in the F_2 population the genotypes were distributed in a 1 : 2 : 1 ratio (¼ A_1A_1, ½ A_1A_2, and ¼ A_2A_2), while the phenotypes were in a 3 : 1 ratio (¾ round, ¼ wrinkled). In other words, the phenotype does not necessarily reflect the genotype, since the A_1A_2 genotype has the same appearance as the A_1A_1 genotype. It was fortunate for Mendel's theory that dominance of the A_1 allele over the A_2 allele was complete. If there had been no dominance, the A_1A_2 phenotype would have been in between the A_1A_1 and the A_2A_2 phenotypes, and it would have appeared as if blending had occurred in the F_1. Different types of gene expression are represented in Figure 3.2. For Mendel's seven pea plant characteristics, there was complete dominance in the sense that the phenotypic value (appearance) of the heterozygote genotype

Box 3.1

Mendel's Classic Paper: Monohybrid Results

Mendel presented the results of eight years of research on the pea plant at a meeting of naturalists in Brunn, Moravia, in 1865, and his paper was published in 1866. Because this paper is the cornerstone of genetics, we have excerpted the following sections, which focus on his monohybrid results.* Box 3.2 describes his dihybrid results.

EXPERIMENTS IN PLANT-HYBRIDIZATION

Experience of artificial fertilisation, such as is effected with ornamental plants in order to obtain new variations in colour, has led to the experiments which will here be discussed. The striking regularity with which the same hybrid forms always reappeared whenever fertilisation took place between the same species induced further experiments to be undertaken, the object of which was to follow up the developments of the hybrids in their progeny. . . .

The paper now presented records the results of such a detailed experiment. This experiment was practically confined to a small plant group, and is now, after eight years' pursuit, concluded in all essentials. Whether the plan upon which the separate experiments were conducted and carried out was the best suited to attain the desired end is left to the friendly decision of the reader. . . .

The Forms of the Hybrids [F$_1$]

. . . in this paper those characters which are transmitted entire, or almost unchanged in the hybridisation, and therefore in themselves constitute the characters of the hybrid, are termed the *dominant*, and those which become latent in the process *recessive*. The expression "recessive" has been chosen because the characters thereby designated withdraw or entirely disappear in the hybrids, but nevertheless reappear unchanged in their progeny, as will be demonstrated later on. . . . Of the differentiating characters which were used in the experiments the following are dominant:

*Translated by William Bateson and the Royal Horticultural Society of London. The original paper was published in *Verh. naturf. Ver. in Brünn*, 1866.

---

was the same as that of one of the homozygote genotypes. The other side of the coin is *recessiveness*. Rather than saying that the A_1 allele dominates the A_2 allele, we could say that the A_2 allele is recessive to the A_1 allele. If, on the other hand, A_1 were completely recessive to A_2, then the A_1A_2 genotype would look the same as the A_2A_2 genotype. These examples consider only complete dominance and recessiveness, but both could be partial. That is, the A_1A_2 heterozygote can lie anywhere in between the observed value for one homo-

1. The round or roundish form of the seed with or without shallow depressions.
2. The yellow colouring of the seed albumen [cotyledons].
3. The grey, grey-brown, or leather-brown colour of the seed-coat, in association with violet-red blossoms and reddish spots in the leaf axils.
4. The simply inflated form of the pod.
5. The green colouring of the unripe pod in association with the same colour in the stems, the leaf-veins and the calyx.
6. The distribution of the flowers along the stem.
7. The greater length of stem. . . .

The Generation from the Hybrids [F₂]

The relative numbers which were obtained for each pair of differentiating characters are as follows:

Expt. 1. Form of seed.—From 253 hybrids 7,324 seeds were obtained in the second trial year. Among them were 5,474 round or roundish ones and 1,850 angular wrinkled ones. Therefrom the ratio 2.96 to 1 is deduced.

Expt. 2. Colour of albumen.—258 plants yielded 8,023 seeds, 6,022 yellow, and 2,001 green; their ratio, therefore is as 3.01 to 1. . . .

Expt. 3. Colour of the seed-coats.—Among 929 plants 705 bore violet-red flowers and grey-brown seed-coats; 224 had white flowers and white seed-coats, giving the proportion 3.15 to 1.

Expt. 4. Form of pods.—Of 1,181 plants 882 had them simply inflated, and in 299 they were constricted. Resulting ratio, 2.95 to 1.

Expt. 5. Colour of the unripe pods.—The number of trial plants was 580, of which 428 had green pods and 152 yellow ones. Consequently these stand in the ratio 2.82 to 1.

Expt. 6. Position of flowers.—Among 858 cases 651 had inflorescences axial and 207 terminal. Ratio, 3.14 to 1.

Expt. 7. Length of stem.—Out of 1,064 plants, in 787 cases the stem was long, and 277 short. Hence a mutual ratio of 2.84 to 1. . . .

If now the results of the whole of the experiments be brought together, there is found, as between the number of forms with the dominant and recessive characters, an average ratio of 2.98 to 1, or 3 to 1.

zygote and that of the other. Overdominance can also occur in which the heterozygote shows a phenotypic value greater than both homozygotes.

Mendel summarized his theory with the *law of segregation:* Inheritance is particulate; that is, there are discrete and inviolable units of inheritance that are transmitted so that offspring have two units, one from each parent. In other words, alleles of a gene pair separate cleanly, with no residual effects on each other. An A_1 allele transmitted from an A_1A_1 parent is no different from

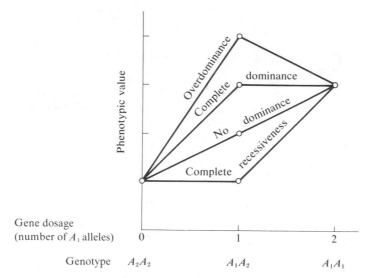

Figure 3.2
Graphical representation of four different types of gene expression.

one transmitted by an A_1A_2 parent. The ratios in the F_2 population are called *monohybrid segregation ratios:* "mono" because they consider only one trait, such as the roundness of the seed; "hybrid" because they are the result of a cross between two truebreeding populations; and "segregation" because they show that the alleles separate out, or segregate, in the F_2 generation. The $1:2:1$ ratio is the monohybrid *genotypic* segregation ratio, and the $3:1$ ratio is the monohybrid *phenotypic* segregation ratio.

Another way of looking at the law of segregation is described in Table 3.1. Each gamete (sperm or egg) contains one allele from each gene pair carried by the parent who produced the gamete. F_1 individuals are all heterozygous, A_1A_2. When they are crossed to breed the F_2 generation, half of the sperm will carry the A_1 allele and the other half will carry the A_2 allele. This is also true for the female gametes. If sperm and eggs unite at random, the offspring produced are those described in Table 3.1. There are three different genotypes in the F_2. Their relative frequencies are ¼ A_1A_1, ½ A_1A_2, and ¼ A_2A_2. If dominance is complete, no difference will be observed between the A_1A_1 and A_1A_2 genotypes, thus yielding a phenotypic segregation ratio of $3:1$.

Table 3.1
F_2 offspring from a monohybrid F_1 cross

		Sperm	
		½ A_1	½ A_2
Eggs	½ A_1	¼ A_1A_1	¼ A_1A_2
	½ A_2	¼ A_1A_2	¼ A_2A_2

Dihybrid Experiments

Mendel also experimented with crosses between varieties of plants differing with respect to two traits, resulting in *dihybrid segregation ratios* in the F_2 population. Box 3.2 contains a short excerpt from Mendel's paper detailing some of these results. As described in Box 3.2, Mendel crossed truebreeding

---

Box 3.2

Mendel's Classic Paper: Dihybrid Results

Box 3.1 contains quotations from Mendel's presentation of his monohybrid results. The following is an excerpt from his discussion of the results when two traits were investigated simultaneously.* The concluding sentence is his statement of the law of independent assortment:

THE OFFSPRING OF HYBRIDS IN WHICH SEVERAL DIFFERENTIATING
CHARACTERS ARE ASSOCIATED

In the experiments above described plants were used which differed only in one essential character. The next task consisted in ascertaining whether the law of development discovered in these applied to each pair of differentiating characters when several diverse characters are united in the hybrid by crossing. . . .

In order to facilitate the study of the data in these experiments, the different characters of the seed plant will be indicated by A, B, C, those of the pollen plant by a, b, c, and the hybrid forms of the characters by Aa, Bb, and Cc.

Expt. 1.—AB, seed parents;
A,   form round;
B,   albumen yellow.
ab,   pollen parents;
a,   form wrinkled;
b,   albumen green.

The fertilised seeds appeared round and yellow like those of the seed parents. The plants raised therefrom yielded seeds of four sorts, which frequently presented themselves in one pod. In all, 556 seeds were yielded by 15 plants, and of these there were:

315 round and yellow,
101 wrinkled and yellow,
108 round and green,
  32 wrinkled and green. . . .

. . . *the relation of each pair of different characters in hybrid union is independent of the other differences in the two original parental stocks.*

*Translated by William Bateson and the Royal Horticultural Society of London. The original paper was published in *Verh. Naturf. Ver. in Brünn,* 1866.

plants that had both yellow cotyledons (the inside of the seed, which Mendel calls albumen) and round seeds with truebreeding plants that had green cotyledons and wrinkled seeds. The F_1 seeds were all round with yellow cotyledons, because roundness and yellowness dominate. The exciting question concerned the F_2. Would yellowness and roundness be transmitted as a package, or would they be inherited independently? When he opened the pods of the F_2, he found that about $3/16$ of the seeds were yellow and wrinkled, and a similar number were green and round. Thus, Mendel concluded that the two traits were inherited independently. The observed ratio was $9:3:3:1$. (See Box 3.2.) That is, $9/16$ were dominant for both traits (yellow and round), $3/16$ were dominant for color and recessive for texture, $3/16$ were recessive for color and dominant for texture, and $1/16$ were recessive for both traits. This ratio is now known as the *dihybrid phenotypic segregation ratio.*

These findings are summarized by Mendel's second law, the *law of independent assortment:* When two traits are inherited, the alleles for each gene assort independently of the other gene. In other words, the alleles of each of the genes segregate as they would have in a monohybrid cross. For example, $12/16$ of the seeds were yellow and $4/16$ were green, which is the $3:1$ monohybrid segregation ratio for color of the cotyledon.

If we generalize the dihybrid cross to eggs and sperm, we see in Table 3.2 that the dihybrid individuals (for traits A and B) of the F_1 generation produce four different kinds of gametes in equal frequencies: $1/4\ A_1B_1$, $1/4\ A_1B_2$, $1/4\ A_2B_1$, and $1/4\ A_2B_2$. Because the two genes, A and B, are transmitted independently, we can determine their joint frequency by multiplying their separate probabilities. For example, the probability of having an A_1 allele in the sperm of an A_1A_2 individual of the F_1 generation is $1/2$. (See Table 3.1.) For the B gene, the probability of having a B_1 allele is also $1/2$. Thus, the probability that F_1 sex cells contain both A_1 and B_1 is the product of their respective probabilities: $1/2 \times 1/2 = 1/4$, as indicated in Table 3.2.

Because the F_2 generation results from crosses or self-pollination of F_1 individuals, we can determine the kinds and expected frequencies of F_2 offspring from a consideration of the sex cell combinations illustrated in Table 3.2. This results in a table with 16 cells, with some of the same

Table 3.2
F_2 offspring from a dihybrid F_1 cross

		Sperm			
		$1/4\ A_1B_1$	$1/4\ A_1B_2$	$1/4\ A_2B_1$	$1/4\ A_2B_2$
	$1/4\ A_1B_1$	$1/16\ A_1A_1B_1B_1$	$1/16\ A_1A_1B_1B_2$	$1/16\ A_1A_2B_1B_1$	$1/16\ A_1A_2B_1B_2$
	$1/4\ A_1B_2$	$1/16\ A_1A_1B_1B_2$	$1/16\ A_1A_1B_2B_2$	$1/16\ A_1A_2B_1B_2$	$1/16\ A_1A_2B_2B_2$
Eggs	$1/4\ A_2B_1$	$1/16\ A_1A_2B_1B_1$	$1/16\ A_1A_2B_1B_2$	$1/16\ A_2A_2B_1B_1$	$1/16\ A_2A_2B_1B_2$
	$1/4\ A_2B_2$	$1/16\ A_1A_2B_1B_2$	$1/16\ A_1A_2B_2B_2$	$1/16\ A_2A_2B_1B_2$	$1/16\ A_2A_2B_2B_2$

genotypes appearing in more than one cell. For example, there are two cells, each of which contains $\frac{1}{16} A_1A_1B_1B_2$. Notice that the genotypes add up to the $9:3:3:1$ dihybrid phenotypic segregation ratio.

Mendel was lucky in his studies of monohybrid ratios to find single-gene, two-allele traits that operate with complete dominance. There are many other traits, such as the size of the seed, that are influenced by more than one gene. And there are many genes with more than two possible alleles. These complications would have made Mendel's search for laws of inheritance much more difficult. His luck held when he considered dihybrid segregation ratios. We know that genes are not just floating around in eggs and sperm. They are carried on *chromosomes*, which literally means "colored bodies" because they stain differently from the rest of the nucleus of the cell. Genes are located at places called *loci* (singular: *locus*) on chromosomes. Eggs contain one chromosome from each pair of the mother's chromosomes and sperm contain one from each pair of the father's. An egg fertilized by a sperm thus has the full chromosome complement — in humans, 23 pairs of chromosomes. If Mendel had studied two genes that were close together on the same chromosome, the results would have surprised him. The two traits would not have been inherited independently. In fact, if they had been very close together on the same chromosome, Mendel would not have found yellow, wrinkled seeds nor round, green seeds. We will discuss this issue in more detail in Chapter 5.

When Mendel wrote the paper about his theory of inheritance in 1865, reprints were sent to scientists and libraries in Europe and the United States, and one even landed in Darwin's office. However, Mendel's findings on the pea plant were ignored by most biologists, who were more interested in the evolution of higher animals. Mendel died in 1884, without knowing the profound impact that his experiments would have during the twentieth century.

Mendelian crosses and expected segregation ratios are widely used today, primarily to determine whether a particular phenotype is influenced by a single gene. Although complex behaviors such as twirling, squeaking, and audiogenic seizures in mice or mental retardation in humans, seem far removed from pea seeds, the laws of heredity discovered by Mendel apply to behavior as well as to peas. We shall now consider single-gene influences for a few behaviors in human and nonhuman animals.

DISTRIBUTIONS

If a particular behavior is primarily influenced by a single gene, it can be expected to display an "either–or" expression. For example, Mendel considered his peas to be either round or wrinkled. This sort of distribution is called qualitative, or discontinuous. However, most behaviors are not distributed in this either–or fashion. Instead, they show smooth, continuous distributions.

This may be because these behaviors are influenced by many genes, each having small effects, as well as by many environmental factors. For such behaviors, simple Mendelian crosses that assume a single gene are not appropriate. New linkage techniques, described in Chapters 5 and 7, can identify single-gene effects even when many genes affect a trait.

One example of an approximate either–or distribution in humans is the ability to taste a bitter substance called phenylthiocarbamide (PTC). It can be tasted by 70 percent of the Caucasians in the United States. However, it is not quite as simple as saying that someone can either taste PTC or not; it depends on the concentration. Taste sensitivity to PTC and related compounds can be assessed by placing a small drop of a test solution on subjects' tongues and asking if they can taste it. A variety of concentrations is usually administered so that a taste threshold can be established for each subject. While some people can taste highly dilute solutions of PTC, others can distinguish only very strong solutions from plain water. Distributions of taste thresholds for English, Africans, and Chinese individuals are presented in Figure 3.3. Individuals who could taste solution number 13, the most dilute solution, and all stronger solutions were most sensitive and thus had the lowest taste threshold. Individuals who could not even taste solution number 1, the strongest solution, were least sensitive.

Unlike most behavioral characters, PTC tasting is not normally distributed with most people scoring at the average. Instead, the distributions, particularly for the English, have two humps (called a bimodal distribution) —one for high thresholds, and one for low thresholds. Variation within the taster and nontaster categories could be due to segregation at loci that have

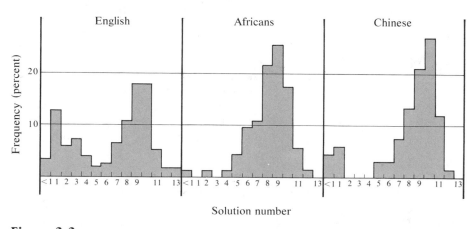

Figure 3.3
Distributions of taste thresholds for PTC in 155 English males, 74 Africans, and 66 Chinese. The strongest solution (1) had a concentration of 0.13 percent in water, the next (2) half this strength, and so on through 13. (From "Taste deficiency for phenylthiourea in African Negroes and Chinese" by N. A. Barnicot. *Annals of Eugenics*, 1950, 15, 248–254. Reprinted by permission of Cambridge University Press.)

relatively small effects on taste sensitivity to PTC, to environmental variation, or to error of measurement. This bimodality suggests that a single major gene may underlie PTC tasting. Simple Mendelian crosses, as described in the next section, support the hypothesis that it is a single-locus, two-allele character, with the allele for tasting PTC dominant. Analyses of PTC tasting have been conducted with mice (Whitney and Harder, 1986); mouse research on tasting another bitter substance is described later. Other examples of discontinuous distributions include genetic diseases with behavioral effects. For example, PKU is a single-gene, recessive condition that causes severe retardation if untreated, as discussed later in this chapter.

Although most behavioral characters show smooth, quantitative variation, rather than qualitative variation, this does not rule out the possibility of single major genes. A major gene may be at work, although its role may be diminished by genetic and environmental effects that cause quantitative variation. A related factor is that genotypes do not always express themselves in the same way in all individuals, due to the complexity of the developmental pathways by which genes are expressed in the phenotypes. Characters that are not expressed in all individuals with the appropriate genotype are said to be *incompletely penetrant*. Those genotypes that are expressed to varying degrees are said to display *variable expressivity*. Although not fully understood, incomplete penetrance and variable expressivity are presumably due to the effects of either environment or other genes. In any case, these phenomena can cloud the effects of single genes.

However, the major difficulty in ascertaining single-gene influences in behavior may be a measurement problem. The phenotypes that behavioral scientists study are often heterogeneous collections of diverse and complicated behaviors. For example, it is unlikely that a single gene can be found for complex behavioral problems such as reading disability or hyperactivity. With cancer research, we no longer talk about the cause or the cure of cancer, but rather the causes and cures of cancers, because we know that there are many different kinds of cancer. It is safe to assume that many of our behavioral phenotypes are similarly heterogeneous.

In summary, bimodal distributions are not necessary to demonstrate a single-gene effect, although it is exciting to find such distributions. A better test of single-gene influence is *segregation analysis,* which essentially looks for Mendelian segregation expected from family data. Earlier segregation analyses assumed qualitative distributions (Morton, 1958). Current sophisticated models can detect major genes for quantitative variables (Elston and Stewart, 1971; Lalouel et al., 1983). For an introduction to segregation analyses and their application to learning disabilities, see Smith and Goldgar (1986). If segregation analysis suggests single-gene influences, linkage analyses, discussed in Chapter 5, are appropriate to pinpoint the gene on a specific chromosome. Indeed, new genetic techniques described in Chapter 7 have so greatly improved the detection of linkage that linkage is now the major way in which single-gene influences are found, even if segregation analysis is unable to show a major gene effect.

MENDELIAN CROSSES IN MICE

With experimental animals, we can conduct behavioral studies parallel to Mendel's studies of pea plant characteristics. Crosses between truebreeding parental populations yield an F_1 generation, and the members of the F_1 generation can be crossed to produce an F_2. One of the favorite subjects of behavioral genetics research is the mouse.

The Genetics of Waltzer Mice

Many studies have been made concerning the effects of a single-gene mutation on mouse behavior. For example, one of the earliest behavioral conditions studied in mammals was that of "waltzing" in mice. In spite of the name, animals exhibiting this behavior are quite ungraceful. They shake their heads, circle rapidly, and are very irritable. Waltzer mice were prized by mouse fanciers and imported to Europe and North America from Asia around 1890 (Gruneberg, 1952). Several different waltzer conditions are known. For example, the Nijmegen waltzer (van Abeelen and van der Kroon, 1967) runs in tight circles in both directions with both horizontal and vertical head shaking. Researchers wondered whether this was the result of a single gene. When waltzer males were crossed with waltzer females, all offspring were waltzers, suggesting that the waltzers are a truebreeding population. Waltzers were then crossed with a nonwaltzing population to produce an F_1 generation. Of the 254 offspring, all were nonwaltzing. This suggests that, if a single gene is operating, the allele for waltzing is recessive to the normal allele.

If a single recessive allele causes waltzing, the F_2 generation should yield the typical Mendelian segregation ratio of ¾ nonwaltzer and ¼ waltzer. As shown in Table 3.3, the data from the F_2 are very close to the expected 3 : 1 ratio.

There are over 300 other examples of inherited neurological defects in mice. Most appear to be due to single-locus, recessive genes. However, some are caused by dominant genes, and a few appear to be due to the combined

Table 3.3
Data observed and expected on the basis of a single-locus recessive model for F_2 Nijmegen waltzer mice

	Normal	Mutant
Observed	124	47
Expected	128.25	42.75

SOURCE: After van Abeelen and von der Kroon, 1967.

effects of genes at several loci. However, the study of behavioral mutants in mice has been eclipsed in recent years by induced mutation and mass screening of behavioral mutations in single-celled organisms, such as bacteria and paramecia, and in invertebrate organisms, such as round worms and fruit flies. These will be discussed in the next chapter because they illustrate the chemical nature of genes as it affects behavior. We shall now consider a couple of additional examples of effects of single-gene mutations on mouse behavior in order to introduce other concepts of Mendelian analysis.

The Genetics of Twirler Mice

Exact Mendelian ratios are not always observed, even for single-locus characters. For example, the "twirler" mouse was first described by Mary Lyon (1958) as shaking its head frequently in a horizontal plane and circling. Evidence indicated that this condition was probably due to a single dominant allele. One test of this hypothesis is to cross twirler males and twirler females. If a single dominant allele is at work, twirlers (born to twirler, crossed with nontwirler, parents) are heterozygotes, and this cross should be like an F_1 cross. Their offspring should show the typical $3:1$ F_2 segregation ratio. However, the results of the cross yielded 84 twirler offspring and 58 normal offspring. Since a $3:1$ segregation ratio would produce 106.5 twirlers and 35.5 normal offspring, the observed ratio departs significantly from these expectations.

However, dominant mutant genes are often lethal in the homozygous condition. If individuals having the homozygous genotype die before investigators observe their behavior, the expected ratio among offspring of a monohybrid cross would be 2 twirlers to 1 normal instead of $3:1$. When this hypothesis is tested, a slight deficiency of twirler offspring is still noted, but the departure from expectation is not sufficient to be statistically significant. Subsequent research supports this hypothesis. About 25 percent of newborn offspring resulting from twirler crosses have a cleft palate or cleft lip and palate. These pups die within 24 hours of birth and consequently would not be observed for behavioral abnormalities. These mice are believed to be the missing homozygotes, accounting for $2:1$, rather than the $3:1$, ratio.

The Genetics of Audiogenic Seizures in Mice

Two other points about single-gene analyses can be made based on the example of sound-induced seizures in mice. Some mice respond to high-frequency sound with wild running, convulsions, and even death; others are apparently unaffected by it. The first point is that behavioral genetics analysis is only as reliable as the behavioral measurements on which it is based. The second point is that Mendelian crosses need not be limited to an analysis of the F_1 and F_2 generations. Predictions can be made to test the hypothesis of a single gene using other crosses, such as a cross between the F_1 generation and the parental population.

Earlier work suggested various modes of inheritance for seizure susceptibility. Researchers suggested dominance (Witt and Hall, 1949), a two-gene model (Ginsburg and Miller, 1963), and a polygenic model (Fuller, Easler, and Smith, 1950). However, a measurement problem, due to the fact that subjects were tested repeatedly, may have obscured some of the earlier results. Animals from a seizure-resistant mouse strain become susceptible to seizures if exposed to a loud sound at an early age (Henry, 1967). Thus, repeated testing of animals may have confounded the response to the first noise and the response to later noises.

When the effects of initial exposure to a loud noise are measured, the situation is greatly clarified (Collins, 1970; Collins and Fuller, 1968). Mice of the C57BL/6J strain, a strain whose members rarely convulse on initial exposure to a loud noise, were crossed with mice of the DBA/2J strain, whose members almost always convulse on initial exposure. As illustrated in Figure 3.4, several additional crosses were conducted in order to provide more extensive analyses. The F_1 animals were crossed back to the parental lines. These crosses are appropriately called backcrosses. B_1 is the backcross to one parental line (P_1, for example, the C57BL strain) and B_2 is the backcross to P_2 (DBA). An F_2 generation was also obtained, as well as crosses between the two backcross generations and the F_1 generation, yielding generations symbolized as B_1F_1 and B_2F_1. The two backcross generations were also crossed yielding B_1B_2.

Mice from these nine groups were individually tested for initial seizure susceptibility at 3 weeks of age. Each subject was placed in a box and exposed to an electric bell that was rung until the onset of a convulsion or for a

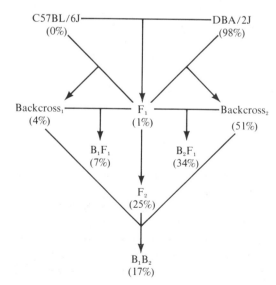

Figure 3.4
Various crosses used to test a single-gene model for audiogenic seizures in mice. Percentages in parentheses refer to the data in Table 3.4.

Table 3.4
Summary of data and genetic analysis for the incidence of initial audiogenic seizures in mice.

Generation	Number of Subjects	Proportion Observed to Have Seizures	Proportion Expected to Have Seizures (Single-Locus Model)
P_1 (C57BL/6J)	45	0.000	
P_2(DBA/2J)	58	0.983	
F_1	89	0.011	
B_1	115	0.035	0.006
B_2	119	0.513	0.497
F_2	105	0.247	0.251
B_1F_1	128	0.070	0.128
B_2F_1	96	0.344	0.374
B_1B_2	185	0.168	0.191

SOURCE: From Collins and Fuller, "Audiogenic seizure prone (asp): a gene affecting behavior in linkage group VIII of the mouse," *Science*, 162, 1137–1139, 1968. Copyright © 1968 by the American Association for the Advancement of Science.

maximum of 1 minute. The proportions of seizures observed and expected, assuming a single-locus model, are presented in Table 3.4. As indicated in this table, C57BL mice had no seizures, almost no F_1 mice had seizures, but almost all DBA mice had seizures. These results suggest that susceptibility to audiogenic seizure on initial exposure to the loud noise may be determined at a single locus by a recessive allele.

In order to test the adequacy of this single-locus, recessive model, it may be hypothesized that (1) DBA mice are homozygous recessive and thus prone to audiogenic seizures, (2) C57BL mice are homozygous dominant, and (3) F_1 mice are heterozygous. The observed proportion of seizures by these geno-types can be used to predict the proportion having seizures in each of the six segregating generations. For example, in the F_2 generation the genotypic segregation ratio should be ¼ homozygous dominant (with a seizure frequency of 0.000) to ½ heterozygous (with a seizure frequency of 0.011) to ¼ homozygous recessive (with a seizure frequency of 0.983). Thus, the expected proportion having seizures in the F_2 generation can be calculated as follows: ¼(0.000) + ½(0.011) + ¼(0.983) = 0.251. It may be seen in Table 3.4 that the results conform closely to those expected on the basis of the model. These results are consistent with the hypotheses that the difference between these two strains in their susceptibility to audiogenic seizures is due to a single gene and that seizure susceptibility is a recessive trait.

Although these results are consistent with a single-gene hypothesis, they are also consistent with other models such as a two-gene model. A two-gene model would also predict these results if the effects of the alleles at the two loci simply added to produce the effect. A two-gene model would hypothesize that the C57BL mice are of genotype $A_1A_1B_1B_1$ and the DBA are $A_2A_2B_2B_2$. The predictions for this two-gene model would be identical to those for the single-gene model. Thus, although the data are compatible with a single-gene model, they also fit other models. One way to provide more definitive support for the single-gene model is to study the prevalence of the behavior in various inbred strains or in recombinant inbred strains. These methods are discussed in the following sections. As discussed in the section on recombinant inbred strains, evidence from a study using this technique suggests that the difference between the C57BL and DBA strains in susceptibility to audiogenic seizures is due to more than one gene.

The Genetics of Tasting Bitterness in Mice

Inheritance of the ability of some mice to taste a bitter substance called sucrose octaacetate (SOA) illustrates a more recent test of a single-gene hypothesis (Whitney and Harder, 1986). C57BL mice do not show a preference between water and SOA-flavored water, whereas mice of the SWR strain strongly avoid drinking SOA. Table 3.5 summarizes the results of the various crosses. F_1 mice from the cross between C57BL and SWR strains nearly all prefer SOA, suggesting complete dominance for tasting SOA. The observed proportions in the backcross and F_2 generations agree closely with the results expected for a dominant single-gene model. Tasting of several other substances, such as strychnine and quinine, also appear to show single-gene influences (Lush, 1984).

Table 3.5
Genetic analysis of SOA tasting in mice

Generation	Number of Subjects	Proportion Observed to Taste	Proportion Expected to Taste (Single-Locus Model)
P_1 (SWR)	97	0.97	1.00
P_2 (C57BL)	73	0.00	0.00
F_1	81	0.98	1.00
B_1	629	0.51	0.50
F_2	199	0.71	0.75

SOURCE: After Whitney and Harder, 1986.

Behavioral Pleiotropism

A general rule is that a single gene affects more than just a single phenotype. *Pleiotropy* is the name given to the multiple effects of a gene. For example, PKU is attributed to a single gene most noted for causing mental retardation. However, PKU individuals also tend to have lighter hair and skin color. Genetic influences on most behaviors are likely to be pleiotropic. In 1915 the first study of behavioral pleiotropism was provided by the geneticist A. H. Sturtevant, inventor of the chromosome map. He found that a single-gene mutation that alters eye color or body color in the fruit fly *Drosophila* also affects their mating behavior.

One of the most often studied examples of behavioral pleiotropism concerns the relationship between albinism in mice and their activity in an open field. The open-field test was first employed by Calvin Hall (1934) to provide an objective index of emotionality. The test involves placing an animal in a brightly lit enclosure and scoring its behavior. An "automated" open field is shown in Figure 3.5. Some animals freeze, defecate, and urinate

Figure 3.5
Mouse in an open field employed by DeFries et al. (1966). The holes near the floor transmit light beams that electronically record an animal's activity. (Courtesy of E. A. Thomas.)

when placed in this presumably stressful situation, whereas others actively explore it. Animals that have relatively low activity and high defecation scores are called "emotional" or "reactive," and those with relatively high activity and low defecation scores are considered "nonemotional" or "nonreactive." The evidence for the validity of this measure has been discussed in some detail (Broadhurst, 1960; Eysenck and Broadhurst, 1964).

It had been known for some time that several albino strains of mice obtain relatively lower activity and higher defecation scores than do pigmented mice. However, because these strains differ from one another at many loci and because some of the albino strains are closely related, it was generally assumed that the differences observed in open-field tests were due to gene differences other than those determining coat color. That is, it was assumed that the observed correlation between albinism and open-field behavior was not causal.

An extensive genetic analysis of open-field behavior initiated by DeFries determined that the greater emotionality of the albino mice was, in fact, due in part to the gene for albinism. In other words, the single gene that determines coat color (called the c locus) has a pleiotropic effect on open-field behavior. But how do we know that the c locus itself, rather than some other locus close to the c locus, is responsible for greater emotionality? One answer is to break up any possible linkages. As we shall see in Chapter 5, nature does this for us through a process called recombination, in which chromosome pairs exchange parts.

One method to ensure that linkages are broken up involves using heterogeneous animals (HS) derived from, for example, an eight-way cross of inbred strains (McClearn, Wilson, and Meredith, 1970). Even using HS animals, it is still possible that certain linkages might not be separated by recombination. The strongest evidence for behavioral pleiotropism is provided by comparisons of individuals with exactly the same genotype except for a newly arisen mutation. When a new mutation arises and is maintained within an inbred strain, mutant and nonmutant subjects within the strain are called *coisogenic* (Green, 1966). An albino mutant from the pigmented C57BL strain was bred to produce a coisogenic albino strain. The two strains have the same genotype except for the single locus determining coat color. If the greater "emotionality" of albino mice is a pleiotropic effect of the c locus itself, then these coisogenic albino mutants should differ from the pigmented C57BL mice. The answer is in favor of the pleiotropism hypothesis: The albino mutants were less active in the open field than the pigmented coisogenic C57BL mice (Henry and Schlesinger, 1967).

Additional analyses suggest that the pleiotropic effect of the c locus on open-field behavior is mediated by the visual system. McClearn (1960) observed that the difference in open-field behavior between an albino strain and a pigmented strain was less under a red light, which reduces visual stimulation. This led to the hypothesis that the albinos are actually afraid of the light in the open field; they may be photophobic. Data from albino and pigmented mice from the same litters support the hypothesis. (See Figure 3.6.) Under the

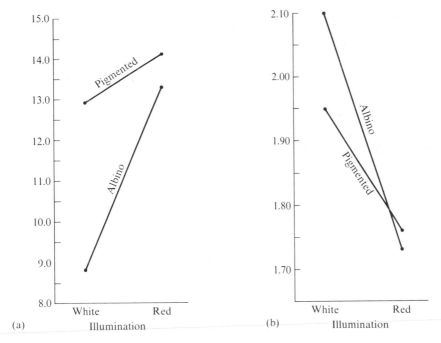

Figure 3.6
Open-field (a) activity and (b) defecation scores of mice tested under different illuminations. (From DeFries et al., 1966.)

red light there is no significant difference between albino and pigmented mice in open-field activity or defecation (DeFries, Hegmann, and Weir, 1966).

The results of this study demonstrate that albino mice have lower activity and higher defecation scores than pigmented mice when tested under white light, and that this difference largely disappears when these subjects are tested under red illumination. It may be concluded that this single-gene pleiotropic effect is mediated through the visual system. If we accept the interpretation that a pattern of low activity and high defecation indicates heightened "emotionality," albinos may be regarded as being more emotional than pigmented animals. Of course, many other genes influence a behavior as complex as open-field behavior, as discussed in Chapter 10.

RECOMBINANT INBRED STRAINS

Another strategy to uncover single-gene effects in behavior is called the recombinant inbred strain method (Bailey, 1971, 1981). As illustrated in Figure 3.7, it begins with a classical Mendelian cross between two inbred strains. An F_1 generation is produced, followed by F_2. In the F_2, genes assort according to Mendel's laws. Recombinant inbred (RI) strains are different

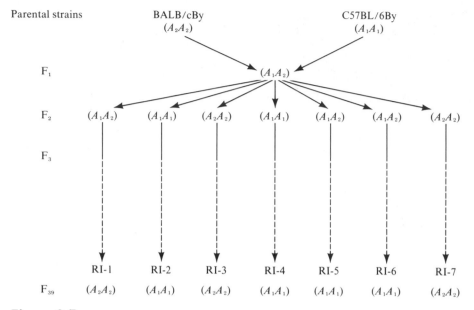

Figure 3.7
The derivation of recombinant inbred (RI) strains by Bailey (1971). In parentheses is an example of a locus, A, for which the parental strains differ. This example illustrates the results presented in Table 3.6, and is explained in the text.

inbred strains that were derived from separate brother–sister pairs from the same F_2 generation. They are called recombinant inbred strains because they are derived from the same F_2 generation after there has been recombination of parts of chromosomes from the parental strains. Like all inbred strains (as described in Chapter 10), RI strains are made homozygous at almost all loci by inbreeding using brother–sister matings over many generations.

How do RI strains derived from the same F_2 generation differ? For those loci with different alleles in the two parental strains, RI strains will show different combinations of alleles. For example, suppose the parental strains had different alleles at the A locus. (See Figure 3.7.) The F_1 will have one allele from each parent, but the F_2 will show segregation. The F_2 consists of homozygous and heterozygous individuals. However, the process of inbreeding increases homozygosity. Thus, each RI strain derived from the F_2 will be either A_1A_1 or A_2A_2. The same will be true for other loci. When we consider two loci at a time (the A and B loci), only ¼ of the RI strains will be $A_1A_1B_1B_1$, ¼ will be $A_1A_1B_2B_2$, ¼ $A_2A_2B_1B_1$, and ¼ $A_2A_2B_2B_2$. In other words, although the RI strains are homozygous at all loci, they represent new combinations of genes at different loci. While each RI strain carries half the genes of each parental strain, there is little chance that the genotype of an RI strain will be the same as that of either parental strain.

These RI strains can be used to determine whether any behavioral differences between the two parental strains are due to single major genes. If a single gene is responsible, each of the RI strains should be just like one parent or the other. In other words, there should be no intermediate phenotypes if just one locus is involved, because each RI strain will be homozygous for the allele of one of the two parental strains. However, if more than one locus is involved in a behavioral difference between two parental strains, more than two kinds of RI strains will be found. Furthermore, if there is just one gene responsible for the behavioral difference, then about half of the RI strains should be like one parent and half like the other, because each RI has an even chance of receiving the allele from either parent.

D. W. Bailey (1971) crossed the BALB/cBy and C57BL/6By inbred strains, which differ for many behavioral traits. Seven highly inbred recombinant strains produced from the F_2 generation of this cross (see Figure 3.7) have since been used in many experiments. One behavioral example concerns a major gene effect for learning to avoid shock in a shuttle box (Oliverio, Eleftheriou, and Bailey, 1973). Mice can be trained to avoid a shock by changing compartments as soon as a light is turned on. C57BL mice perform this task poorly, avoiding the shock only 14 percent of the time. The seven RI strains are either low performers like the C57BL mice or high performers like the BALB mice, as indicated in Table 3.6. These results suggest that a single gene may underlie the marked difference in avoidance learning between BALB and C57BL mice. If the RI strains had been intermediate to the two progenitor strains, *polygenic* (multigene) influences would have been implicated. For example, this was the result found in an RI study of susceptibility to audiogenic seizures (Seyfried, Yu, and Glaser, 1979). Using 21 RI strains derived from C57BL and DBA progenitors, 13 of the 21 RI strains were intermediate to the C57BL and DBA progenitor strains. This finding suggests susceptibility to audiogenic seizures for these two strains is due to polygenic influences, even though the classical Mendelian crosses were consistent with a single-gene model.

The SOA-tasting classical Mendelian results mentioned earlier have been supported by RI analysis: For six RI lines derived from nontasting and tasting strains, four lines showed nearly complete aversion to SOA and two

Table 3.6
Percent avoidance over 250 trials (50 trials per day for 5 days) for BALB and C57BL mice and 7 recombinant inbred strains

Low Performers		High Performers	
Strain	Avoidance (%)	Strain	Avoidance (%)
C57BL	14	BALB/c	46
RI-2	13	RI-1	60
RI-4	17	RI-3	36
RI-5	7	RI-7	58
RI-6	8		

SOURCE: After Oliverio et al., 1973.

lines were unable to taste SOA, as illustrated in Figure 3.8 (Whitney and Harder, 1986).

Although there are other ways to search for single-gene effects, the RI method is useful in screening behavioral differences between two progenitor strains. In addition to the RI strains, *congenic* strains have been created that are the same as one of the progenitor strains except for a small segment of chromosome derived from the other. This is accomplished by repeated backcrossing of mice with a particular gene to an inbred strain. The resulting substrain is called *congenic* because it is genetically identical to the inbred strain except that it contains the transferred gene and genes closely linked to it (Bailey, 1981).

The use of congenic strains and other comparisons using RI strains facilitate locating single-gene influences on specific chromosomes (Eleftheriou and Elias, 1975). The pattern of single-gene characteristics whose location on chromosomes is known can be compared to the pattern obtained for a particular behavioral trait in RI and congenic strains, called a *strain distribu-*

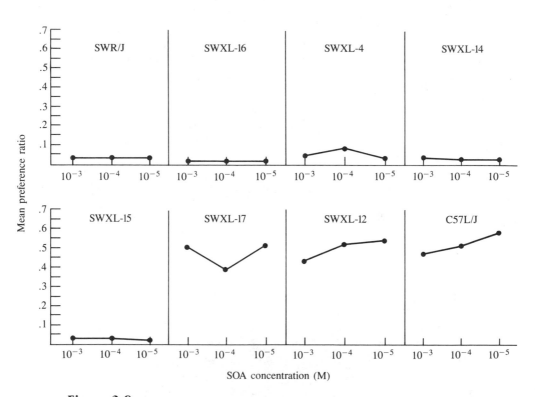

Figure 3.8
Mean preference ratios for six recombinant inbred strains, plus the SWR/J and C57L/J parental strains, across three SOA concentrations. (From Whitney and Harder, 1986).

tion pattern, in order to map the chromosomal location of the gene that affects the behavioral trait. For example, a major gene difference between the two progenitor strains for learning to avoid shock appears to be on chromosome number 9 (Oliverio, Eleftheriou, and Bailey, 1973). As we might expect, most of the behaviors submitted to RI analysis suggest polygenic effects (Broadhurst, 1978).

MENDELIAN ANALYSIS IN HUMANS

In many respects, the human species is an unfavorable population for genetic analysis: Experimental crosses cannot be performed, environmental control cannot be imposed, the generation interval is relatively long, and the number of offspring per family is relatively small. However, these problems are not insurmountable. Because the human population is large, a rich store of genetic information is potentially available. Although planned Mendelian crosses are not feasible, data can be collected from families in which particular types of mating have occurred. Although environmental control cannot be imposed, it is possible to study members of different families who have been reared in more or less similar environments. Although the human generation interval is long, data from several generations can nonetheless be obtained. Finally, although the number of children in human families is relatively small, data from many families can be pooled to provide adequate samples for statistical analysis. The fact that many important advances have occurred in human genetics within the past two decades demonstrates that human populations—like those of mice, fruit flies, bread mold, and colon bacteria —can also be subjected to genetic analysis.

Single-gene analyses of human behavior are principally concerned with testing the adequacy of single-locus hypotheses. Researchers first note familial transmission of some character of interest and then determine whether the observed pattern of transmission conforms to that expected on the basis of a simple Mendelian model. Pedigrees, or family trees, are frequently used to depict hereditary information. An example of a pedigree is given in Box 3.3

A pedigree analysis of many families can be used to determine whether a particular behavioral character conforms to Mendelian expectations. As in the examples of mouse behavior, we can test for the adequacy of a single-gene model, as well as the mode of transmission. The following sections consider the basic expectations for certain crosses for dominant and recessive transmission. The expectations are different for genes on the sex chromosomes, as compared to those on the other 22 pairs of human chromosomes, called *autosomes*. We shall first discuss autosomal dominant and recessive expectations and then consider transmission of single genes on the sex chromosomes. Because specific predictions can be made about single-gene characters, these expectations are of practical use in genetic counseling. (See Box 3.4.) Useful

Box 3.3

Pedigree Analysis

A sample *pedigree* (from *pié de grue*, "crane's foot," a three-line mark denoting succession) is illustrated below. Affected individuals—those manifesting the condition under study—are designated by solid symbols. Individuals who were probably affected are shown by a cross. Nonaffected individuals are indicated by open symbols. Females are represented by circles and males by squares. A diamond is used to represent an individual whose sex is unknown. Parents are joined by a horizontal *marriage line*, and offspring are listed below. Members of a *sibship* are connected to a horizontal line that is joined by a perpendicular line to their parents' marriage line. The *siblings* (brothers and sisters) are listed from left to right in order of birth.

In this hypothetical pedigree each generation is designated by a Roman numeral, and each individual within a generation is denoted by an Arabic numeral. For example, individuals II-5 and II-6 were siblings, but information concerning their parents was not included in the pedigree. The marriage of individuals II-2 and II-3 resulted in no children. In order to save space, a number enclosed in a large symbol can be used to indicate the number of siblings of like condition. Thus, III-1, III-2, and III-3 were unaffected males in a family of seven children. Twins are indicated by two symbols that are connected either at or just below the sibship line. Individuals III-4 and III-5 were *monozygotic* (identical) twins, indicated by the short vertical line that descends from the sibship. Individuals III-6 and III-7 were *dizygotic* (fraternal) twins. The finding of a family of interest frequently comes only after the discovery of a particular affected individual. This specific individual who first comes to the attention of the investigator is referred to as the *index case* (or the *propositus* or *proband*). The index case, individual III-6, is indicated by an arrow.

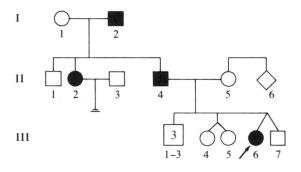

Box 3.4

Genetic Counseling

The need for genetic counseling is clear. About 20 percent of infant deaths are now attributed to genetic defects (Porter, 1977), and 3 percent of all newborns have some genetic birth defect. About 1 in 200 have a chromosomal abnormality (see Chapter 6), about 1 in 250 have a single-gene defect, and at least 1 to 2 percent have problems that involve polygenic influences (Epstein and Golbus, 1977). Genetic counseling is becoming an important tool in preventive medicine (Thompson and Thompson, 1986).

There are three equally important steps in genetic counseling. First, a precise diagnosis must be made. Second, detailed pedigrees must be obtained. Third, the expected risk and burden must be carefully explained. The risk may be great but the burden small, as in the case of color blindness. But even when all the tools of single-gene analysis are used to make the best possible estimate of risk, predictions are often hampered by incomplete information and reduced penetrance.

An exciting new area is prenatal diagnosis involving amniocentesis (described in Box 6.1). This technique permits an analysis of enzyme deficiencies for as many as 75 single-gene diseases, as well as a diagnosis of any chromosomal abnormalities of the fetus. Another promising approach to genetic counseling involves screening populations for particular single-gene problems. In the past, most people who sought genetic counseling did so because they or someone in their family already had an affected child. Now, for some single-gene, recessive diseases, it is possible to detect carriers as well as to diagnose the disease prenatally. It is important to identify carriers, since 25 percent of the offspring will be affected in matings between two carriers. A number of cities have established programs to identify carriers of Tay-Sachs disease (infantile amaurotic idiocy, discussed in the next chapter). In California, 33,000 carriers were detected and 350 couples were identified who were both carriers (Epstein and Golbus, 1977). For those couples, prenatal diagnosis can determine whether a particular pregnancy will result in a Tay-Sachs child. Sickle-cell disease is another example of a single-gene, recessive condition that can be screened on a large scale, and can be subjected to prenatal diagnosis.

Another preventive approach is to screen infants, as has been discussed in the case of PKU. PKU screening laws exist in over 40 states and, as a result, 90 percent of PKU babies are detected before they leave the hospital. Twenty million people have been screened for PKU, and about 1,500 infants with PKU have been discovered (Scriver, 1977). In addition, 14 states have programs for screening newborns for diseases other than PKU (Culliton, 1976). In 1974, New York passed a bill that mandates that hospitals must screen for abnormalities if they have the know-how. There are arguments against such screening, however. For example, screening for sickle-cell anemia at birth is questionable because there is no known treatment. On the other hand, carriers can be detected in this way.

Important new information often creates ethical dilemmas (Omenn, 1978). Genetic counseling abounds with both new information and ethical problems. Discussions of the ethical issues are complex in that they require a balance between the rights of a fetus to be born healthy, the rights of parents, and the rights of society. A negative tone often pervades such discussions because of the specter of abortion. However, in the vast majority of cases, genetic counseling provides positive information for parents. For example, 98 percent of the pregnant women who go through amniocentesis are relieved to find that their fetus has no chromosomal abnormality. Although many parents fearing a genetic disorder in their families might choose not to have children, genetic counseling and prenatal diagnosis may be able to relieve them of that fear.

introductions to the psychological and medical aspects of genetic counseling are available (Kessler, 1979; Thompson and Thompson, 1986).

Dominant Transmission

Testing particular hypotheses about the transmission of single genes is really a matter of logic. Table 3.7 lists the expected outcomes of certain crosses if a character is influenced by a single dominant gene. By definition, only one allele (*A*) will cause an individual to be affected. Thus, the only way an offspring can be affected is if at least one of the parents is affected. In other words, we should not find affected offspring with unaffected parents (Table 3.7[1]). Also, about half of the offspring of an affected parent should be affected, because each offspring has a 50-50 chance of inheriting that allele from the affected parent. We can determine whether an affected parent is a heterozygote or a homozygote by studying the parent's parents. If an affected individual has only one affected parent, then the individual must be heterozygous. As indicated in Table 3.7(2), half of the offspring of matings between affected heterozygotes and unaffected individuals are likely to be affected. Three quarters of the offspring will be affected in matings between two affected heterozygotes. All the offspring of an affected homozygote parent will be affected. In order to determine the mode of inheritance, pedigrees such as the simple ones in Figure 3.9 must be analyzed.

Many genetic defects are single-gene, recessive characters, but dominant ones are rare. They are quickly eliminated by natural selection, as discussed in Chapter 8. However, one well-known neural disorder, *Huntington's chorea*, is caused by a single dominant gene. Huntington's disease is characterized by loss of motor control and progressive deterioration of the central nervous system that begins with personality changes, forgetfulness, and other behavioral problems. When this condition was traced through many generations of pedigrees, a consistent pattern of transmission was observed. Most afflicted individuals had a parent who was also afflicted, and approximately half of the children of an affected parent eventually develop the disease. The persistence of this insidious dominant lethal gene in the population is due to the fact that the disease is not usually expressed until after the childbearing years. This late age of onset illustrates the principle that hereditary conditions are not always manifested at birth.

Because Huntington's disease is so clearly inherited as a single gene and because of the availability of large extended families with the disorder, it is the first disease whose gene was mapped to a specific chromosome using new recombinant DNA techniques discussed in Chapter 7. This discovery will eventually make it possible to isolate the Huntington's gene, determine the gene's product, and, it is hoped, intervene to prevent the onset of this horrible genetic disease.

Huntington's disease also illustrates the fact that the age of onset of a given condition may differ among individuals. The distribution of age of

Table 3.7
Offspring expected from various crosses for a completely dominant single autosomal gene (*A* refers to the dominant allele; thus, *Aa* and *AA* individuals are affected.)

1. Matings Between Two Unaffected Individuals

	Unaffected	
	a	*a*
Unaffected *a*	*aa*	*aa*
a	*aa*	*aa*

2. Matings Between a Heterozygote and an Unaffected Individual

	Affected	
	A	*a*
Unaffected *a*	*Aa*	*aa*
a	*Aa*	*aa*

(Shaded area indicates affected offspring.)

3. Matings Between Two Heterozygotes

	Affected	
	A	*a*
Affected *A*	*AA*	*Aa*
a	*Aa*	*aa*

4. Matings Between an Affected Homozygote and an Unaffected Individual

	Affected	
	A	*A*
Unaffected *a*	*Aa*	*Aa*
a	*Aa*	*Aa*

(a)   (b)

Figure 3.9

Hypothetical pedigrees. The condition indicated in (a) could be due to a recessive autosomal gene, but not a dominant; whereas that in (b) could be due to a dominant, but not a recessive.

onset of Huntington's chorea in 762 patients is plotted in Figure 3.10. The usual age of onset is between 30 and 50 years. Data of this type are very useful for obtaining age-corrected incidence data. As an oversimplified example, assume that only one half of individuals genetically predisposed to develop a disease actually express the disease by age 40. In such a case, the observed incidence of the disease among individuals at age 40 should be doubled to obtain age-corrected incidence data. Another condition with variable age of onset is schizophrenia, discussed in Chapters 11 to 13.

Recessive Transmission

In the case of single-gene, recessive conditions, affected individuals are homozygous recessive. Carriers are heterozygotes, and thus have only one of each allele. Table 3.8 presents the expectations for offspring of different

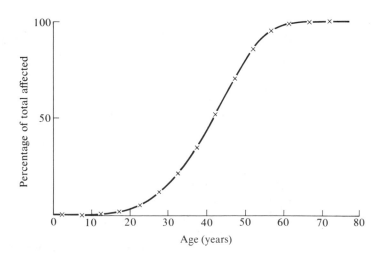

Figure 3.10

Huntington's chorea. Distribution of age of onset in 762 patients. (From *Principles of Human Genetics*, 3rd ed., by Curt Stern. W. H. Freeman and Company. Copyright © 1973.)

Table 3.8
Offspring expected from various crosses for a completely recessive, single autosomal gene (*a* refers to the recessive allele; thus, *aa* individuals are affected.)

1. Matings Between Two Carriers

		Carrier	
		A	*a*
Carrier	*A*	*AA*	*Aa*
	a	*Aa*	*aa*

(Shaded area indicates affected offspring.)

2. Matings Between an Affected Individual and an Unaffected Homozygote

		Affected	
		a	*a*
Unaffected	*A*	*Aa*	*Aa*
	A	*Aa*	*Aa*

3. Matings Between an Affected Individual and a Carrier

		Affected	
		a	*a*
Carrier	*A*	*Aa*	*Aa*
	a	*aa*	*aa*

4. Matings Between Two Affected Individuals

		Affected	
		a	*a*
Affected	*a*	*aa*	*aa*
	a	*aa*	*aa*

types of matings. The catch is that we often cannot tell whether an individual is an unaffected homozygote or a carrier. Natural selection cannot distinguish them either, which is the reason why recessively inherited problems persist. Sometimes pedigree information can distinguish carriers. At a biochemical level, carriers sometimes differ from unaffected dominant homozygotes and can be discriminated through the use of "carrier tests," as described in the next chapter.

For recessive characters, one commonly finds affected offspring with unaffected (carrier) parents (Table 3.8[1]). One quarter of the offspring of two carriers are affected. Also, as indicated in Table 3.8, none of the offspring of an affected parent is affected unless the spouse is affected or is at least a carrier. All the offspring of two affected individuals are affected. Another mark of a recessive character is that it occurs more frequently in offspring of matings between related individuals—that is, children of consanguineous marriages. This is called *inbreeding*, discussed in Chapter 8. Marriages between related individuals, such as cousins, are more likely to result in offspring with a double dose of recessive alleles that had been carried in the heterozygous state by the parents.

As mentioned previously, PKU is a type of severe mental retardation caused by a single, recessive gene. A pedigree for PKU is shown in Figure 3.11. Individuals marked with a cross were probably affected, but their condition was not known for certain because they died young. A cousin marriage is indicated in the middle of the pedigree. As these individuals all lived in an isolated group of small islands in Norway, there may actually have been more inbreeding than indicated by this pedigree. Although PKU may manifest incomplete penetrance, the pattern of transmission indicated in Figure 3.11 nonetheless conforms to that expected of an autosomal recessive gene, because affected subjects have normal parents and an increased incidence accompanies inbreeding. Of the 18 children in generation IV, 4 were definitely

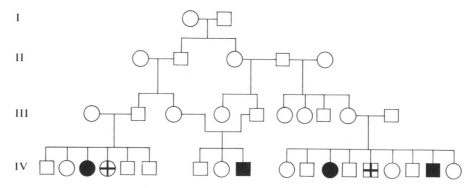

Figure 3.11
Pedigree of phenylketonuria and associated mental deficiency. For key to symbols, see Box 3.3. (From *Principles of Human Genetics*, 3rd ed., by Curt Stern. W. H. Freeman and Company. Copyright © 1973. After Følling, Mohr, and Ruud, 1945.)

affected and 2 were probably affected. If both unaffected parents of each sibship were carriers, as must be the case for a recessive condition, then only one fourth of the children would be expected to be affected. Although this departure from expectation (6 versus 4.5) is not significant, an excess of affected individuals is frequently observed in pedigree data because of a bias called *truncate selection.* Couples who are at risk of producing affected off-spring but who have not actually done so are excluded from pedigree data. This results in more affected individuals than would be expected from Mendelian calculations. PKU will be discussed in greater detail in the next chapter.

Sex-Linked Transmission

Pedigrees are also useful in detecting single genes located on the sex chromosomes. As discussed in Chapter 5, females have two large X chromosomes, whereas males have only one X and a small chromosome called Y. Recessive genes on the X chromosome express themselves in males when they have one such allele on their single X chromosome. Females, however, show the condition only when they have the allele on both of their X chromosomes. Thus, one of the first indications of sex-linked (meaning X-linked) recessive transmission is a greater incidence in males.

The transmission of the sex chromosomes from one generation to the next is portrayed in Figure 3.12. The figure shows that daughters inherit their father's X chromosome but sons do not. Thus, sons cannot inherit sex-linked conditions from their father. Daughters inherit sex-linked genes from their

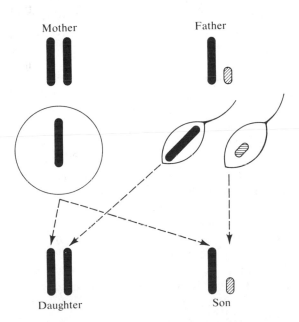

Mother Father

Daughter Son

Figure 3.12
Transmission of the X chromosomes from one generation to the next. (From *Principles of Human Genetics,* 3rd ed., by Curt Stern. W. H. Freeman and Company. Copyright © 1973.)

father, but they do not express a recessive condition unless they receive another such allele on the X chromosome from their mother.

Table 3.9 presents the expectations for offspring of certain matings involving a single recessive gene on the X chromosome. Affected fathers produce no affected offspring unless the mother is a carrier. However, all the daughters of an affected male are carriers (1). Half the sons of a carrier mother will be affected (2 and 3). Half of her daughters will be carriers if the spouse is a normal male (2), and half will be affected if the spouse is affected (3). Finally, all the sons of an affected woman will be affected, and all of her daughters will be carriers if the father is normal (4). Another peculiarity of recessive, sex-linked inheritance is the alternate-generation phenomenon, in which the maternal grandfather and his grandsons are affected, but none of those in the intermediate generation. However, half of the grandfather's daughters' sons are affected. Because the affected grandfather's daughters are all carriers, their sons have a 50-50 chance of being affected.

Five pedigrees depicting the transmission of color blindness are presented in Figure 3.13. Although a number of different forms of color blindness are known, the more common types have a similar genetic basis and thus will not be differentiated here. The pattern of transmission evident in Figure 3.13 conforms closely to that expected of a sex-linked, recessive gene. Recombinant DNA techniques, described in Chapter 7, have been used to isolate the genes on the X chromosome that are responsible for color-blindness (Natans et al., 1986).

In the 1960s some data pointed to the possibility that a major gene on the X chromosome influences a specific cognitive trait called spatial ability. Tests of spatial ability involve tasks such as deciding what a two-dimensional form will look like in three dimensions and mentally rotating two- or three-dimensional objects to determine whether they are the same or different from a standard shape. Of all the specific cognitive abilities, spatial ability shows the most consistent sex difference. Males tend to score higher on tests of spatial ability.

One possible explanation for sex differences is sex linkage. Suppose that a major recessive gene enhancing spatial ability is on the X chromosome. Males would express such an allele more frequently than females, and would thus have higher scores. Moreover, because sons never receive their father's X chromosome, fathers and sons should be less similar than mothers and daughters, mothers and sons, or fathers and daughters if X-linked genes are involved. Earlier data suggested that fathers and sons are less similar than other relatives and revealed other expected patterns of familial similarity for spatial scores. These data seemed to support the hypothesis of a major X-linked gene for spatial ability. However, data from larger studies do not confirm the hypothesis of sex-linkage (Bouchard and McGee, 1977; DeFries et al., 1976; DeFries et al., 1979; Guttman, 1974; Loehlin, Sharan, and Jacoby, 1978). For example, in the study by DeFries and associates (1979), the correlation for spatial ability scores in 672 father–son pairs was 0.33 in a Caucasian group and 0.26 in 241 father–son pairs in an Oriental sample. The average correla-

Table 3.9
Expected frequencies of offspring for a sex-linked recessive character (X_a represents the X chromosome with the particular recessive allele; thus, X_aY males and X_aX_a females are affected.)

1. Matings Between an Affected Father and a Homozygous Normal Mother

2. Matings Between a Normal Father and a Carrier Mother

(Shaded area indicates affected offspring.)

3. Matings Between an Affected Father and a Carrier Mother

4. Matings Between a Normal Father and an Affected Mother

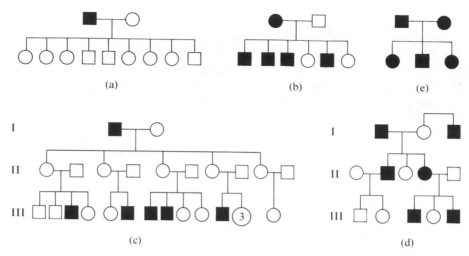

Figure 3.13

Pedigrees of color blindness. (a) Part of pedigree no. 406, Nettleship. (b) Pedigree no. 584. (c) Part of Horner's pedigree. (d) Pedigree, Whisson, 1778, the first known pedigree of color blindness. (e) Pedigree, Vogt. (From *Principles of Human Genetics*, 3rd ed., by Curt Stern. W. H. Freeman and Company. Copyright © 1973.)

tion for other family relationships was 0.33 for the Caucasian sample and 0.25 for the Oriental sample. Thus, these results do not support the hypothesis that spatial ability is influenced by a sex-linked, recessive gene.

Scope of Single-Gene Effects in Human Behavior

Only a few of the many known single-gene behavioral effects in humans have been discussed in this chapter. Although classical Mendelian crosses are still used to demonstrate single-gene effects, molecular genetic techniques described in Chapter 7 offer much more power for identifying single-gene effects even when many genes and environmental influences are operating. The scope of single-gene effects in human behavior can be gleaned from Victor McKusick's *Mendelian Inheritance in Man* (1986a). This catalogue lists 1,172 established autosomal dominant, 610 autosomal recessive, and 124 X-linked characteristics of human beings. In addition, there are over 2,000 tentatively identified single-gene effects. More than 100 of the established single-gene effects list mental retardation as a clinical symptom; it is likely that many of these genes affect other less obvious aspects of behavior as well.

It should be noted that many of these single-gene effects were isolated because they produced a disorder of some kind. Normal variation in behavior is likely to be influenced by many genes: Any one of many genes can disrupt development, but the normal range of behavioral variation is likely to be orchestrated by a system of many genes, each with small effect, as well as by environmental influences. Three results suggest that many genes affect behavioral variation: (1) As noted earlier, most behaviors are distributed in a smooth, continuous distribution rather than the dichotomous, either–or traits studied by Mendel; (2) artificial selection studies of nonhuman animals (Chapter 10) show a gradual response to selection over generations rather than the rapid response expected if a single gene is at work; (3) various strategies to isolate single-gene effects on normal behavioral variation in human as well as nonhuman animals have generally failed, as in the previous example of spatial ability. This is the reason why behavioral genetics requires quantitative genetic approaches as a first step towards understanding the etiology of behavioral variation. However, each of the many genes that affect behavior operate according to the laws of hereditary transmission discovered by Mendel.

As discussed in Chapter 9, a dispute arose during the early twentieth century between the so-called Mendelians and biometricians. Mendelians looked for single-gene effects as shown by Mendelian segregation ratios for qualitative, either–or traits; biometricians knew that most traits are continuously distributed and for this reason argued mistakenly that Mendelian inheritance is wrong. The dispute was resolved when it was understood that a continuous distribution will be observed if a trait is influenced by several genes, each having small effects, as well as environmental factors. This is the essence of quantitative genetics, the topic of Chapter 9. As discussed in the following chapter, the resolution of the dispute between Mendelians and biometricians has taken on new significance, because molecular genetic techniques have advanced to the point where it is possible to isolate and identify the effects of single genes even when many genes and environmental factors affect a trait, thus bridging the gap between single-gene approaches and quantitative genetics.

SUMMARY

In each body cell of higher organisms, there are pairs of chromosomes that carry hereditary factors, now called genes. A gene may exist in two or more different forms (alleles). One allele can dominate the other. Genotype refers to alleles considered two at a time as they exist in individuals. Mendel's monohybrid experiments showed that alleles are not contaminated by the presence of an alternate allele in the genotype, and that the two alleles separate (segregate) completely during hereditary transmission. Mendel's dihybrid and trihybrid experiments demonstrated that the alleles for one trait assort independently of the alleles for other traits. In this way, Mendelian genetics provides a mechanism for the maintenance of vast amounts of genetic variability.

Mendelian crosses can be used to determine the mode of inheritance for behavioral characters. There are many examples of single-gene abnormalities in mice. Genes with well-known morphological effects, such as albinism, may also affect behavior (pleiotropism). The use of recombinant inbred strains provides another strategy to uncover single-gene effects in animals. Mendelian analysis of human pedigrees can be used in a similar manner as illustrated for Huntington's disease, PKU, and color blindness. The most powerful approach for finding single-gene effects on behavior is the use of recombinant DNA techniques, which are discussed in Chapter 7.

CHAPTER·4

Mechanisms of Heredity and Behavior

Mendel was convinced that his "elements" were material units located in the gametes. The state of knowledge at the time, however, made it impossible for him to specify their physical nature in any greater detail. It was fortunate that the basic Mendelian laws could be established by treating the "elements," or genes, as hypothetical constructs, without precise knowledge of their location or structure.

In this chapter the story will be made more complete by describing the chemical structure and function of genes. In the second part of the chapter we will consider examples of the mechanisms by which genes influence behavior and how mutations can be used to "dissect" behavior. In subsequent chapters, the packaging of genes in chromosomes is discussed (Chapter 5), chromosomal abnormalities are considered (Chapter 6), and the new genetic tools of recombinant DNA are described (Chapter 7).

MECHANISMS OF GENE ACTION

In 1902 A. E. Garrod discussed a rare human condition, called alkaptonuria, in which affected individuals display the remarkable symptom of excreting urine that turns black when exposed to air. Garrod concluded that the condition was inherited, and that it obeyed the newly discovered Mendelian laws. As significant as it was then to have an example of Mendelian inheritance in humans, the conclusion Garrod drew about the physiological basis of the disorder was of even greater importance. Analysis revealed that the urine of the affected individual contained large quantities of homogentisic acid instead of the usual compound, urea. Garrod knew that homogentisic acid is normally converted into urea. He thus suggested that the usual metabolic route that converts the acid into urea had been blocked and called it an "inborn error of metabolism." Garrod (1908) also proposed that other defects in humans, such as albinism, may be due to similar metabolic blocks. William Bateson (1909) suggested that inborn errors of metabolism might be due to the failure of the enzymes that control the normal reactions.

Other work distributed over the next 20 years on the inheritance of pigmentation in plants and animals (see Sturtevant, 1965, for a review) supported the hypothesis that genes produce some sort of biochemical substance. Another line of investigation lent support to this general proposition. The existence of human blood groups was described in 1900, and their Mendelian inheritance was shown in 1929, establishing another link between genetics and a biochemical process. While these results showed that genes could influence the physiological functioning of an organism, it was not clear whether these were typical or unusual situations.

In the 1930s the common bread mold, *Neurospora*, was introduced into genetic research. This organism has extremely simple nutritional requirements and normally can survive on a simple medium containing salts, glucose, and a compound called biotin. From this simple diet, the organism is capable of metabolizing all the complex chemicals required for life. George Beadle and E. L. Tatum (1941) x-rayed spores of the fungus, and found that some of the organisms had undergone mutation and were no longer able to survive on this simple medium. By analyzing the nutritional requirements of these mutants, they were then able to describe the normal metabolic sequence and show that each enzymatic step in the sequence is under the control of a single gene. These results gave rise to the "one-gene, one-enzyme hypothesis," and it became increasingly reasonable to assume that the basic mechanism of gene action operates through the production of enzymes.

Later in this chapter we describe behaviorally relevant single genes. The main point is that the basic function of genes was understood by the 1940s. It was known that genes control the production of enzymes. However, the chemical structure of genes was not understood until the 1950s. Knowledge of the chemical structure of genes led to an understanding of how genes faithfully replicate themselves and how they control enzyme production.

THE CHEMICAL NATURE OF GENES

A great deal of research has been done in an attempt to understand the chemical nature of the gene itself. The chemical hereditary substance would have to meet several requirements:

1. It would have to be found in the nucleus of the cell, because chromosomes, which are the carriers of the genes, are found within nuclci.
2. The substance would have to be capable of self-duplication, because the genes have this ability.
3. The chemical would have to be capable of existing in various forms, because it was known that there are a large number of genes and that they occur in different allelic forms. In other words, it would have to be able to carry a variety of genetic information.

The successful synthesis of all of the available data was accomplished by James Watson and Francis Crick (1953a, 1953b). They hypothesized that deoxyribonucleic acid (DNA) is the fundamental component of the hereditary material, and proposed a molecular structure that could account for its biological properties. This structure was confirmed by subsequent research, and its confirmation signaled the explosive growth and development of molecular biology. Briefly, and in necessarily oversimplified form, the basic features of the molecular structure of the gene and its action are as follows. A DNA molecule consists of two strands, each composed of phosphate and deoxyribose sugar groups. The strands are held a fixed distance apart by pairs of bases (nitrogenous compounds). There are four bases involved: adenine, thymine, guanine, and cytosine. Due to the structural properties of these bases, adenine always pairs with thymine and guanine always pairs with cytosine (Figure 4.1). The strands coil around each other to form a double helix (Figure 4.2).

The double nature of the helix and the restrictions on base pairing make possible the self-duplication of the DNA molecule. In the process of cell division, the helices of the DNA molecule unwind, the base pairs separate, and one of each pair remains attached to each strand (Figure 4.3). Within the nucleus of the cell, the raw materials necessary for the construction of new DNA are in the form of nucleotides, which consist of one of the four bases, a deoxyribose sugar, and a phosphate. Nucleotides pair with the exposed bases of the unwound strands and ultimately form a complementary strand paired with each of the originals. By this process, two molecules of DNA are produced where there was previously but one.

GENES AND PROTEIN SYNTHESIS

One function of DNA is self-duplication. In fact, the process just described may be very similar to the way the original replicating cells duplicated themselves billions of years ago. The other major function of DNA is translation from the genetic code to enzyme production. This occurs in several steps, illustrated schematically in Figure 4.4. First, the information of one strand of the DNA molecule is transcribed onto a different sort of nucleic-acid molecule. This single-stranded molecule, ribonucleic acid (RNA), is composed of a ribose sugar, a phosphate, and the same bases as DNA, with the exception

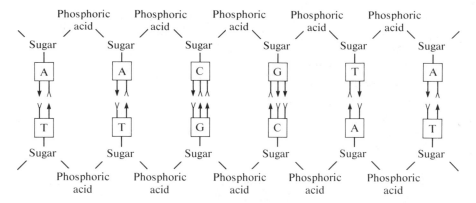

Figure 4.1

Flat representation of a DNA molecule. A = adenine; T = thymine; C = cytosine; G = guanine. (From *Heredity, Evolution, and Society* by I. M. Lerner. W. H. Freeman and Company. Copyright © 1968.)

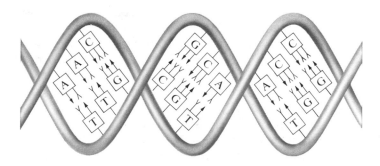

Figure 4.2

A three-dimensional view of a segment of DNA. (From *Heredity, Evolution, and Society* by I. M. Lerner. W. H. Freeman and Company. Copyright © 1968.)

that uracil substitutes for thymine. By a process of base pairing similar to that of the duplication of DNA, a complementary RNA strand is formed using the DNA strand as a template. In Figure 4.4, the dark DNA strand is being transcribed, although recent research indicates that in some cases both strands are transcribed (Adelman et al., 1987). This RNA molecule, called messenger RNA, enters the cytoplasm where it connects with ribosomes, the sites of protein synthesis. Within the cytoplasm is another form of RNA, transfer RNA. This RNA, which has a helical structure, exists in a variety of forms, each of which attaches to a specific amino acid. The transfer RNA molecules,

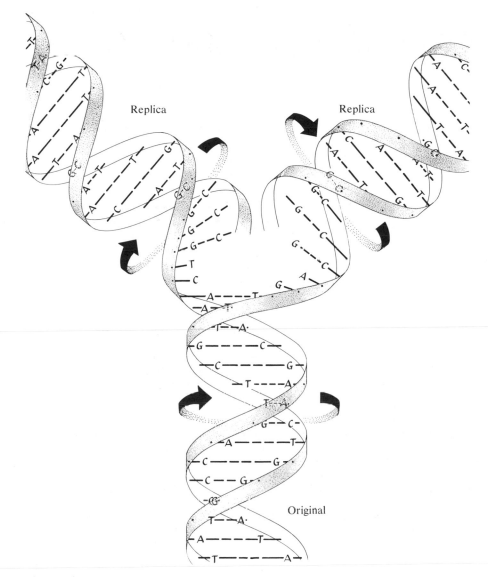

Figure 4.3
Replication of DNA. (After *Molecular Biology of Bacterial Viruses* by G. S. Stent. W. H. Freeman and Company. Copyright © 1963.)

with their attached amino acids, pair up with the messenger RNAs in a sequence dictated by the rules of base pairing.

Amino acids are incorporated into growing polypeptide chains at the incredible rate of about 100 per second. The polypeptide chains in turn constitute enzymes and other proteins. In this way, the genetic information of DNA becomes expressed in the production of specific enzymes.

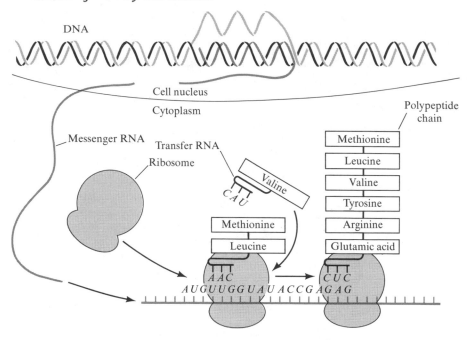

Figure 4.4

The "central dogma" of molecular genetics states that genetic information flows from DNA to messenger RNA to protein. Genes are relatively short segments of the long DNA molecules in cells. The DNA molecule comprises a linear code made up of four types of nucleotide base: adenine (*A*), cytosine (*C*), guanine (*G*), and thymine (*T*). The code is expressed in two steps. First, the sequence of nucleotide bases in one strand of the DNA double helix is transcribed onto a single complementary strand of messenger RNA (which has the same bases as DNA except that thymine is replaced by the closely related uracil, or *U*). The messenger RNA is then translated into protein by means of complementary transfer RNA molecules, which add amino acids one by one to the growing chain as the ribosome moves along the messenger RNA strand. Each of the 20 amino acids found in proteins is specified by a "codon" made up of three sequential RNA bases; the sequence of codons that code for a specific polypeptide is called a cistron. (From "The mechanism of evolution" by F. J. Ayala. Copyright © 1978 by Scientific American, Inc. All rights reserved.)

The basic unit of the genetic code has been shown to be triplet sequences of bases, with each succeeding triplet specifying an amino acid. (See Table 4.1.) For example, three adenines in a row on the DNA molecule (AAA) will be transcribed in the messenger RNA as three uracils (UUU). When on the ribosome, this messenger RNA triplet will attract transfer RNA with the triplet sequence AAA. This particular transfer RNA is the one that "carries" the amino acid phenylalanine. Although there are 64 possible triplet codes, there are only 20 amino acids; some amino acids are coded by as many as six triplet codes. Since the same genetic code applies to all living organisms, "breaking" this code has been one of the great triumphs of molecular biology.

Table 4.1

The genetic code

First letter	Second letter				Third letter
	A	G	T	C	
A	Phe	Ser	Tyr	Cys	A
	Phe	Ser	Tyr	Cys	G
	Leu	Ser	chain end	chain end	T
	Leu	Ser	chain end	Try	C
G	Leu	Pro	His	Arg	A
	Leu	Pro	His	Arg	G
	Leu	Pro	Gln	Arg	T
	Leu	Pro	Gln	Arg	C
T	Ile	Thr	Asn	Ser	A
	Ile	Thr	Asn	Ser	G
	Ile	Thr	Lys	Arg	T
	Met	Thr	Lys	Arg	C
C	Val	Ala	Asp	Gly	A
	Val	Ala	Asp	Gly	G
	Val	Ala	Asp	Gly	T
	Val	Ala	Asp	Gly	C

NOTE: Each amino acid is coded by a triplet of three bases, as shown in the table, which is a compact way of setting out the 64 possible triplets.

The four bases are denoted by the letters A, G, T, and C. In DNA the four bases are: A = Adenine; G = Guanine; T = Thymine; C = Cytosine.

The 20 amino acids are identified as follows:

Ala = Alanine	Leu = Leucine
Arg = Arginine	Lys = Lysine
Asn = Asparagine	Met = Methionine
Asp = Aspartic acid	Phe = Phenylalanine
Cys = Cysteine	Pro = Proline
Glu = Glutamic acid	Ser = Serine
Gln = Glutamine	Thr = Threonine
Gly = Glycine	Try = Tryptophan
His = Histidine	Tyr = Tyrosine
Ile = Isoleucine	Val = Valine

SOURCE: After Cavalli-Sforza and Bodmer, *The Genetics of Human Populations.* W. H. Freeman and Company. Copyright © 1971. Data from Crick, 1966.

A NEW VIEW OF MENDEL'S "ELEMENTS"

These developments have provided a dramatically new view of Mendel's hypothetical "elements." A gene is a functional unit that codes for a particular polypeptide. Structurally, it is composed of a stretch of nucleotide bases of DNA that is only 2 nanometers in diameter and can be visualized only with an electron microscope. It is estimated that the DNA of the bacterium

Eschericia coli, which has a single chromosome, has nearly 4 million nucleotide base pairs. DNA of our species, which has 23 pairs of chromosomes, involves about 3 billion nucleotide base pairs if we just consider one member of each pair of chromosomes. If human beings have about 3 billion nucleotide base pairs and if the average gene is several hundred bases long, one might guess that there are well over a million genes. However, the number of genes is far less; recent findings concerning the complexity of DNA make it difficult to estimate the number of genes in the human genome, although several methods converge on an answer of 50,000 to 100,000 genes (McKusick, 1986b). In recent years, new molecular genetic techniques have greatly advanced our knowledge of Mendel's elements and have special relevance to behavioral genetic analyses. These advances are described in Chapter 7, following two chapters on chromosomes.

The focus of behavioral genetics is on DNA variations among individuals as they affect behavior. DNA variation is due to two types of mutation, or copying error. Chromosomal rearrangements can lead to inversions, duplications, and deletions of large segments of DNA. Far more common are copying errors at a single base, called a point mutation, in which one base is deleted, an extra base is inserted, or one base is substituted for another. Deletions and insertions are called frameshift mutations because they change the reading frame of all subsequent codons in the gene. If three bases or a multiple of three are changed, the effect may be less drastic because only one amino acid will be changed in the protein product. New techniques discussed in Chapter 7 can harness such DNA variation in order to assess its role in behavioral variation among individuals.

An important aspect of the molecular view of Mendel's "elements" is its demystification of the process by which genes influence behavior. Genes do not code for behavior per se; structural genes code for sequences of amino acids that form the thousands of proteins of which organisms are made. Enzymes are proteins that serve as organic catalysts. They speed biochemical reactions that would otherwise be sluggish or would not occur at all under the conditions of temperature and pressure prevailing within an organism's body. These reactions and their timing are fundamental to the development of all the systems of an organism and to the functioning of the various organs. The influence of genes, therefore, is not exerted through some mysterious mechanism. The pathways from genes to behavior run through the skeletal system; the muscles; the endocrine glands; the digestive, respiratory, and excretory systems; the immune system; and the autonomic, peripheral, and central nervous systems. Investigations of these pathways, therefore, involve studies in molecular biology and biopsychology.

Although the connection between genes and behavior can be understood as a general theoretical proposition, the specific details have been elucidated in only a limited number of cases. However, this area of research is becoming increasingly popular, and we can predict that increased knowledge of the biochemical, physiological, and anatomical mechanisms of genetic influence on behavior will be very rapid in the near future.

PHENYLKETONURIA

Phenylketonuria (PKU) illustrates the way in which a single gene can influence behavior. It is the earliest and best-understood condition of genetic involvement in mental retardation. The trail of research that led to our current understanding of PKU had its beginning in Norway in 1934. A dentist with two retarded children was distressed because they exuded a peculiar odor that so aggravated his asthmatic condition that he was unable to stay with them in a closed room. He had them examined by Asbjörn Fölling, who began the search for the cause of the peculiar odor by analyzing the urine of the children. His search quickly paid off in the isolation of phenylpyruvic acid. Fölling postulated that the disease was inherited as a single recessive gene, and that phenylpyruvic acid was present in the urine because of a disturbance in the metabolism of phenylalanine, an essential amino acid. Somehow the excess of phenylpyruvic acid was related to mental retardation.

The disease came to be known as Fölling's disease, or phenylketonuria, and it became the subject of research in a number of laboratories around the world. This research revealed that the metabolic problem is caused by the absence or inactivity of a particular enzyme, phenylalanine hydroxylase. This enzyme converts phenylalanine to tyrosine. If this conversion is blocked, phenylalanine levels increase in the blood and phenylpyruvic acid accumulates in the urine. Apparently, the high level of phenylalanine in the blood depresses the level of other amino acids, depriving the developing nervous system of needed nutrients. Behavioral pleiotropic effects, especially hyperactivity and irritability, are also part of this syndrome.

This knowledge concerning the biochemical origin of PKU made possible a search for rational therapies. If a particular enzyme is deficient, it might be possible to provide the necessary amounts of that enzyme. Although that approach may soon be technically feasible (Ambrus et al., 1978), an approach that has been used successfully involves minimizing the need for the enzyme. If the mental retardation of PKU is caused by a buildup of phenylalanine, the amount of phenylalanine in the diet can be reduced. Phenylalanine is found in a wide variety of foods, particularly meats. In 1953 a special diet was prepared that was very low in phenylalanine. Although it normalized levels of phenylalanine, it did not improve intelligence of older PKU children. However, when the diet was administered to very young PKU children, retardation apparently was prevented (Hsia et al., 1958). As a result of this work, routine screening programs have been established to identify PKU infants at birth.

Through these programs and related programs to test relatives of affected persons, it became possible to assay the intelligence of individuals identified on the basis of the chemical defect. Previously, most research had been conducted on individuals biochemically identified as phenylketonurics from a population of individuals already determined to be mentally retarded. The assumption had been that all phenylketonurics were probably institutionalized for mental retardation. On this assumption, calculations had been made that the mean IQ of phenylketonurics was approximately 30. With the new screening procedure, a surprising number of individuals were discovered who were biochemically phenylketonurics, but whose intelligence was in the normal range. This discovery necessitated a reevaluation of the efficacy of dietary treatment. If some of the individuals treated with the special diet would have developed normal intelligence in any case, then the report of IQs of treated subjects in the range of 80 to 90 could hardly be taken as evidence that the diet prevented retardation.

A great deal of research has subsequently been devoted to the problem of the efficacy of the low phenylalanine diet, and it remains somewhat controversial. A straightforward experiment comparing the outcomes for treated and untreated groups, both identified at birth, would provide a critical test. However, this would involve withholding a potentially effective treatment from a group of patients, which is ethically impossible and is unlikely ever to be done. An approach to this type of comparison can be made, however, by comparing the IQs of treated patients with those of older siblings, who were untreated because the dietary therapy had not yet been invented. The results of such a comparison by Hsia (1970) are given in Figure 4.5. With late treatment or no treatment at all, the distribution of IQs ranges from 10 to 110; siblings treated from an early age have IQs ranging from 65 to 120. Furthermore, the distribution of IQs in this latter group indicates that most cases are at the higher end of the distribution. These results constitute reasonable evidence that the diet is in fact a useful therapy.

This discovery of normal IQs in untreated phenylketonurics also raises the issue of the possible heterogeneity of the condition. A number of variant forms have now been described (Murphey, 1983; Scriber, Kaufman, and Woo, 1988). Hsia (1970) recommends as a working definition that patients with persistent blood phenylalanine levels in excess of 25 milligrams percent be diagnosed as "classical" phenylketonurics. Of these, approximately one fourth may achieve normal intellectual functioning without dietary treatment. It is difficult to be sure of the diagnosis of patients with levels between 15 and 25 milligrams percent, but those with levels between 2 and 15 milligrams percent probably exhibit one of the variants of phenylketonuria that do not cause retardation. In fact, data from the newborn-screening programs indicate that for every two PKU cases, there is one case of elevated phenylalanine levels not caused by the PKU gene.

Important new information concerning PKU continues to be reported. For example, it has long been known that PKU heterozygotes are less able to convert phenylalanine to tyrosine than are normal homozygotes. But only recently have researchers discovered that heterozygotes may have lower IQs than normal homozygotes (Bessman, Williamson, and Koch, 1978). This same study and others also suggest that the retardation effect of PKU may occur to some extent prenatally, rather than solely after birth. Now that PKU individuals grow up essentially normal, a new problem has arisen: PKU mothers must be restarted on the special low-phenylalanine diet as early as possible during pregnancy to prevent toxic levels of phenylalanine during fetal development.

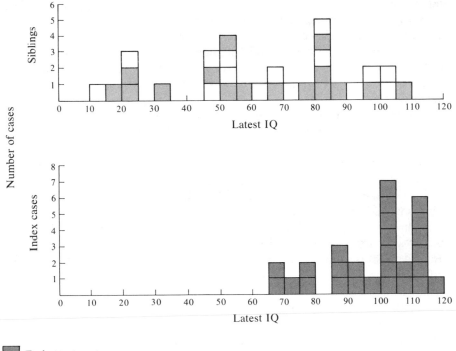

Early treatment

Late treatment

No treatment

Figure 4.5
Frequency histograms of IQ scores of early-treated index cases versus late-treated or untreated siblings. (From "Phenylketonuria and its variants" by D. Y.-Y. Hsia. In A. G. Steinberg and A. G. Bearn, eds., *Progress in Medical Genetics*. Grune & Stratton, Inc., 1970. Used with permission.)

    The considerable success in elucidating the biochemical mechanisms of PKU has inspired an intensive research effort directed toward identifying and analyzing other conditions of mental retardation associated with abnormal metabolism of amino acids. At least a dozen have now been identified, but most are much more rare than PKU and none is understood nearly as well.

OTHER RECESSIVE METABOLIC ERRORS

In addition to amino acidurias such as PKU, there are single-gene metabolic errors involving carbohydrates and lipids (fats). Galactosemia is an example of a carbohydrate metabolic defect. Individuals homozygous for the autosomal recessive allele lack an enzyme to convert galactose to glucose. In many cases, early death results, and affected individuals who do survive are severely mentally retarded. However, as in the case of PKU, a rational therapy has been developed. Early identification of affected infants, coupled with replacement of milk with galactose-free substances, is quite successful. Heterozygotes have half the normal enzyme activity, which is apparently sufficient to metabolize galactose.

Second only to PKU as a cause of mental retardation is Lesch-Nyhan syndrome. First described in 1964, it is caused by a recessive gene on the X chromosome that effects an enzyme involved in purine metabolism. Development is normal until the second half of the first year, when hypertonia and severe spasticity begin to emerge. In addition to these severe motor defects that make speech difficult, the most dramatic feature of this syndrome is aggressive self-mutilation that often begins when teeth erupt, seen early as biting lips and fingers, with consequent tissue loss. No reduction in sensitivity to pain appears to occur—patients scream in pain as they bite themselves— and physical restraint is necessary. The link between purine metabolism and self-mutilating behavior remains a mystery; neurotransmitters, especially dopamine, have also been implicated and may explain the bizarre behavioral manifestations of the syndrome (Kopin, 1981).

Tay-Sachs disease (infantile amaurotic idiocy) is a recessive condition caused by the absence of an enzyme that breaks down a lipid. Such individuals are apparently normal at birth but begin to show symptoms of *nystagmus* (spasmodic movement of the eyes) and paralysis when a few months old. The condition steadily worsens to a state of profound idiocy, paralysis, and blindness. Death usually intervenes before two years of age. Autopsy has shown that nerve cells of the brains of affected individuals contain abnormal amounts of a lipoid substance and the neurons show degenerative changes. A carrier test has been developed, and it is also possible to diagnose affected fetuses prenatally by sampling the amniotic fluid. (See Box 6.1.)

The point of these examples is to indicate that genes affect behavior the same way that they affect other phenotypes. When we talk about single-gene influences on behavior, it is simply a convenient way to indicate that we are considering the effects of DNA production and regulation of proteins as described earlier in this chapter. Sometimes students react to these examples of single-gene influences by saying, "But genes didn't *really* cause these behavioral disturbances. The genes caused metabolic problems and the behavioral effect was only a by-product of the enzyme deficiency." However, that is just the point. Genes are not magical elements that somehow blossom into behavior patterns, as when the puppeteer pulls a puppet's strings. Genes are segments of DNA that code for protein production. In that sense, all aspects of ourselves—our bones as well as our behavior—are by-products of this process.

REGULATOR GENES

Until now, we have considered the physical basis for genes that code for particular proteins, known as *structural genes*. However, it is clear that genes are not completely active from the moment of conception. Since genes turn on and off throughout development, after birth as well as before, some mechanism must be responsible for regulating the timing and quantity of protein production. The hypothesized mechanism, suggested by Jacob and Monod (1961), is called the *operon model.* In addition to structural genes that serve as templates for protein production, there are *operator* genes and *regulator* genes. The operator gene is a short segment of DNA next to the structural gene that serves as an on–off switch, determining whether the structural gene will be transcribed for protein production. Together, the operator gene and structural gene are called an *operon.* Regulator genes switch the operator gene on and off by producing a repressor that binds with the segment of DNA referred to as the operator gene. This model suggests a mechanism for the regulation of genes. (See Figure 4.6.) The repressor produced by the regulator gene will bind with the operator gene, shutting down the structural gene. If this were the end of the story, all such structural genes would be permanently turned off. However, the repressor can also bind with a *regulatory metabolite,*

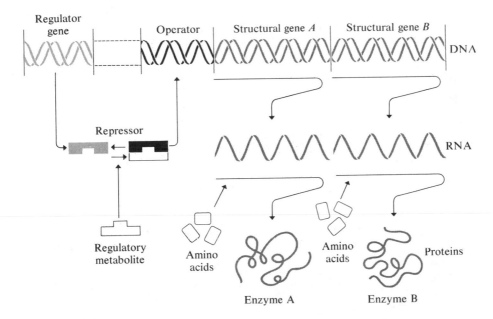

Figure 4.6
The operon model. When the regulatory metabolite binds with the repressor (produced by the regulator gene), the repressor does not bind with the operator. The structural gene will then be switched on and transcribe RNA, thus producing its enzyme. (From "The control of biochemical reactions" by J.-P. Changeux. Copyright © 1965 by Scientific American, Inc. All rights reserved.)

a product of other operons or substances from outside the cell. When this happens, the repressor is, in a sense, inactivated, because it will bind with the regulatory metabolite rather than with the operator gene. The result is that the operator gene, no longer bound to the repressor, will permit transcription of messenger RNA so that the structural gene is, in effect, turned on.

Suppose that this structural gene produces an enzyme that breaks down a regulatory metabolite and that the metabolite binds with the gene's repressor. When a structural gene is shut down by its repressor, a regulatory metabolite builds up. As the metabolite builds up, it begins to bind with the repressor until the repressor is inactivated and the structural gene turns on. The structural gene produces an enzyme that metabolizes the regulatory metabolite. As the metabolite is broken down, the repressor again begins to bind with the operator gene and eventually shuts down the operon. In summary, presence of the regulatory metabolite turns on an operon, and its absence turns an operon off.

The operon model has been demonstrated with the single-celled bacterium *E. coli*. Discussion of the details of the operon in bacteria will make the concept of regulatory genes clearer. (See Figure 4.7.) *E. coli* produces about 700 different enzymes. The enzyme that we know the most about is β-galactosidase, which metabolizes lactose into glucose and glactose. In a normal *E. coli* cell there are about 3,000 β-galactosidase molecules, because a normal cell lives in a lactose-rich environment. The cell needs the enzyme to metabolize lactose, its major energy source. However, when *E. coli* are placed in environments without lactose, there are as few as three β-galactosidase mole-

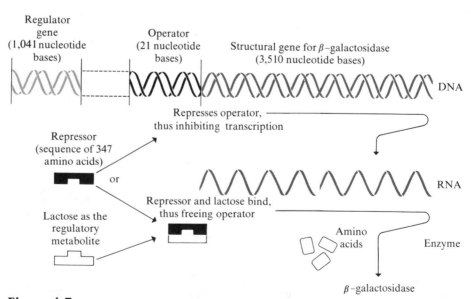

Figure 4.7
The operon model for β-galactosidase in *E. coli*. (Adapted from description by J. D. Watson, 1976.)

cules. This is a good example of how enzyme production is rigidly controlled, not simply pumped out at a constant rate.

Synthesis of messenger RNA is critical to this control. We know this because messenger RNA exists for only 2 to 3 minutes, and variation in its synthesis is related to protein production. For example, in a lactose-rich environment, an average cell will have 35 to 50 β-galactosidase messenger RNA molecules at any given time. In the absence of lactose, the same cells will have less than one messenger RNA molecule on the average. Thus, the presence of the short-lived messenger RNA that is translated into the enzyme is dependent on the need for that enzyme. Synthesis of messenger RNA is controlled by the operon system.

The regulator gene for β-galactosidase involves 1,041 nucleotide bases, and the repressor that it codes for is a sequence of 347 amino acids. This repressor binds to another segment of DNA (the operator gene), which is only 21 nucleotide bases long. When the repressor binds to the operator, it blocks transcription of messenger RNA for the structural gene (3,510 bases) that codes for the β-galactosidase enzyme. In this way, the operon for the enzyme is switched off. However, the repressor is inactivated when lactose binds with it. When lactose builds up, the repressor no longer shuts down the operon, and the β-galactosidase enzyme is again produced. This enzyme metabolizes lactose and, when its job is done, it is again repressed. As a sidelight, it is known that the repressor for β-galactosidase also represses two neighboring structural genes that produce enzymes crucial to the metabolism of lactose.

Interactions among operons can provide mechanisms for relatively permanent changes in gene functioning. Figure 4.8 shows two operons, each containing one structural gene that is, in fact, a regulatory gene (RG) for the other operon. Thus, if operon 1 is active, the product RG_1 will turn off operon 2. If some inducer from outside the system, I_1, a regulatory metabolite for the repressor coded by RG_1, appears even briefly, operon 2 can be switched on. G_2 will produce enzyme 2 and RG_2 will produce its repressor, which switches off operon 1. Operon 2 is now locked on and operon 1 is locked off until such time as an inducer for operon 2 appears. Far more complex types of interactions could be postulated. But the foregoing should make it clear that operon-

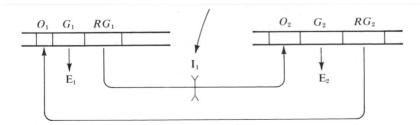

Figure 4.8
An operon circuit that would switch the production of one enzyme off and another on. (From *General Genetics*, 2nd ed., by A. M. Srb, R. D. Owen, and R. S. Edgar. W. H. Freeman and Company. Copyright © 1965.)

type systems could be responsible for most of the sequential regulation of genes and the selective workings of genes in different tissues that accompany developmental processes.

To this classical operon model has been added two DNA sequence elements necessary for the regulation of structural genes in complex organisms: promoters and enhancers. The promoter is required for accurate and efficient initiation of transcription; enhancers increase the rate of transcription from promoters. A promoter is a short DNA sequence of about 100 base pairs that is the spot at which RNA polymerase binds to DNA to begin transcription. The promoter gene lies next to the operator gene, the site at which the repressor molecule binds.

For bacterial β-galactosidase the promoter consists of two blocks of base pairs. One block of six base pairs is located 10 base pairs from the start of the structural gene; the other block is 25 base pairs further upstream and consists of 10 base pairs. When the operator gene is free from repression, RNA polymerase binds to the promoter segment farthest upstream from the structural gene. The other segment of the promoter then opens up into its single strands, exposing the DNA template to transcription of RNA. Mutations in either of these DNA promoter sequences can alter the rate at which messenger RNA is synthesized by a factor of 1,000.

In addition to the inhibitory control of the operon model, many other types of regulatory control are being explored. For example, the enhancer sequences, which have been found in many organisms, including humans, substantially increase transcription of any gene near them (Gluzman and Shenk, 1984). An interesting aspect of enhancers is that their activity can be induced by environmental events such as heat shock, exposure to heavy metals, and viral infection. Operator, promoter, and enhancer sequences only affect genes on the same chromosome (called *cis* acting). Genetic regulation across chromosomes is referred to as *trans*—for example, the regulatory product of operons can act on chromosomes other than the one in which they are found.

Newly discovered features of DNA of higher organisms, such as split genes and transposable genes—described in Chapter 7—are also thought to play a role in gene regulation, as are factors that affect the rate at which messenger RNA molecules are translated into proteins. Genetic regulation is likely to hold many surprises of relevance for behavioral analyses. For example, it appears that experience, including learning, can alter gene expression in the nervous system (Black et al., 1987).

GENETIC REGULATION OF DEVELOPMENT

Any of these control processes can regulate long-term developmental change, as indicated in the previous example of the operon model (Figure 4.8). Two regulatory processes specific to development are also under study. Temporal genes have been discovered that affect the timing of developmental events. For example, in the nematode round worm, described later, a dominant mutation slows differentiation by causing cells to repeat the lineage of their ancestors; a recessive mutation accelerates differentiation by causing cells to express the fates normally expressed by their descendants (Ambros and Horvitz, 1984). In other organisms, temporal genes have been studied by identifying mutations that change enzyme levels developmentally. The first and most thoroughly studied temporal gene is the so-called *Gus* locus that controls an enzyme, β-glucuronidase, involved in glucose metabolism. In some inbred mouse strains — for example, the C3H strain — a temporal gene causes an abrupt decline in activity of this enzyme at about 12 days of age. Prior to this age, production of the enzyme is normal, indicating that the control elements and the structural gene for β-glucuronidase are intact.

There are also temporally controlled genes in humans, such as the gene that controls the expression of Huntington's disease in midlife (see Chapter 3), although these are difficult to study because the gene product is not known. The best-known example in humans is adult lactose intolerance. In nearly all mammals except humans, intestinal lactase that hydrolyzes lactose in milk is high at birth and then declines after weaning until less than 10 percent of its peak activity remains. In humans, the enzyme often persists at high levels into adulthood, presumably as an adaptation to the use of milk products as a staple food. However, some adults do not show persistence of high lactase levels and are therefore intolerant to milk. Another classic example involves developmentally regulated genes that code for components for hemoglobin unique to embryonic development, fetal development, and postnatal development. Such temporally controlled events may be genetically programmed (Paigen, 1980).

A second regulatory process specific to development involves homeoboxes. Discovered in 1984, homeoboxes are DNA segments of 180 base pairs that appear in several gene complexes important for the timing of development and that are surprisingly similar in many species (Gehring, 1987). Homeoboxes were first discovered as part of homeotic genes in *Drosophila* in which mutations can produce normal structures in the wrong places, such as legs where antennae should be. *Homeotic* is from the Greek meaning *similar*, describing misplaced but similar body parts. Over 20 homeboxes have been found in *Drosophila*, usually within developmentally relevant genes; 12 similar sequences have been located in mice and six in humans. Homeoboxes code for a protein that binds to DNA, presumably regulating the transcription of genes to which it binds. Although its function is not yet understood, it is known that genes containing homeoboxes are active very early in development — for example, by 10 weeks of age for the human embryo.

Clearly, genetic mutations can affect the course of development. However, it is generally agreed that, although there are some temporal genes, development is not hard-wired. Although genetic control of development is often alluded to as a blueprint, a circuit diagram, or a computer program, development is not coded in the genes in this preformationist, scale-model sense. A recipe is a better analogy than a blueprint, because a recipe does not imply a one-to-one mapping between genes and their outcomes:

> A recipe is not a scale model, not a description of a finished cake, not in any sense a point-for-point representation. It is a set of *instructions* which, if obeyed in the right order, will result in a cake. . . . There is no simple one-to-one mapping, then, between genes and bits of body, any more than there is mapping between words of recipe and crumbs of cake. The genes, taken together, can be seen as a set of instructions for carrying out a process, just as the words of a recipe, taken together, are a set of instructions for carrying out a process. (Dawkins, 1982, pp. 295–296)

However, even the recipe analogy falls short of conveying the dynamic nature of genetic effects in development.

Sydney Brenner, who initiated developmental genetic research on the nematode, stated the issue bluntly:

> At the beginning it was said that the answer to the understanding of development was going to come from a knowledge of the molecular mechanisms of gene control. . . . I doubt whether anyone believes that anymore. The molecular mechanisms look boringly simple, and they don't tell us what we want to know. We have to try to discover the principles of organization, how lots of things are put together in the same place. I don't think these principles will be embodied in a simple chemical device, as it is for the genetic code. (In Lewin, 1984, p. 1327)

For example, a thousand different molecules must be synthesized in a specific sequence during the half-hour life cycle of bacterium. It used to be assumed that this sequential synthesis was programmed genetically—each gene is programmed to turn on at the right moment. However, the sequence of steps is not efficiently programmed in DNA; transcription of DNA depends on the substrate, the products of earlier DNA transcriptions. In other words, there is no code for the specific sequence of enzymes needed for development; rather, it is a case of "this leads to that." The reason for this complexity is that natural selection builds upon what is available and what works. As François Jacob (1982, p. 35) phrased it, "Evolution proceeds like a tinkerer who, during millions of years, has slowly modified his products, retouching, cutting, lengthening, using all opportunities to transform and create." The frequency of a mutant gene increases in a population if it has a positive effect on reproductive fitness in the complex context of all of the developmental interactions among other gene products of the organism. Using the recipe analogy, slight differences in the ingredients or the cooking process can drastically alter the whole cake, and we can study the effect of such recipe mutations on the outcome even though there is no one-to-one correspondence between recipe and bits of cake. The point is that development is not programmed step by step in DNA; it is the jerry-built result of millions of small experiments to sculpt an efficient and effective reproducing organism. Nonetheless, the fact that genetically identical animals such as inbred mice and identical twins are so similar phenotypically suggests that, despite the complexities of the mechanisms involved in development, genes guide the process with a sure hand.

Although behavioral genetics is usually phrased in terms of structural genes, its methods are just as appropriate for detecting genetic variability that arises from gene regulation. For example, individuals within an inbred strain of mice will develop identically in terms of all genetic regulatory processes that are coded in DNA at conception. To the extent that changes in DNA expression or rearrangements are brought about by environmental factors not controlled ultimately by DNA, the mice could differ. This is as it should be because changes of this type, although they involve DNA, are initiated by environmental factors. As we shall see in later chapters, an important feature of behavioral genetics is that it also takes into account nongenetic influences that are clearly involved in behavioral development.

MUTATION

Although DNA replication is highly reliable, mistakes sometimes occur. Some of the early work in genetics concerned mutations induced by x-rays, a procedure used by Hermann J. Muller in 1927 with *Drosophila*. Since this discovery of the *mutagenic* effect of x-rays, other agents have been discovered for experimentally inducing mutations, including some chemical compounds and extreme temperatures. Thus, certain environments can produce changes in genes. It should be noted that this phenomenon differs greatly from the old notion of inheritance of acquired characters. Mutations are random and most are harmful. Such mutations are rare in nature, perhaps only one in several million DNA replications. Given the large number of genes in an individual, however, it is clear that such mutations are also an important source of genetic change on an evolutionary time scale.

As we have seen, the average protein is a sequence of about 300 amino acids, each of which is coded by a triplet sequence of nucleotide bases. A change in any one of these bases can radically change the functioning of the gene's product through a change in the amino acid sequence. For example, sickle-cell anemia involves a change of just one of the 146 amino acids linked to form the β-chain of hemoglobin.

As we shall see in the following section, the capability of experimentally inducing mutations has proved to be of marked value in genetic research and in behavioral genetics analysis. It has contributed to our understanding of the molecular structure of genes, as well as the biochemistry of gene action. It has also been important to evolutionary theory. Recall that Darwin took great pains in considering the possible sources of heritable variation. Somewhat reluctantly, he concluded that Lamarckian mechanisms are an important source of variability. Contemporary evolutionary theory views mutation as the ultimate source of the genetic variability on which natural selection depends.

USING MUTATIONS TO DISSECT BEHAVIOR

As noted earlier, Beadle and Tatum (1941) used mutations in the common bread mold to study the operation of genes. They found mutants that were unable to convert certain substances into more complicated compounds. Their work revealed the sequence of normal metabolism, and demonstrated that each step in the sequence is controlled by a single gene—the "one-gene, one-enzyme" hypothesis.

In 1967 Seymour Benzer suggested that a similar approach could be used to dissect the physiological events underlying behavior. In the decades since Benzer's paper, hundreds of behavioral mutants have been selected in organisms as diverse as bacteria, fruit flies, crickets, and mice. Nearly all of these studies involved inducing mutations through the use of chemicals, screening individuals for behavioral abnormalities, and then searching for specific genetic causes.

In this section we shall look at attempts to perform genetic dissection of behavior in single-celled organisms, such as the bacteria and paramecia, as well as in more complex organisms, such as the round worm and fruit flies. The point of this section is to show how genes influence behavior. Genes affect behavior in the same way that they affect any phenotype—by controlling the production of enzymes.

Bacteria

Although the behavior of bacteria was studied in detail in the early part of this century, until recently this research area has been neglected. To be sure, the behavior of bacteria is by no means attention grabbing—in fact, most researchers ascertain if bacteria are dead or alive only by determining whether they reproduce. However, they do behave. They move toward or away from many kinds of chemicals (called *chemotaxes*) by rotating their four to eight tiny propellerlike flagella. When the flagella rotate in a clockwise direction, the single-celled organisms swims forward smoothly for about a second. When the flagella rotate counterclockwise, the organism tumbles and changes direction. The reason for the renewed interest in this simple behavior is that so much is known about the genetics of bacteria. This knowledge has led to rapid advances in isolating genes and the proteins responsible for various aspects of this behavior (Adler, 1976; Parkinson, 1977).

Great progress has been made since the first behavioral mutant in bacteria was isolated in 1966 (reviewed by Koshland, 1980). Behavioral

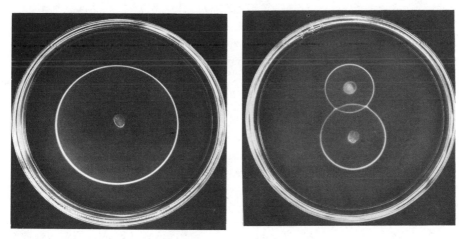

Figure 4.9
Bacteria on the move in normal and mutant strains. (a) Normal bacteria placed in the center of a plate with galactose radiate outward, forming a ring. (b) This plate has both galactose and ribose, and two mutant strains—one that cannot metabolize galactose and one that cannot metabolize ribose. Each strain moves independently of the other, forming two separate rings. (Courtesy of Julius Adler.)

mutants are isolated by screening huge numbers of bacteria for deviant behavior. Specific behavioral mutants that can accomplish only certain aspects of a response are especially important. For example, some mutant bacteria can swim well but cannot recognize certain chemical stimuli. One mutant strain can swim and recognize chemicals but cannot metabolize galactose. Another strain's only shortcoming is that it cannot metabolize ribose. When these strains are put on a plate with ribose and galactose, each consumes the sugar it can metabolize, and each strain moves independently of the other, as shown in Figure 4.9.

The normal swimming of bacteria involves steps much like those in more complex behaviors of higher organisms. The bacterium cell must detect a stimulus and then transmit this information to produce an appropriate response. Genetic dissection using the mutant strains indicates that the first step, recognizing the chemical stimulus, probably involves about 20 genes and at least nine genes are needed just to detect different types of sugars. The genes produce proteins that bind with particular substances, such as sugars and amino acids. The second step, signaling, involves using the products of at least three genes that recognize the extent of binding that has occurred. Finally, there are many genes involved in rotating the flagella. Some genes influence the rotor, other genes determine whether the flagella rotate clockwise or counterclockwise, and still others control the duration of the behavior. These genes have been mapped on the single circular chromosome of *E. coli*, as illustrated in Figure 4.10. This figure emphasizes the genetic complexity of an apparently simple behavior in a simple organism.

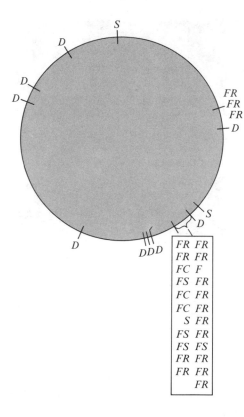

Figure 4.10

The single, circular chromosome of *E. coli* indicating the position of the genes that influence swimming. *D* refers to genes that produce proteins that bind with certain substances, thus detecting chemical stimuli. *S* are loci that signal a response by detecting the proportion of binding sites occupied. *F* loci are mutants that influence the rotor itself (*FR*), the switch that determines the direction that the rotor turns (*FS*), and genes that control the switch (*FC*). (From "Behavioral genetics in bacteria" by J. S. Parkinson. Reproduced, with permission, from *Annual Review of Genetics*, 11, 397–414. Copyright © 1977 by Annual Reviews, Inc. All rights reserved.)

This illustrates the fact that most normal behavior is influenced by many genes. As pointed out in the previous chapter, there are many single genes that can seriously disrupt normal behavioral development. However, even if many genes influence a particular behavior, a single gene can still disrupt the behavior. An automobile, which requires thousands of parts for its normal functioning, makes a good analogy. If any one part breaks down, the car may not run properly. In the same way, single genes can drastically affect behavior that is normally influenced by many genes.

Paramecia

Paramecia, like bacteria, are one-celled organisms, but they are larger and their movement is more obvious. They are covered by cilia that propel the cell forward or backward. Paramecia avoid certain chemical and thermal stimuli by backing up and then swimming forward in a new direction. Paramecia are also interesting because of their unusual mode of reproduction. They are diploid (that is, they have a pair of chromosomes), unlike bacteria, which have only one chromosome. Diploidy would complicate molecular genetic analysis in paramecia, except for the fact that they can reproduce by a process called *autogamy*. Autogamy is self-sexual reproduction, in which identical gametes are produced and fertilize each other, producing completely homozygous organisms in just a few hours.

A mutagenic agent is fed to paramecia and they are allowed to reproduce by autogamy, thus passing on mutations to their offspring. Behavioral mutants are screened from thousands of such organisms. Over 300 behavioral mutants have been isolated, and 20 genes have been implicated in the avoidance behavior of paramecia (Kung et al., 1975). For example, in some solutions, wild paramecia show repeated avoidance reactions. Some mutants, however, cannot swim backward; they are called *pawn* mutants (after the chess piece that can only move forward). Over 150 different pawn mutants have been identified, involving 62 different mutants at only three loci. Other mutants include: *paranoiac* (prolonged backward movement), which involves mutations at any of five different loci; *spinner* (spins in place in a certain solution), involving one locus; and *sluggish* (very slow mover) involving one locus. Figure 4.11 shows tracks of the movements of these strains during periods of about 10 seconds.

Analyses of avoidance behavior in paramecia have focused on the membrane and its electrical properties. For example, the pawn mutants have a defect in the permeability of the membrane for a particular chemical involved in the electrical response. Although researchers have done some molecular analyses of specific gene products (enzymes) underlying these behavioral mutants, such studies have not advanced as far as those of bacteria.

Round Worms

The nematode round worm *Caenorhabditis elegans* is intermediate in complexity between single-celled organisms and complex metazoans, such as fruit flies, mice, or human beings. It is about 1 millimeter long, and it has 959

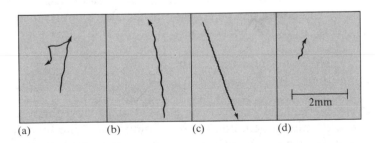

(a)   (b)   (c)   (d)

Figure 4.11
Movements of various paramecia strains. (a) Wild-type: spontaneous avoidance reactions, which cause turns. (b) Pawn: forward motion with no avoidance reactions. (c) Paranoiac: sustained, rapid backward motion. (d) Sluggish: very slow movement. (Adapted from "Genetic dissection of behavior in *Paramecium*" by C. Kung et al. *Science*, 188, 898–904. Copyright © 1975 by the American Association for the Advancement of Science.)

somatic cells, 302 of which are nerve cells. Most are self-fertilizing, and about 100,000 descendants can be produced in one week. Genetic dissection of the nematode's nervous system has given this lowly worm the distinction of being the only organism for whom we know the entire design of the nervous system (Kenyon, 1988).

The nematode is also the first multicelled organism for which the complete developmental fate map and timing of each cell has been determined. Genetic information about the nematode is accumulating rapidly; it has a total of about 2,000 genes (80 million DNA bases), and as many as a quarter of these are essential during embryogenesis (Cassada et al., 1980). Behavioral analyses have focused on chemotaxis mutants and dozens have been isolated; about half of these have observable differences in the anatomy of their nervous systems (Ward, 1977). For a recent behavioral genetic analysis using this organism, see Johnson (1986).

Drosophila

Benzer studied behavioral mutants in *Drosophila* because of the great store of available genetic information. Long lists of *Drosophila* behavioral mutants have been compiled (e.g., Grossfield, 1976). They include: *sluggish* and *hyperkinetic* mutants; *wings-up*, which cannot fly because of a defect in the muscle; *easily shocked*, in which jarring produces behavior similar to an epileptic seizure; *paralyzed*, which collapses whenever the temperature goes above 28° C; and *drop dead*, which walks and flies normally for a couple of days and then suddenly falls on its back and dies in just a few hours. More complex behaviors are also being studied. One such group involves courtship behavior in male *Drosophila*. Males follow other flies and then vibrate their wings prior to mounting for copulation. Behavioral mutants for aspects of courtship have been found. One male mutant, called *fruitless*, courts males as well as females and does not copulate with the females. Another male mutant cannot disengage from the female after copulation and is given the dubious title *stuck* (Hall, 1977a).

Benzer used an ingenious trick to narrow the search for mutant genes to the X chromosome. His trick also made possible the use of the important tool referred to as genetic mosaics, which we will discuss in the next section. Although the method is somewhat complicated, the goal is simply to create *Drosophila* with mutant genes on the X chromosome. Like human beings, male *Drosophila* have one X and one Y chromosome, and females have two X chromosomes. However, unlike the human situation, flies with two X chromosomes *and* one Y chromosome are fertile females. We will consider the special case of two *attached* X chromosomes that are inherited as a pair.

Suppose that we induce mutations among male *Drosophila* and cross them with nonmutated females with a double-X chromosome and a Y chromosome. Table 4.2 shows the types of offspring from such a cross. Two kinds of offspring do not survive—ones that receive the double-X from the mother

and an X from the father and offspring that receive the Y from the mother
and the Y from the father. This leaves only two kinds of offspring. One type
includes XY males with the X chromosome from the father, which thus show
any recessive mutations induced on the X chromosome of their father. Note
that this particular male receives the Y from his nonmutated mother. The
other type of offspring is a female like her mother (double-X and Y). She
received the double-X from her mother and cannot express mutations on the
X chromosome because she has not received an X chromosome from her
father. However, such a female can express mutations on the other chromosomes from her father. But these characters will be expressed only if they are
dominant because she receives only one autosome of each pair from her
mutated father. In summary, the males from such a cross express all of the
recessive mutations on the X chromosome and the females express only the
dominant mutations on the autosomal chromosomes.

In his early work, Benzer (1967) screened these offspring for their
response to light. Normal, wild-type *Drosophila* are positively *phototactic*—
that is, they move toward light. Two mutant male offspring from the above
cross were found to be nonphototactic (did not move toward light). Analyses
indicated that a single gene on the X chromosome was responsible for this
strange behavior. For example, when these males were mated to double-X
and Y females, all of their sons exhibited the same behavior. The mechanism
here is the same as that illustrated in Table 4.2. The only viable male offspring
from such a cross received the X chromosome from their father and express
the recessive mutated gene on the X chromosome. Females do not receive the
X chromosome from their father. These results indicated that nonphototaxis
was due to a gene mutation on the X chromosome. Hotta and Benzer (1970)
extended this work by describing a series of nonphototactic mutants with
problems of the retina. All of the mutant genes were located at five loci on the
X chromosome. Mutant analysis has revealed more than 100 different genes
involved in the structure of the *Drosophila* eye (Ready, Hanson, and Benzer,
1976). Other behavioral mutants have also been studied. For example, 48
mutants defective in flight behavior have been found (Homyk and Sheppard,
1977). The mutations were mapped to 34 different sites on the X
chromosome.

Table 4.2
Offspring from mating of normal male and attached-X
female *Drosophila*

		Male Gametes	
		X	Y
Female Gametes	$\widehat{XX}$ Y	$\widehat{XX}$Y (nonviable) XY (normal male)	$\widehat{XX}$Y (attached-X female) YY (nonviable)

Genetic Mosaics

Another method developed by Benzer to dissect behavior more precisely is the analysis of *genetic mosaics*. Genetic mosaics are individuals with different genes in various cells of the body. Normally, of course, life begins as a single cell, and the genetic material in that cell is replicated in every other cell of the body. In addition to the female with double-X and Y chromosomes, *Drosophila* have another genetic curiosity that permits mosaic analysis. There is an unstable, ring-shaped X chromosome that is frequently lost soon after fertilization. (See Figure 4.12.) Females with one normal X and one unstable X become mosaics because some cells lose the unstable X, and thus have only one X chromosome, compared to other cells with two X chromosomes. The cells with a single X chromosome will express all the genes on that chromosome. Moreover, in *Drosophila*, individuals with a single X chromosome are male and those with two X chromosomes are female. Thus, these mosaics are composites of male and female cells and are called *gynandromorphs* (from the Greek, meaning "characteristics of both sexes"), or XX-XO mosaics.

The XX and XO cells of gynandromorphs are not randomly intermingled; large, continuous areas with many cells are of the same genotype. By comparing many such XX-XO mosaics, it is possible to isolate the parts of the body responsible for certain behaviors. For example, work by Hotta and Benzer (1970) with 477 mosaics indicates that if the head is male, courtship behavior is male. A mosaic with a male head will follow other flies and vibrate its wings regardless of the sex of other parts of its body. Subsequent work indicates that only half of the brain must be male to produce male courtship behavior (Hall, 1977b). Hotta and Benzer also examined 130 mosaics that were successful in following and vibrating their wings and in copulation. The mosaic analysis suggested that successful copulation requires a male thorax (the part of an insect's body between the head and abdomen), as well as a male head. Of course, sex isn't all in the head; male genitals are also required for successful copulation.

The final step is to mate the females with the ring-shaped X chromosome to the mutant males described above. (See Figure 4.12.) These males pass on their X chromosome, with its mutant gene, to the female offspring. Half of the time, these female offspring also receive one of the "disappearing" X chromosomes from their mother. In this way, the mosaics are not just male-female mosaics. They are also mosaic for the recessive mutated genes on the X chromosome. "Marker" genes for body color, for example, can also be inserted on the X chromosome to permit easy identification of the particular "male" and "female" cells.

This method has been applied to many behavioral mutants to analyze which body parts must be mutant for the mutant behavior to occur. In an early study, mosaics were created using a gene that causes a defective visual response to light. Although there is considerable variability among flies, they normally move away from gravity (negative geotaxis) and toward light (posi-

Step one: Establish males with mutagenized X chromosome. (See Table 4.2.)

Step two: Establish female mosaics from females with unstable $X(X_u)$ chromosome.

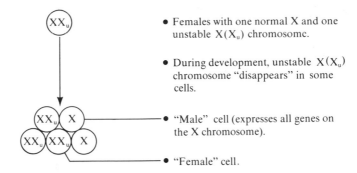

- Females with one normal X and one unstable $X(X_u)$ chromosome.

- During development, unstable $X(X_u)$ chromosome "disappears" in some cells.

- "Male" cell (expresses all genes on the X chromosome).

- "Female" cell.

Step three: Mate males with mutagenized X and females with unstable $X(X_u)$.

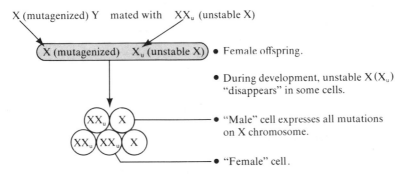

- Female offspring.

- During development, unstable $X(X_u)$ "disappears" in some cells.

- "Male" cell expresses all mutations on X chromosome.

- "Female" cell.

Figure 4.12
Mosaic analysis using *Drosophila.*

tive phototaxis). Thus, in a dark tube, most normal flies will climb straight up the tube whether or not there is a light at the top. A certain mutant gene produces a diminished response to light. When this mutant gene is present in mosaics, some flies will have one mutant eye and one normal eye. When these mosaics are put in a dark tube, they also climb straight to the top because their geotaxis is not affected. (See Figure 4.13a.)

However, when a light is placed at the top of the tube, the mosaic flies attempt to equalize the light coming into the two eyes, and thus keep their defective eyes toward the light. As a result, the mosaic flies climb up the tube in a spiral, keeping their mutant eyes closer to the light (Figure 4.13b). Spiral climbing occurs regardless of the amount of normal female tissue present, as long as one eye is normal and the other is mutant. Thus, the mutation is specific and isolates the defective visual response to the eye itself. As Benzer has indicated, phototaxis is a complex behavioral response: "Light is absorbed

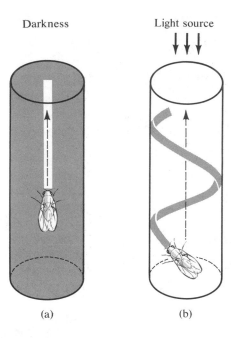

Darkness

Light source

(a)

(b)

Figure 4.13
(a) Behavior in darkness displayed by a mosaic fly with normal vision in one eye and defective vision in the other. Since the fly is negatively geotactic, it climbs straight up. (b) When light is shining from above, the same fly turns its defective eye toward the light and climbs in a helical path. (From Hotta and Benzer, 1970.)

by a pigment in the receptor cell, producing neural excitation, transmission at synaptic junctions, integration in the central nervous system involving comparison with other inputs, and generation of appropriate motor signals such that the fly walks in a particular direction" (1967, p. 1118). A mutation resulting in a defect in any of these structures or processes could lead to a change in phototaxis. As indicated earlier, mutants for more than 100 different genes affecting the eye have been isolated. These are now used in mosaic analyses to dissect the visual system and to provide information about the normal sequence of events.

Mutant and mosaic analysis has also been applied to learning. A *Drosophila* mutant defective in avoidance learning has been isolated (Dudai et al., 1976). Flies were taught to avoid a particular odor that had been coupled with shock. Although flies avoided the odor only 30 percent of the time, the learning lasted a day (Quinn, Harris, and Benzer, 1974). After screening about 500 lines of flies with mutagenized X chromosomes, one mutant line was found that could not learn to avoid the odor associated with shock even though it had normal sensory and motor behavior. In addition to this mutant, called *dunce*, other learning mutants called *rutabaga, cabbage,* and *turnip* have been found, as well as a memory mutant, *amnesiac* (Aceves-Pina et al., 1983), and interesting behavioral research has been conducted to clarify the learning processes involved in these mutants (Dudai, 1983; Duerr and Quinn, 1982; Kyriacou and Hall, 1984). Recent reviews of mutant research on learning and courtship behaviors are available (Quinn and Greenspan, 1984; Tully, 1984).

Genetic Dissection and Development
(Fate Mapping)

The method of genetic dissection of behavior promises to assume even greater importance in behavioral genetics (Hall, 1984). Although mice are much slower breeders than bacteria, paramecia, round worms, and *Drosophila*, genetic dissection of the nervous system of mice has also begun. One interesting technique involves *chimeras*, animals derived by putting together cells from different embryos. If mutant and normal mouse embryos are fused, the resulting chimera is a composite of normal and mutant cells. Chimeras can be used to identify the specific tissue involved in gene action because chimeric mice that show the mutant defect should have the same physical defect (McLaren, 1976). Although the chimeric technique has been primarily used to study neurological mutants, behavioral research has also been conducted using chimeras between two inbred strains of mice. In general, this work indicates that behavior is affected by more than one cell population (Nesbitt et al., 1981; Nesbitt, Spence, and Butler, 1979). The more fine-grained mosaic analysis has been limited primarily to *Drosophila*, however, because it is easy to generate XX-XO mosaics (Hall, 1977a).

A review of genetic dissection research concludes that the next phase of research will begin to unravel the developmental interaction between genes and behavior: "Many of the mutants that have been selected for behavioral alterations are altered in the development of parts of the nervous system. This means that eventual understanding of the effects of these mutants must be sought by understanding the gene control of development. This is the most challenging problem confronting the analysis of neurological mutants" (Ward, 1977, p. 444).

One step in this direction is mosaic fate mapping. In 1929 Alfred Sturtevant (see Benzer, 1973) proposed that mosaics could be used to "map" what happens to the cells of the *blastoderm*—the surface of the blastula, which is a hollow ball consisting of a single layer of cells surrounding the yolk in the young embryo. In XX-XO mosaic blastoderms, there is a dividing line between the areas of XX and XO cells. The orientation of this boundary is random, which means that the likelihood that this boundary will pass between two cells is proportional to the distance between them. We also know from histological (tissue) analysis that the location of a cell in the blastoderm determines its fate throughout development (Baker, 1978). Thus, in the adult mosaic, body parts that are usually of different genotypes (XX versus XO) are likely to be farther apart in the blastoderm. Sturtevant scored pairs of body parts in 379 mosaics for the frequency with which one part was XX and the other XO. The more often this happened, the farther apart were the ancestor cells in the blastoderm, as determined by histological analysis (Gehring, 1976).

Benzer and Hotta extended this approach to behavior (Benzer, 1973). First, using 703 mosaics, they constructed an anatomical fate map, as shown in Figure 4.14. The distances between the origins of various parts are scored in

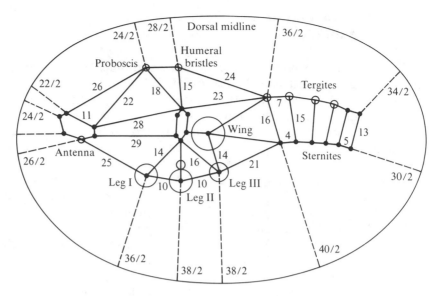

Figure 4.14
A blastoderm fate map of the external body parts of *Drosophila*, based on the probability that two parts of the body are of different genotypes in mosaic individuals. (From "Genetic dissection of behavior" by S. Benzer. Copyright © 1973 by Scientific American, Inc. All rights reserved.)

sturts, in honor of Sturtevant. One sturt represents a 1 percent probability that the two structures will be of different genotypes. For example, legs I and II were of different genotypes 10 percent of the time, legs II and III differed 10 percent of the time, and legs I and III differed 20 percent of the time.

Benzer and Hotta then began to relate behavioral characteristics to this fate map. For example, the hyperkinetic mutant shakes all six of its legs when anesthetized. They showed that each leg's shaking was independent of that of the other legs. When they mapped the shaking of the legs in the hyperkinetic flies, they found that the location of the original cells in the blastoderm is near the location for the appropriate leg, but always below it, in a region that is the origin of part of the nervous system. (See Figure 4.15.) The embryonic locations of the *drop-dead* and the *wings-up* mutant genes mentioned earlier are also shown in Figure 4.15. Mutant and mosaic analyses have been conducted for other behaviors, such as flying (Homyk, 1977).

Benzer has summarized this approach as follows:

> In tackling the complex problems of behavior the gene provides, in effect, a microsurgical tool with which to produce very specific blocks in a behavioral pathway. With temperature-dependent mutations the blocks can be turned on and off at will. Individual cells of the nervous system can be labeled genetically and their lineage can be followed during development. Genetic mosaics offer the equivalent of exquisitely fine grafting of normal and mutant parts, with the

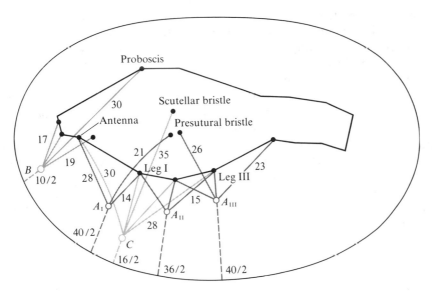

Figure 4.15

A fate map showing the sites on the blastoderm at which mutant genes (*A*) affect leg shaking in the hyperactive mutant. A similar map exists for the other side of the blastoderm, which affects shaking of the three legs on the other side of the body. The blastodermal site of the (*B*) drop-dead and (*C*) wings-up mutant genes are also shown. (From "Genetic dissection of behavior" by S. Benzer. Copyright © 1973 by Scientific American, Inc. All rights reserved.)

entire structure remaining intact. What we are doing in mosaic mapping is in effect "unrolling" the fantastically complex adult fly, in which sense organs, nerve cells and muscles are completely interwoven, backward in development, back in time to the blastoderm, a stage at which the different structures have not yet come together. Filling the gaps between the one-dimensional gene, the two-dimensional blastoderm, the three-dimensional organism and its multidimensional behavior is a challenge for the future. (Benzer, 1973, p. 15)

These attempts to use mutations to dissect behavior, including developmental dissections, will profit immensely from the tools of the "new genetics" described in Chapter 7.

GENES AND INDIVIDUAL DIFFERENCES

In the previous chapter, discussing Mendelian genetics, a basic message has been genetic variability. The knowledge that genes are sequences of DNA nucleotide bases also adds to our understanding of variability. The ultimate source of genetic variability is mutation, random changes in nucleotide base sequences. Although most mutations are harmful, they are the price paid for genetic variability and the possibility of evolutionary change. We also know that all genes are not simply templates of protein production. Regulator genes control the production of proteins, and mutations of these genes can result in additional genetic variability.

Studies of enzyme differences indicate that more than a third of structural genes have two or more alleles. This is likely to be just the tip of the iceberg of genetic variability. It now appears that most genetic variability lies with genetic regulation rather than with structural genes, that is, with regulation of levels of expression of DNA rather than with differences in the amino acid sequences coded by structural genes. For behavioral geneticists, the point is that genetic variability is the rule rather than the exception; the task of behavioral genetics is to determine the extent to which genetic variability accounts for behavioral differences among individuals.

SUMMARY

Perhaps the most exciting advance in biology in this century has been in understanding Mendel's elements. Early in the century, inborn errors of metabolism suggested the "one-gene, one-enzyme hypothesis." Knowledge of the double-stranded DNA molecule followed. The structure of DNA relates to its dual functions of self-replication and protein synthesis. Genes are sequences of nucleotide pairs long enough to code for a specific sequence of amino acids in a particular polypeptide. In addition to structural genes that serve as templates for protein production, there are also regulator genes that control the transcription of structural genes.

Mutations are changes in the nucleotide bases of DNA. Much of our early knowledge of genes was the result of experimentally induced mutations. Mutations are now being used to dissect behavioral sequences in many organisms, such as bacteria, paramecia, round worms, and *Drosophila*. The combination of mutants and mosaics in *Drosophila* permits powerful analyses of the fine structure of behavioral development.

CHAPTER·5

Chromosomes

$\mathbf{A}$dvances in the field of cytology—the study of the cell and its contents—led to a major breakthrough in understanding the physical nature of heredity. In the mid-nineteenth century, it was generally accepted that cells are the basic units of living organisms. Aided by new knowledge about the chemistry of dyes, cytologists were able to stain the contents of cells to make them more visible for study. It was soon found that a portion of the cell, the *nucleus*, contains a number of small rod-shaped bodies called *chromosomes* ("colored bodies") because they can be stained by particular dyes. The number of chromosomes is the same in all cells of an organism, except for the sex cells (the sperm and eggs, which are called *gametes*). In the non-sex cells (called *somatic cells*), chromosomes come in pairs. The total number of chromosomes in each somatic cell is called the *diploid number*. In sex cells, only half of the chromosomes—one member of each pair—are represented. The number of chromosomes in each sex cell is referred to as the *haploid number*. All normal individuals of a species have the same number of chromosomes in each cell, although the number varies widely from one species to another. For example, the pea plant has 7 pairs of chromosomes, wheat has 21, mosquitoes have 3, fruitflies have 4, carp have 52, mice have 20, dogs have 39, and humans have 23. It gradually became clear that chromosomes were involved in heredity.

MENDEL AND CHROMOSOMES

Mendel did not know that his elements were parts of chromosomes. His law of independent assortment states that the elements, or alleles, for two different traits are inherited independently. However, in Chapter 3 we noted that Mendel would have been in for a surprise if he had happened to study two traits affected by genes located close together on the same chromosome. In this case, the traits would have been inherited together, and he would not have found independent assortment. The dihybrid segregation ratio of $9:3:3:1$ would have been closer to a $3:1$ ratio. As indicated in Figure 5.1, the ratio would be $3:1$ if the two loci are so closely linked that no recombination occurs. If the A and B loci were less tightly linked, some recombinants (such as $A_1A_1B_2B_2$) would occur in the F_2 progeny.

It is now known that the pea plant has seven pairs of chromosomes. So given that Mendel studied seven traits, the probability is very small (0.006) that all seven genes were located on different chromosomes pairs. Why, then,

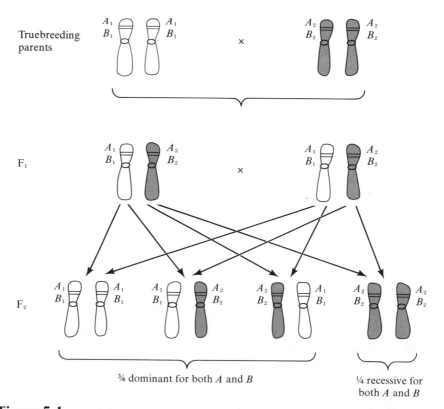

Figure 5.1
Dihybrid segregation ratio if genes are very closely linked on the same chromosome. (A_1 and B_1 are dominant.)

didn't Mendel find dihybrid segregation ratios that indicated linkage? The answer to that question may be found in Box 5.1.

CELL DIVISION

Cell division involves two processes. The first involves duplication of cells. We begin life as a single cell and end up with about 10^{14} cells in our adult bodies. Second, a sample of half of our genes is represented in each of our sex cells (sperm and eggs). The two types of cell division involved in these processes are *mitosis* and *meiosis*, respectively. (See Figure 5.2.)

Box 5.1

Why Didn't Mendel Find Linkage?

Mendel studied seven single-gene traits of the pea plant. The pea plant has only seven chromosomes. The probability that each of the seven genes selected at random would be on a different chromosome is only 6 in 1,000. Therefore, why didn't Mendel observe any dihybrid segregation ratios other than 9:3:3:1?

The answer to the paradox is *not* that the seven genes that Mendel studied are on seven different chromosomes (Blixt, 1975). All seven genes have now been mapped to specific chromosomes, and we know that two of them are linked closely on chromosome 4. These loci are linked sufficiently close to yield a significant departure from a 9:3:3:1 ratio. However, given seven traits, there are 21 possible dihybrid crosses. Only one of these would have shown this departure from the expected 9:3:3:1 dihybrid segregation ratio. Mendel apparently did not perform that particular dihybrid cross.

In addition to the two tightly linked genes, Mendel reported dihybrid ratios for some other genes that were on the same chromosome. However, unless genes are close together on the same chromosome, they will recombine due to the process of *crossing over*, in which chromosomes exchange parts (as discussed in detail later in this chapter). The greater the distance between loci, the greater the frequency of recombination. The genes that Mendel happened to study are, in fact, so far apart on their chromosomes that they would not have shown a significant departure from the expected dihybrid segregation ratios for unlinked genes.

We should note that departures from expected dihybrid segregation ratios are not just a nuisance in Mendelian genetics. They provide an important tool for determining whether genes are, in fact, located on the same chromosome. Departures from expected segregation ratios suggest that the genes are linked. The less recombination that occurs, the closer together are the genes on the chromosome.

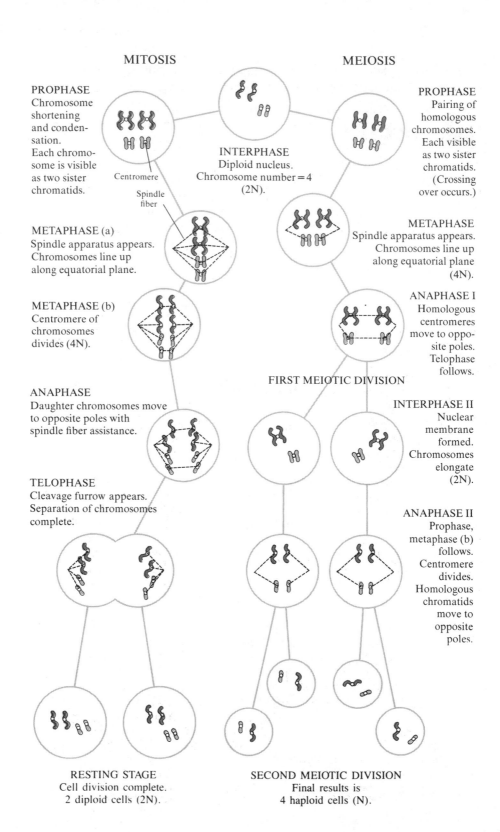

MITOSIS MEIOSIS

PROPHASE
Chromosome
shortening
and conden-
sation.
Each chromo-
some is visible
as two sister
chromatids.

Centromere

Spindle
fiber

INTERPHASE
Diploid nucleus.
Chromosome number = 4
(2N).

PROPHASE
Pairing of
homologous
chromosomes.
Each visible
as two sister
chromatids.
(Crossing
over occurs.)

METAPHASE (a)
Spindle apparatus appears.
Chromosomes line up
along equatorial plane.

METAPHASE
Spindle apparatus appears.
Chromosomes line up
along equatorial plane
(4N).

METAPHASE (b)
Centromere of
chromosomes
divides (4N).

ANAPHASE I
Homologous
centromeres
move to oppo-
site poles.
Telophase
follows.

ANAPHASE
Daughter chromosomes move
to opposite poles with
spindle fiber assistance.

FIRST MEIOTIC DIVISION

INTERPHASE II
Nuclear
membrane
formed.
Chromosomes
elongate
(2N).

TELOPHASE
Cleavage furrow appears.
Separation of chromosomes
complete.

ANAPHASE II
Prophase,
metaphase (b)
follows.
Centromere
divides.
Homologous
chromatids
move to
opposite
poles.

RESTING STAGE
Cell division complete.
2 diploid cells (2N).

SECOND MEIOTIC DIVISION
Final results is
4 haploid cells (N).

Box 5.2

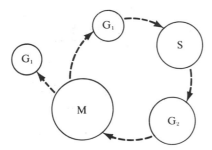

Typical Cell Cycle

G_1 is a period of initial cell growth, lasting about eight hours.

S   is a period (about six hours) in which chromosome replication occurs. Since genes are carried on chromosomes, and since DNA is the genetic material, this is a period of active DNA synthesis.

G_2 is a second period of growth, which lasts about five hours.

M   is the final phase, in which cell division occurs. This is the shortest phase (about one hour) and yields two identical daughter cells that begin the cycle over again. During this phase, the chromosomes become visible under the light microscope and various stages of mitosis may be observed.

Mitosis

In the process of growth, cells divide into two "daughter cells," each of which then later divides into two more, and so forth. Many cells in our bodies go through a typical cell cycle lasting about a day. (See Box 5.2). On the other hand, some cells, such as nerve cells, rarely divide. Others (e.g., liver cells) divide only when injured; cancer cells divide extremely rapidly.

Study of the chromosomes revealed that a remarkable series of changes takes place during cell division. The major features of this process of mitosis are illustrated in the left half of Figure 5.2. Prior to the splitting of the cell, the chromosomal material is replicated and spindle fibers become attached to the centromere of each chromosome (a genetically inactive chromosome region).

Figure 5.2

(A comparison of mitosis and meiosis. N is the number of chromosomes of each type. (From James D. Watson, *Molecular Biology of the Gene*, 3d ed., Benjamin/ Cummings Publishing Company, Menlo Park, CA. Copyright 1976.)

During cell division, half of the material goes into one daughter cell, and half into the other. Since different chromosomes are somewhat distinctive in shape and size, researchers were able to determine that each daughter cell receives an equivalent chromosomal complement. This distinctiveness of chromosomes also resulted in the conclusion that chromosomes exist in pairs. In Figure 5.2, one chromosome of each pair is of paternal origin; the other is of maternal origin. Thus, two pairs of homologous chromosomes, or matched sets, are shown in the figure. Artificial chromosomes that have recently been constructed in yeast are beginning to elucidate some of the complexities of mitosis (Murray and Szostak, 1987).

Meiosis

The process by which one set of chromosomes is contributed by each parent is due to *meiosis*. Meiosis essentially consists of the splitting of a cell into two, without the prior doubling of chromosomal number that occurs in mitosis. (See Figure 5.2.) More precisely, meiosis begins with doubled chromosomes. Each cell divides twice, so each original diploid cell yields four gametes, each with a haploid number of chromosomes. One member of each pair of homologous chromosomes is drawn into each haploid daughter cell before the division is complete.

The haploid set of chromosomes included in any one gamete, however, is not necessarily the same haploid set that the individual received from its mother or father. A reshuffling occurs, so that an individual transmits to its offspring some of the chromosomes received from its mother and father. This reshuffling of chromosomes can create considerable genetic variability. For example, your gametes (sperm or eggs) can have any of over 8 million possible haploid combinations (2^{23}) of your 23 pairs of chromosomes. Any of these haploid combinations can fertilize, or be fertilized by, the 8 million possible combinations of gametes of your mate. This means that you and your mate can potentially create 64 trillion chromosomally different zygotes. Meiosis results in even more genetic variability by means of the process of crossing over, discussed in the following section.

In short, mitosis occurs in almost all cells of the body. Its function is to duplicate cells. Meiosis occurs only in the ovaries and testes, which produce the sex cells. Its function is to shuffle chromosomes and produce gametes with single chromosomes rather than chromosome pairs.

CROSSING OVER

Understanding of the chromosomal basis of heredity helped to delineate exceptions to the law of independent assortment, as mentioned in Box 5.1. It was evident long ago that there must be more genes than there are chromosomes. In other words, each chromosome must contain a number of loci. If two characteristics are determined by loci on the same chromosome, the alleles at the two loci may not assort independently.

Genes on the same chromosome are said to be *linked* if their alleles do not assort independently. However, linkage does not mean that the traits occur together in a population. Linkage refers to loci, not to alleles for segregating loci. For example, hemophilia and color-blindness are both determined by recessive alleles of genes on the same (X) chromosome. However, a hemophiliac is no more likely than anyone else to be color-blind, because the alleles for hemophilia and color-blindness are not inherited together even though their loci are on the same chromosome.

Alleles for two loci on the same chromosome may be separated in a population by the process of *crossing over*. During the first stage of meiosis, homologous chromosomes line up pair by pair. (See Figure 5.2.) Each member of each pair duplicates, and the duplicates (chromatids) separate, except at the centromere. The chromosomes frequently come into contact and exchange parts or "cross over." Crossing over may occur within a locus, as well as between loci. That is, crossing over can break up DNA at any point, not just at some point that divides two loci. This is another important source of genetic variability, because the sequence of nucleotide bases can change if chromosomes trade bases within a locus (thus creating new alleles).

The "life expectancy" of a genetic unit is related to its size. A whole chromosome will survive intact only one generation if the chromosome, on the average, undergoes one crossover every time a gamete is formed. A smaller genetic unit, say 0.002 of a chromosome (close to the size of the average gene), has a 0.9998 chance of surviving intact from either parent.

One way of understanding the important contribution that crossing over makes to genetic variability is to consider your parents. Your father had the same 23 pairs of paternal and maternal chromosomes in every cell of his body, except his sperm. When he formed sperm through meiosis, some of the homologous chromosomes from his mother and father crossed over, so chromosomes in his sperm were a unique patchwork of his mother's and father's chromosomes. The patchwork chromosomes in the sperm that fertilized the egg that produced you are now half the chromosomes in every cell of your body. Your gametes, in turn, are recombined chromosomes from your father and mother.

Figure 5.3 is an illustration of this process for one pair of chromosomes only. It should be remembered that the same events may be occurring at the same time for all other chromosome pairs. The maternal chromosome, carrying the alleles A_1, B_2, and C_1, is represented in white; the paternal chromosome, with A_2, B_1, and C_2, is gray. When each homologous chromosome duplicates to form sister chromatids, these chromatids may cross over one another. During this stage, the chromatids can break and rejoin. Each of the chromatids will be transmitted to a different gamete. Consider only the A and B loci for the moment. As shown at the bottom of Figure 5.3, one gamete will carry the genes A_1 and B_2, as in the grandmother, and one will carry A_2 and B_1, as in the grandfather. The other two will carry A_1 with B_1 and A_2 with B_2. For the latter two pairs, recombination has taken place.

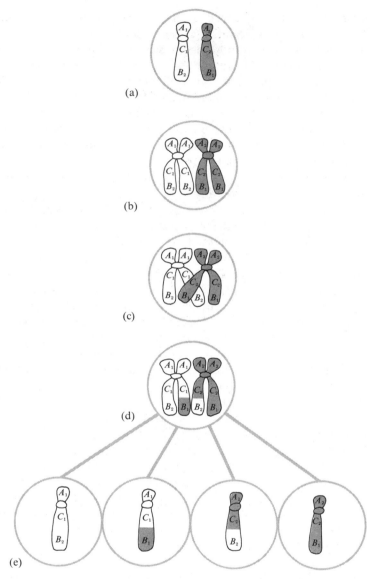

Figure 5.3
Diagrammatic illustration of crossing—the mutual exchange of material by homologous chromosomes. (After "The inheritance of behavior" by G. E. McClearn. In L. G. Postman, ed. *Psychology in the Making*, Copyright © 1963. Used with permission of Alfred A. Knopf, Inc.)

Crossing over of this kind can occur at any place along the chromosome, and the probability of recombination is a function of the distance between the particular genes. In Figure 5.3, for example, crossing over has not affected the relationship between the A locus and the C locus. All gametes are either A_1C_1 or A_2C_2, as in the grandparents, since the crossover did not occur between these loci. Crossing over could occur between the A and C loci, but that would be less frequent than between A and B. Because of this, the crossover gametes occur less often than the noncrossover. Genes located on different chromosomes do, of course, assort at random. These facts have been used as a tool to map genes on chromosomes.

Linkage: Mapping Genes

If two genes are closely linked on the same chromosome, they will not assort according to the $9:3:3:1$ segregation ratio predicted by Mendel's law of independent assortment. Moreover, the closer the genes are on a chromosome, the less likely they are to be separated by crossing over. Thus, we can map the distance between two genes on a chromosome by determining the extent to which recombination occurs. For example, in Figure 5.3 the A locus must be closer to the C locus than it is to the B locus, if recombination is less frequent between A and C than between A and B. For plants and nonhuman animals, controlled test crosses can be obtained to determine the extent of recombination. For example, consider genes A and B, each with two alleles. A test cross involves matings between "doubly heterozygous" individuals and "doubly homozygous" ones. For example, consider the cross between $A_1A_2B_1B_2$ (doubly heterozygous) and $A_2A_2B_2B_2$ (doubly homozygous) individuals. When the A and B loci are on different chromosomes, there will be four types of offspring, each with equal frequency:

Frequency of Gametes Produced by Doubly Heterozygous Parents

	$\frac{1}{4}\ A_1B_1$	$\frac{1}{4}\ A_1B_2$	$\frac{1}{4}\ A_2B_1$	$\frac{1}{4}\ A_2B_2$
	Offspring			
Gametes from Doubly Homozygous Parents A_2B_2	$\frac{1}{4}\ A_1A_2B_1B_2$ +	$\frac{1}{4}\ A_1A_2B_2B_2$ +	$\frac{1}{4}\ A_2A_2B_1B_2$ +	$\frac{1}{4}\ A_2A_2B_2B_2$

However, if A and B were close together on the same chromosome, the law of independent assortment would not prevail. Suppose that the doubly heterozygous parent had the A_1 and B_1 alleles on one chromosome and the A_2 and B_2 alleles on the other. The doubly homozygous parent has A_2B_2 on both chromosomes. Without crossing over, we would not find any of the "mixed" offspring, $A_1A_2B_2B_2$ or $A_2A_2B_1B_2$. If an offspring has an A_1 allele, then that offspring should also have a B_1 allele (that is, *not* B_2B_2) because the A_1 allele is on the same chromosome as the B_1 allele from the doubly heterozygous parent. Similarly, if the offspring has a B_1 allele, that offspring should also have an A_1 allele, not A_2A_2. However, crossing over could create these recombinations. *Map distance*, the relative distance between genes on a chromosome, is simply the percentage of recombination. For instance, where A and B are on different chromosomes, the map distance (percent recombination) is 50 — meaning that 50 percent of the offspring are recombinant types. This suggests that there is no linkage, and thus independent assortment. However, suppose the percentage of the two recombinant genotypes were 40 percent instead of 50 percent. The map distance between A and B genes would be 40 map units. If there is less recombination, the map distance is smaller and the genes are closer together on the chromosome. Map units are called centimorgans (cM) in honor of Thomas Hunt Morgan, who pioneered work on gene mapping in *Drosophila*. One centimorgan corresponds roughly to about 1 million base pairs.

In 1906, Bateson found two genes in the sweet pea that did not assort independently. By 1915, T. H. Morgan and colleagues summarized results for 85 genes in the fruit fly *Drosophila* and showed that they fall into four linkage groups. We now know that *Drosophila* has four chromosomes.

For humans, the problem is more difficult because naturally occurring test crosses must be found. Expectations for recombination must be estimated for a given set of family data. Linkage for the sex chromosome is more easily studied because males have an X and a Y chromosome and females have two X chromosomes. Recessive traits on the X chromosome show a greater incidence in males. Because the gene is recessive, a woman will show the trait only if she is homozygous for those alleles. In males, however, there is no corresponding locus on the Y chromosome, so a single recessive gene on the X chromosome is expressed. If a gene is on the X chromosome, we would expect to find no father-to-son inheritance because fathers give their sons only the Y chromosome. Other predictions can be made concerning familial resemblances for an X-linked trait. (See Chapter 3.) The first X-linkage to be established for humans was color blindness, and several dozen other X-linked traits were demonstrated during the next half century.

Once sex linkage is demonstrated, map distance can be determined by the "grandfather method." Females who are doubly heterozygous for two genes on the X chromosome can be detected because their fathers will show the alleles present on one of their X chromosomes. Recombination can then be studied for the sons of the doubly heterozygous females, and map distance can be determined.

In 1936 Haldane established the distance between the genes for color blindness and hemophilia on the X chromosome. Over 100 genes have now been identified as being located on the human X chromosome. In 1951 the first linkage between two human autosomal genes (located on a chromosome other than the sex chromosomes, X and Y) was demonstrated (Mohr, 1954), and in 1967 M. C. Weiss and H. Green localized the first human gene to a specific chromosome. Linkage studies are most powerful when large pedigrees are studied; three generations with at least eight individuals (at least two affected) constitutes the minimum. Chapter 7 discusses linkage in greater detail and uses as a concrete example the famous Venezuelan pedigree that led to the discovery that the gene for Huntington's disease is located on chromosome 4.

Rather than using large pedigrees, another approach to linkage is the sib-pair method, which also relies on classical genetic analyses of segregation (Solomon and Bodmer, 1979). The sib-pair method examines the co-segregation of a behavioral trait and a genetic marker in pairs of siblings, both of whom show the particular behavior. If the behavioral trait is linked to a genetic marker, Mendel's second law of independent assortment should be violated. The advantage of this approach is that the mode of inheritance does not need to be known and it can be used to isolate genetic effects on complexly inherited rare traits. For example, Hodgkin's disease is a form of cancer that affects 1 in 10,000 individuals and shows no familial inheritance in over 90% of cases. Nonetheless, a rare dominant gene that causes Hodgkin's disease in only about 5% of individuals with the disease has been detected. Despite its rarity, when this genetic form of the disease occurs in sibling pairs, it shows linkage to a marker on chromosome 6.

The probability of linkage is expressed as an odds ratio: the probability that the observed data would arise under one hypothesis (e.g., linkage at 10 percent recombination between the marker and the trait) as a ratio to the probability of nonlinkage (i.e., 50 percent recombination). Odds ratios from independent pedigrees can be multiplied together, and when the odds become overwhelming, linkage is considered proven. By convention, an odds ratio of 1,000:1 in favor of linkage is used to demonstrate linkage, and an odds ratio of 100:1 against linkage is considered sufficient to reject linkage. If linkage is established, the recombination fraction at which the odds ratio is largest, referred to as θ, is used as the recombination fraction.

In place of the odds ratio, human geneticists use the logarithm to the base ten of the odds ratio, which is called a LOD (logarithm of odds) score. LOD scores can be summed across independent pedigrees. A LOD score of 3 signifies linkage, and a -2 LOD score indicates nonlinkage. The advantage of the LOD score approach with large pedigrees is that a single pedigree can demonstrate linkage, which means that even linkage for a rare disease that occurs only in a particular pedigree can be detected.

Linkage and mapping research continues to progress rapidly. At the 1986 Human Gene Mapping meeting, a total of 376 autosomal and 124 X-chromosome loci were mapped; 222 other assignments were provisional. In part, this rapid progress is due to the development of a mapping technique called somatic cell hybridization, described in Box 5.3, which makes it possible to study linkage experimentally rather than analyzing segregation of traits in pedigrees. The major reason, however, for the great expansion in linkage results is the use of recombinant DNA techniques that have resulted in over 1,000 genetic markers that can be used in linkage studies, as discussed in Chapter 7.

Box 5.3

Somatic Cell Hybridization

A completely different method—one that sounds like science fiction—has greatly increased our knowledge of human chromosomes. The technique is called *somatic cell hybridization* because it fuses human cells with the cells of other mammals. *Somatic* refers to cells of the body, as opposed to sex cells. The resulting hybrid cells have different assortments of human chromosomes and chromosomes from the other mammals. The figure on the facing page shows (a) the 40 chromosomes of a mouse cell, (b) the 46 chromosomes of a human cell, and (c) a man–mouse hybrid cell of 73 chromosomes, only eight of which are human chromosomes. All of the mouse and human chromosomes are active in these hybrid cells. By comparing enzymes from replicates, or clones, of such hybrid cells, we are able to determine which chromosomes are responsible for specific enzymes. The last panel of the figure (d) shows hypothetical results for four human enzymes and three human chromosomes in five hybrid clones. Linkage is determined by noting which human enzymes and chromosomes appear simultaneously in the different hybrids. For example, in the figure, enzymes I and III appear together in all five clones and they are correlated with the presence of chromosome 2. Thus, genes for these two enzymes must be linked on chromosome 2.

In recent years, recombinant DNA techniques have led to two direct mapping methods that have greatly assisted mapping efforts: *in situ hybridization* and *high-resolution chromosome sorting*. These techniques are described in Chapter 7.

It should be noted that somatic cell hybridization has a severe limitation: The genes must be expressed in the hybrid cell if they are to be detected. Nonetheless, well over 100 human genes have been mapped using this technique. Moreover, once a segment of DNA has been cloned, using the recombinant DNA techniques described in Chapter 7, the clone can be used as a probe to identify the same DNA sequences on a chromosome in a hybrid cell. This obviates the need for gene expression in somatic cell hybridization, an important advance, because many human DNA segments are not expressed in man–mouse hybrid cells. Localization of the linked gene in hybrid cells takes advantage of naturally occurring reciprocal translocations in which segments of two different chromosomes have been exchanged. Over 300 human reciprocal translocations have been identified, and cells from the affected individuals have been stored; if a gene is absent from chromosomes with a particular translocation, it must normally reside in the region spanned by the translocated segment. Over 100 human genes have been assigned to particular subchromosomal regions using this technique.

Somatic cell hybridization. (a) The 40 chromosomes of a mouse cell. (b) The 46 chromosomes of a human cell. (c) Mouse and human chromosomes are both present in a hybrid cell formed by the fusion of two cells like those whose chromosomes are pictured here. There are 73 chromosomes in this cell, only eight of which are human (marked with arrows; courtesy of F. H. Ruddle and R. S. Kucherlapati.) (d) Hypothetical results for somatic cell hybridization. (From "Hybrid cells and human genes" by F. H. Ruddle and R. S. Kucherlapati. Copyright © 1974 by Scientific American, Inc. All rights reserved.)

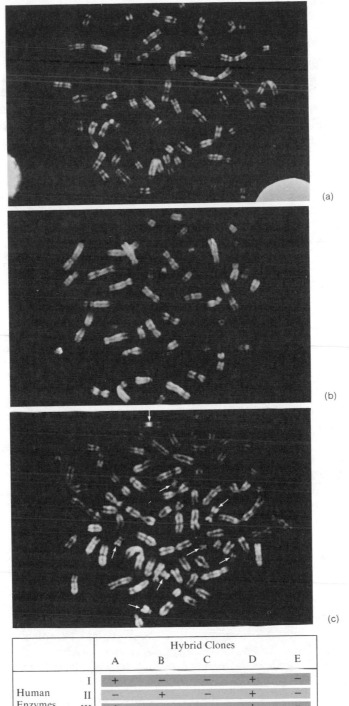

		Hybrid Clones				
		A	B	C	D	E
Human Enzymes	I	+	−	−	+	−
	II	−	+	−	+	−
	III	+	−	−	+	−
	IV	+	+	+	−	−
Human Chromosomes	1	−	+	−	+	−
	2	+	−	−	+	−
	3	−	−	−	+	+

(a)

(b)

(c)

(d)

HUMAN CHROMOSOMES

Although the chromosomes of *Drosophila* and other organisms were subjected to detailed analyses as early as the 1930s, human cytogenetics lagged far behind. Prior to 1956, students were taught that the number of chromosomes in humans was 48 (24 pairs). However, after using improved techniques, it was reported in 1956 that the normal diploid chromosome number in humans is 46, not 48 (Tjio and Levan, 1956). Since that date, important developments in human cytogenetics have occurred with great rapidity.

Karyotypes

In order to study the chromosomal complement, or *karyotype*, of an individual, a sample of white blood cells (*leukocytes*) is usually obtained and cultured in the laboratory for two or three days. A chemical (phytohemagglutinin) is added to the culture to stimulate growth and cell division. Dividing cells are then exposed to colchicine, a chemical that inhibits the separation of doubled chromosomes. This results in the accumulation of cells at the stage of mitosis, illustrated in Figure 5.2, in which the duplicated chromosomes have shortened and thickened and are still attached at the centromeres. The cells are then washed with a saline solution, resulting in swelling of the cells and spreading out of the chromosomes. When these cells are squashed or air dried, the chromosomes tend to lie in the same optical plane. The cells are then stained and photographed under high-power magnification. The chromosomes in the photograph may then be cut out and rearranged according to their size and the location of the centromeres. The karyotype of a normal male is shown in Figure 5.4.

An international conference was held in Denver, Colorado, in 1960 for the purpose of standardizing the classification of human chromosomes. The resulting "Denver classification" is based on both chromosome length and location of the centromere. If the centromere divides a chromosome into arms of approximately equal length, the chromosome is said to be *metacentric*. If the centromere is very close to one end of the chromosome, the chromosome is referred to as being *acrocentric*. If the centromere is located somewhere between the middle and one end, the chromosome is described as being *submetacentric*.

As indicated in Figure 5.4, the 23 pairs of human chromosomes are classified into seven distinct groups. Group A includes chromosome pairs 1, 2, and 3. These are large chromosomes that can be distinguished from others on the basis of their size and the central location of the centromere. Group B includes chromosome pairs 4 and 5, which are large submetacentric chromosomes. Group C, the largest group, includes chromosome pairs 6 through 12 and the X chromosome, all of which are medium sized and submetacentric. Group D includes the medium-sized acrocentric chromosome pairs 13, 14, and 15. Group E chromosome pairs (16, 17, and 18) are relatively short and

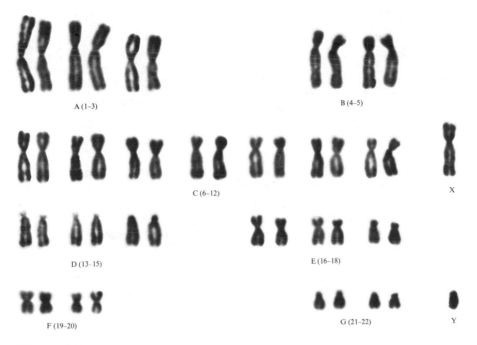

A (1–3)

B (4–5)

C (6–12)

X

D (13–15)

E (16–18)

F (19–20)

G (21–22)

Y

Figure 5.4
Male karyotype. (Courtesy of Dr. Margery Shaw.)

metacentric or submetacentric, and those in Group F (19 and 20) are metacentric and shorter. Group G chromosomes are very short and acrocentric. This group includes chromosomes 21 and 22, as well as the Y chromosome.

A symbolic system has also been devised for describing the chromosomal complement of an individual. The total chromosome number is indicated first, followed by the sex-chromosome constitution and any autosomal abnormality. A plus or minus sign before a chromosome number or letter indicates that an entire autosome is represented an extra time or is missing, and a question mark indicates uncertainty. This system of nomenclature is illustrated in Table 5.1.

Banding

In 1968, another significant advance was made in the technology of human chromosome identification. In that year, T. Caspersson and coworkers reported that metaphase (middle stage of mitosis) chromosomes can be stained by fluorescent DNA-binding agents to yield a pattern of up to 320 light and dark bands when viewed with a fluorescence microscope. Other banding techniques have since been found that yield generally similar results.

Table 5.1
Nomenclature for human chromosome complements, including aberrant ones

Abbreviation	Description
46, XY	Normal male
46, XX	Normal female
45, X	22 pairs of autosomes, one X chromosome; one sex chromosome missing
47, XXY	22 pairs of autosomes; one extra sex chromosome
45, XY, −C	Male; one chromosome missing in group C
47, XX, +21	Female; one extra chromosome number 21
45, XX, −?C	Female; one autosome missing, probably in group C
45, X/46, XX	A mosaic, some cells like those of a normal female and some missing an X chromosome.

SOURCE; After Hsia, *Human Developmental Genetics.* Year Book Medical Publishers. Copyright © 1968. Adapted from Chicago Conference Standardization in Human Cytogenetics, Birth Defects Original Article Series II: 2. New York: The National Foundation, 1966.

The most widely used techniques are those that do not require special equipment such as fluorescence microscopes. In 1976 it was discovered that many more bands can be detected when chromosomes are stained during earlier stages of mitotic division than in metaphase (Sanchez and Yunis, 1977, Yunis, 1976). Figure 5.5 compares the patterns of human chromosomes at metaphase and prophase. In the left half of each chromosome (the left chromatid) we see the banding picture as of 1972. The chromatids at right depict the banding pattern using techniques available in 1976.

It is not clear what the bands represent. At first, they were thought to be active genes, but it is now known that the average band has 30,000 nucleotide bases, which is over 20 times more than necessary to code for most proteins. It was then suggested that the bands represent related proteins, but this view is now considered unlikely. Thus, we are left with no good explanation. Nonetheless, the bands are extremely useful in identifying chromosomes. Previous techniques often resulted in no more than the assignment of individual chromosomes to groups on the basis of their size and the location of the centromere. Now it is possible to identify each chromosome on the basis of its characteristic banding pattern. In addition, the improvement of chromosomal identification has facilitated analyses, such as a gene-mapping technique known as somatic cell hybridization (see Box 5.3) and the identification of minor chromosomal abnormalities.

Banding has progressed to the point that it is possible to compare human chromosomes to those of other species. The chromosomes of the great apes (chimpanzee, gorilla, and orangutan) and humans are very similar (deGrouchy, Turleau, and Finaz, 1978). Although the great apes have 48 chromosomes and humans have 46, banding analyses have shown that the differ-

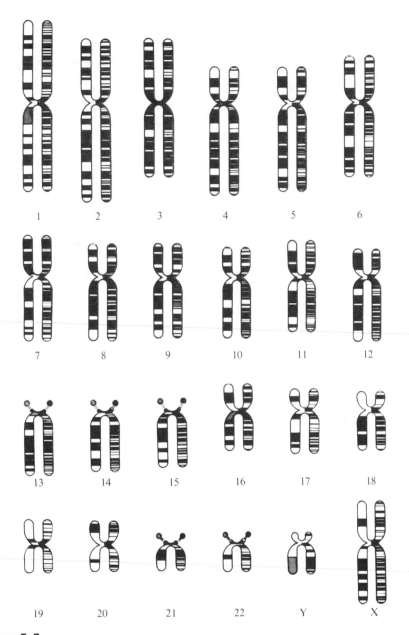

Figure 5.5
Representation of human chromosome bands. For each chromosome, the left chromatid is the banding pattern (320 bands) observed at metaphase. The right chromatid is the banding pattern (1,256 bands) in prophase. (From J. J. Yunis, "High resolution of human chromosomes," *Science*, 191, 1268. Copyright © 1976 by the American Association for the Advancement of Science.)

ence in chromosome number is due to the fact that two short chromosome pairs in the great apes fused to form a large chromosome pair (number 2) in the human species. As shown in Figure 5.6, humans are not very different chromosomally from chimpanzees. In fact, in both species the banding patterns are similar for chromosome pairs 3, 6, 7, 8, 10, 11, 19, 20, 21, and 22 and for the X chromosome. The other chromosomes usually differ by a single inversion.

CHROMOSOMAL ABNORMALITIES

Although meiosis is usually a very orderly process, irregularities occasionally occur. Chromosomes can break and result in a *deletion* of part of a chromosome. Loose, broken pieces of chromosomes tend to stick to other chromo-

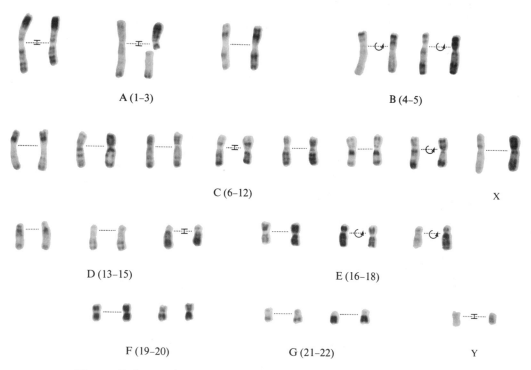

Figure 5.6
Comparison of the karyotype of man to that of the chimpanzee (*Pan troglodytes*) after R-banding. For each pair, the chromosome on the left is a human chromosome. The chromosome on the right is the best chimpanzee homologue. The symbol ○ indicates an inversion. The symbol I indicates another rearrangement, such as a fusion. (From J. deGrouchy, C. Turleau, and C. Finaz, "Chromosomal phylogeny of the primates." Reproduced, with permission, from *Annual Review of Genetics*, 12, 289–328. Copyright © 1978 by Annual Reviews, Inc. All rights reserved.)

somes. A deleted segment of chromosome can stick onto the end of the homologous chromosome, creating a *duplication* of that segment. Sometimes the pieces end up on the same chromosome, but in an *inverted* position. Inversions can change the location of the centromere. In other cases, the pieces may stick to a nonhomologous chromosome, which is known as a *translocation*. These types of chromosomal anomalies are illustrated in Figure 5.7.

Sometimes the chromatids of duplicated chromosomes do not separate properly during meiosis (*nondisjunction*); as a result, some gametes end up with extra or missing chromosomes. This results in two abnormal cells, one with an extra chromosome and one missing a chromosome, as indicated in Figure 5.8. These conditions are referred to as *aneuploidy*. When a gamete with an extra chromosome unites in fertilization with a normal gamete, the resulting zygote will have three of that particular chromosome. This is called a *trisomy* ("three bodies"). When a gamete missing a particular chromosome unites with a normal gamete, the result is a *monosomy* ("single body").

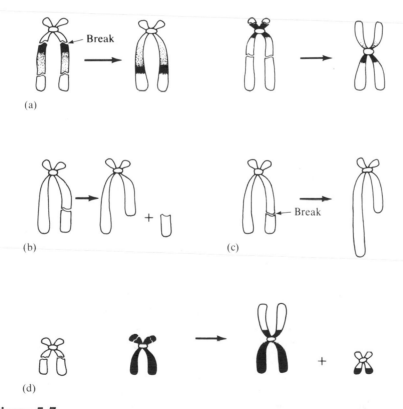

Figure 5.7
Common types of chromosomal rearrangements. (a) Inversions. (b) Deletion. (c) Duplication. (d) Translocation between nonhomologous chromosomes.

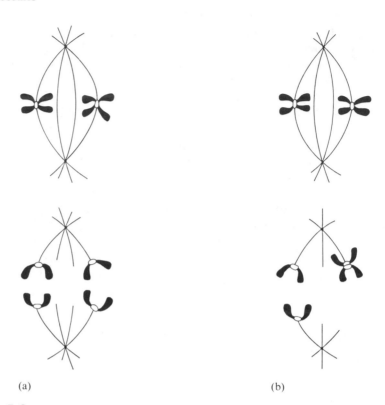

(a)   (b)

Figure 5.8
Diagrammatic representation of (a) normal cell division and (b) nondisjunction.
(After Nadler and Borges, 1966.)

      Chromosomal abnormalities involve gross genetic imbalance and usually result in multiple defects, including behavioral problems like mental retardation. J. G. Boué (1977) presented some surprising statistics concerning chromosomal abnormalities. Half of all fertilized human eggs have a chromosomal abnormality, and most of these result in early spontaneous abortions (miscarriages). At birth, about 1 in 200 babies has an obvious chromosomal abnormality. About half of these are abnormalities of the sex chromosome.

      The newer banding techniques have led to the discovery that deleterious chromosomal abnormalities may occur for each of the 23 human chromosomes (Sanchez and Yunis, 1977). In addition, there are many minor chromosomal anomalies with no apparent manifestations. Partial trisomies and minor deletions are most common and have been discovered in nearly every chromosome. Most complete trisomies are found only in spontaneous abortions, although cases of trisomy in live births have been discovered for chromosome pairs 8, 9, 13, 18, 20, 21, and 22 and for the sex chromosomes. Monosomies are found only in early spontaneous abortions, except for chromosome pair 21 and the X chromosome. For humans as well as *Drosophila*, the excess or missing chromosomal material in live births is almost always less than 5 percent of the total DNA. More severe abnormalities are presumably lethal.

SEX CHROMOSOMES AND THE LYON HYPOTHESIS

In 1949 it was observed that normal males and females have a striking difference in the nuclei of their cells (Barr and Bertram, 1949). In normal females, a small body, called a *chromatin* (or *Barr*) *body*, lies near the inner surface of the membrane of the nucleus. Staining indicates that the chromatin consists of DNA. In normal males, however, there is no Barr body. it was subsequently found that the number of these chromatin bodies is one less than the number of X chromosomes, as indicated in Figure 5.9. A male with an extra X (XXY) would thus stain positively for one chromatin body.

Later research showed that the chromatin body is a single condensed X chromosome. In 1961 Lyon hypothesized that this X chromosome is genetically inactive. Inactivation of an X chromosome in females compensates in part for the different amount of genetic material in the sex chromosomes of normal males and females. Males have only one large X and the Y chromosome, which is very small, while females have two large X chromosomes. Lyon also hypothesized that the inactive X chromosome could be either maternal or paternal in origin, even in different cells of the same individual. If the X chromosome that inactivates is of maternal origin, all daughter cells in the cell line resulting from subsequent mitotic divisions will also have an

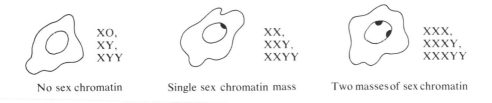

XO, XY, XYY   No sex chromatin

XX, XXY, XXYY   Single sex chromatin mass

XXX, XXXY, XXXYY   Two masses of sex chromatin

XXXX, XXXXY   Three masses of sex chromatin

XXXXX   Four masses of sex chromatin

Figure 5.9
Correspondence between sex-chromatin patterns and sex-chromosome complements. (After Hsia, 1968.)

inactivated maternal X chromosome. Thus, females are mosaics for the X chromosome: in some of their cells, the X chromosome from their fathers is active; in other cells, the active X chromosome is from their mothers. This inactivation of the X chromosome occurs in the embryo by the third week of life. Considerable evidence has accumulated in support of Lyon's hypothesis. It has recently been suggested that inactivated X chromosomes may be reactivated as part of the aging process (Wareham et al., 1987).

The sex-chromatin test has greatly facilitated the determination of individuals with X-chromosome anomalies. This test is much simpler and less expensive than karyotype analysis. Cells may be easily obtained for examination by lightly scraping the inside of the cheek. These cells are then spread on a slide, stained, and examined microscopically for the presence of sex-chromatin masses. Because of the economy of this test, large-scale surveys have been undertaken. For example, among 8,621 mentally defective, institutionalized males and mentally handicapped schoolboys, about 0.8 percent were found to have sex-chromatin anomalies (Hsia, 1968). This is approximately twice the incidence observed in the general population.

Abnormalities of the sex chromosomes and autosomes and their effect on human behavior are the topic of the next chapter.

CHROMOSOMES AND GENETIC VARIABILITY

We now know that Mendel's elements of inheritance are carried on chromosomes and they assort independently through the shuffling process of meiosis. Crossing over contributes even more to genetic variability by providing for recombination of genes at different loci on the same chromosome. Meiosis is responsible for the fact that we are somewhat like our parents in that we receive one chromosome of each homologous pair from each of them. However, meiosis and crossing over guarantee that we are genetically unique. Our particular combination of chromosomes and genes has never occurred before and never will occur again.

SUMMARY

Genes do not assort in a completely independent manner because they are associated on chromosomes. Crossing over breaks up linkages between alleles on chromosomes, and the rate at which this occurs has been used to locate different genes on the same chromosomes. Chromosomes replicate during the process of mitosis, creating duplicate cells. Meiosis shuffles chromosome pairs and results in haploid gametes. The 23 pairs of human chromosomes can be identified because of unique banding patterns. Abnormal karyotypes, occurring as frequently as in 1 of 200 births, are caused by deletions, translocations, inversions, and nondisjunction.

CHAPTER·6

Chromosomal Abnormalities and Human Behavior

As noted in the previous chapter, human chromosomal abnormalities are quite common. As many as half of all human fertilizations involve such abnormalities, and most of these result in early spontaneous abortion. About 1 in 200 fetuses have chromosomal anomalies and survive until birth. However, some of these babies die soon after they are born. For example, only 10 percent of trisomy-18 individuals (incidence: about 1 in 5,000 births; see Figure 6.1) live for more than one year (Gorlin, 1977). Death ensues in the first month of life for 50 percent of the individuals with trisomy-13 (incidence: about 1 in 6,000 births; see Figure 6.2). Other chromosomal abnormalities are such that the individuals survive, but result in behavioral as well as physical manifestations. One of these involves the deletion of 15 to 80 percent of the short arm (p) of chromosome 4 or 5 (Figure 6.3). The syndrome is called *cri du chat* (cry of the cat) because of a monotone cry, nearly one octave higher than usual, during the first month or two of life. This chromosomal abnormality usually results in severe retardation and accounts for about 1

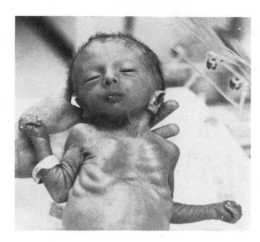

Figure 6.1
Patient with trisomy-18
syndrome. (Courtesy of George F.
Smith, M.D.)

percent of institutionalized retardates with IQs less than 35. However, an even more important cause of mental retardation is trisomy-21, which will be discussed along with other behaviorally related chromosomal abnormalities later in this chapter.

Although behavioral effects of chromosomal abnormalities have been studied in species other than *Homo sapiens*, most research has focused on human beings (Borgaonkar, 1977). This research has shown that the effects of most abnormalities are broad and general. This should not be surprising, considering that many genes are involved in a major chromosomal anomaly. Because the effect is general, we typically find no specific biochemical or morphological abnormalities to characterize a particular chromosomal abnormality (Smith, 1977). Behaviorally, the result of this general effect is that nearly all major chromosome abnormalities influence cognitive ability, which would be expected if cognitive ability is affected by many genes (Lewandowski and Yunis, 1977).

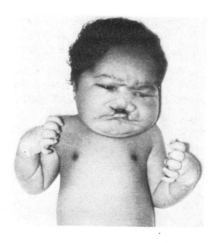

Figure 6.2
Patient with D-trisomy (trisomy-13)
syndrome. (Courtesy of George F. Smith,
M.D.)

(a)

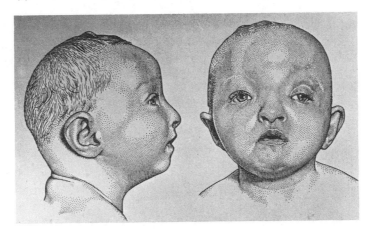

(b)

Partial deletion of B short arm

46,XX,*Bp*−

Figure 6.3
(a) Patient with *cri du chat*. (Courtesy of George F. Smith, M.D.) (b) Karyotype of
a patient affected with *cri du chat*, showing partial deleton of the short arm (*p*−)
of number 5. (Courtesy of Arthur Robinson, M.D.)

Recent improvements in banding techniques have led to the identification of many new chromosomal anomalies (Borgaonkar, 1984). For example, extra chromosome 15 material has long been reported, but during the past decade it has been shown that this material often involves a duplicated copy of the short arm of chromosome 15 (designated inverted duplicated 15p), and an extra copy of a portion of chromosome 22 ("cat eye" syndrome, so named because the pupil appears slit). The inverted duplicated 15p syndrome often involves delays in motor development and later seizures; the cat eye syndrome frequently leads to mental retardation. A deletion of a portion of chromosome 15 appears to be involved in the Prader-Willi syndrome, which is characterized by an uncontrollable appetite in childhood, short stature, and mental retardation. Most of the new chromosomal anomalies involve unique deletions and insertions of bits of chromosomal material.

However, in this chapter, we shall focus on the classical chromosomal syndromes in humans involving entire additional or deleted chromosomes. Down's syndrome, which is caused by the presence of an extra autosomal chromosome, was discovered first. A disproportionate number of chromosomal abnormalities involve the sex chromosomes rather than autosomes. About half of all surviving individuals with chromosomal abnormalities have a problem with the sex chromosome. We shall discuss four of the best-known sex chromosomal anomalies: Turner's females with only one X chromosome, females with one or more extra X chromosomes, Klinefelter's males with one or more extra X chromosomes, and males with one or more extra Y chromosomes.

DOWN'S SYNDROME

The first human autosomal abnormality was discovered in 1959. Patients with *Down's syndrome*, or *mongolism*, were found to have 47 chromosomes instead of the normal 46. One of the small chromosomes in group G is present in triplicate rather than in duplicate, yielding the karyotype shown in Figure 6.4. Another name for this condition is trisomy-21, because the trisomy was thought to involve the next-to-the-smallest autosome (number 21 by the Denver system of enumeration). We now know that the smallest autosome is the one in triplicate. Thus, trisomy-21 should really be called trisomy-22, but it has become too firmly entrenched in the literature to change the numbering system. We shall also continue to refer to the condition as trisomy-21.

Down's syndrome is so common (an incidence of about 1 in 700 newborns) that its general features are probably familiar to everyone. Infants affected with Down's syndrome are often quiet and uncrying during the early weeks of life. Although more than 300 abnormal features have been reported for Down's syndrome children, a handful of specific physical disorders are diagnostic because they occur in over 95 percent of Down's individuals, such

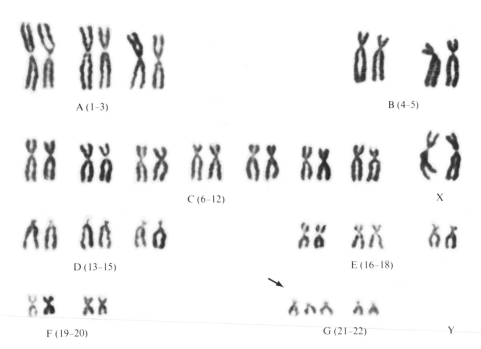

Figure 6.4
Karyotype of a Down's (trisomy-21) female. (From *Slide Guide for Human Genetics* by R. A. Boolootian. Copyright © 1971 John Wiley & Sons, Inc. Reprinted by permission of John Wiley & Sons, Inc.)

as a wide space between first and second toes. Other common disorders include increased neck tissue (85 percent), muscle weakness (80 percent), Brushfield spots (speckled iris of eye, 75 percent), mouth kept open (65 percent), protruding tongue (60 percent), and epicanthal folds, or slanting eyes (60 percent) (Pueschel, 1982). (See Figure 6.5.) Some symptoms, such as the abundant neck tissue and epicanthal folds, become less prominent as the child grows, whereas others, such as mental retardation and short stature, are noted only as the child grows. About two-thirds have hearing deficits, and one-third have severe congenital heart defects. In the past these problems have resulted in high mortality during the first few months of life and an average life span of only 20 years, although modern medical intervention has decreased the mortality rate. The pervasive effects of trisomy-21 make it unlikely that we will discover any single drug treatment to intervene in the development of Down's syndrome. There is no effective treatment.

One of the most striking features of Down's syndrome is its resultant mental defects. The average IQ among institutionalized Down's syndrome patients is below 50, and 95 percent have IQs between 20 to 80 (Connolly, 1978). The traditional cutoff criterion for retardation is an IQ of 70, and the average IQ in the population as a whole is 100. The older literature apparently underestimated the mental capabilities of individuals with Down's syndrome

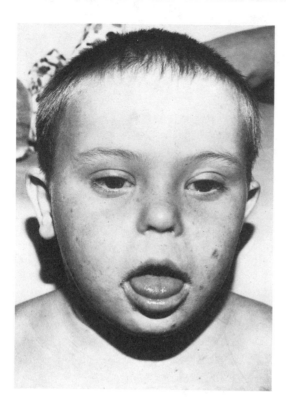

Figure 6.5
Patient with Down's syndrome.
(Courtesy of George F. Smith,
M.D.)

by focusing on institutionalized individuals. The majority of Down's syndrome children are in the mild-to-moderate range of mental retardation — some have IQs in the average range, and only a few are severely retarded. Earlier characterizations of the temperament of Down's syndrome children were also overly stereotypic. In personality as well as social behavior, Down's syndrome children are much more similar to normal children than they are different (Cicchetti and Serafica, 1981).

For 95 percent of individuals with Down's syndrome, the trisomy is a result of nondisjunction during meiosis, as described in the previous chapter (see Figure 5.8). Essentially, the chromatids fail to separate, so that one gamete ends up with both chromatids while another is missing one. In the case of Down's nondisjunction, the gamete with an extra chromosome 21 frequently unites with a normal gamete containing one chromosome 21, thus producing the trisomy. No individuals have been found with only one chromosome 21, which would occur if the gamete with no chromosome 21 united with a normal gamete. Therefore, we can assume that this monsomy is lethal. Too little genetic material is usually more damaging than extra material. Nondisjunction occurs during meiosis; however, in about 2 percent of Down's syndrome children, mosaicisms are found in which some cells are normal, suggesting nondisjunction following one of the first mitotic cell divisions after conception.

Translocation Down's Syndrome

Sometimes the extra chromosome 21, or a large part of it, attaches to another chromosome, usually 14, 21 or 22 (see Figure 5.7). It will lead to trisomy-21 if the individual has the normal pair of chromosome 21 in addition to the translocated chromosome 21. Such Down's syndrome individuals appear to have 46 rather than 47 chromosomes. The translocation version of Down's is different from that originally described, in that it is often inherited. Nondisjunction trisomy-21 is rarely inherited because few Down's individuals reproduce. In fact, in almost all cases, nondisjunction Down's individuals have normal parents. In contrast, Down's individuals with the translocation usually have a parent with a translocation. As illustrated in Figure 6.6, the parent's translocation is balanced in the sense that there is a normal amount of chromosomal material, and the individual is phenotypically normal. However, the gametes produced by this parent include balanced and unbalanced translocations, as well as normal gametes. When these gametes are fertilized by normal gametes, the zygotes may be of four types: inviable, normal, balanced translocation, and translocation with trisomy-21. For genetic-counseling purposes, it is important to determine whether the siblings of a child with translocation Down's syndrome carry a balanced

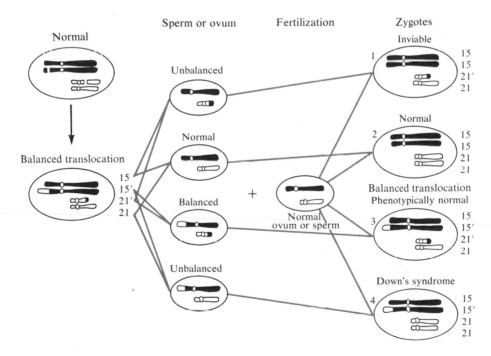

Figure 6.6
Diagram showing suggested origin of translocation and its genetic consequences. (After Polani et al., 1960.)

translocation. Those who do would have a substantial risk of bearing a child with this syndrome.

Relation to Maternal Age

Down's syndrome, first described by Langdon Down in 1866, the same year that Mendel published his classic paper, defied explanation for many years. Although it was occasionally found to be familial, it was clearly not due to a simple dominant or recessive gene. Another factor that confounded researchers was that Down's syndrome is related to the age of the mother. Its higher incidence among children of older mothers gave rise to environmental explanations, such as reproductive exhaustion. However, we now know that children of older mothers have more chromosomal anomalies in general, including trisomy-21.

All the immature eggs of a female mammal are present at birth. These have a diploid number of chromosomes, and the meiotic process that produces haploid gametes occurs periodically throughout the female's fertile years. Nondisjunction during this meiotic process is more likely to occur as the female grows older.

The prevalence of Down's syndrome increases significantly in children born to mothers 35 years of age and older, especially in surveys from the 1960s and 1970s. In Table 6.1, the percentage of Down's infants (9,441 cases) born to mothers of various ages is compared with the percentage of normal babies born to women in the same age-groups. The column on the

Table 6.1
Distribution of Down's syndrome by mothers' ages

Mother's Age	% Down's Infants Born	% of Normal Births	% Down's Births / % Normal Births
−19	1.9	4.9	0.39
20–24	10.5	26.1	0.40
25–29	14.5	30.9	0.46
30–34	16.6	22.1	0.75
35–39	27.0	12.0	2.25
40–44	25.2	3.7	6.81
45+	4.3	0.3	14.33
Total	100.0	100.0	
Mean age	34.43	28.17	

SOURCE: After Penrose and Smith, 1966.

far right is the ratio of Down's births to normal births for each age-group. Up to 30 years of age, the ratio does not change much. From 30 to 34 years of age, the ratio increases slightly. After 35, it rises abruptly. For example, women from 35 to 39 years of age produce only 12 percent of all normal births but 27 percent of Down's infants. Even more striking, women from 40 to 44 produce only 4 percent of all normal births but 25 percent of Down's infants.

In recent analyses of data for live births as well as large studies of prenatal diagnoses, the rate of Down's syndrome continues to show a strong relationship to maternal age: The risk for Down's is about 1 in 150 for women 35 to 39 years old, 1 in 50 for women 40 to 44, and nearly 1 in 10 for women over 44 (Hook, 1982). During the past two decades, many women delayed childbearing, and it is likely that an increase in reproduction will be seen by women in the higher-risk age categories.

Nonetheless, the incidence of Down's syndrome is declining: Twenty years ago, the incidence of Down's syndrome was about 1 per 600, but now it is closer to 1 per 1,000. This decrease is due to fewer Down's births by older women; 20 years ago, half of the children with Down's syndrome were born to women over 35 years of age, but now this has been reduced to 20 percent (Hook, 1982). The incidence is the same in all races and socioeconomic classes. Although the effect of maternal age is clear, research during the past few years indicates that paternal age also has some effect: In about a quarter of the cases of Down's syndrome, the extra chromosome 21 is of paternal origin (Magenis and Chamberlin, 1981). Multiple causality of the nondisjunction is likely, including genetic predispositions, maternal radiation, and diseases.

This is an important social problem. As many as 10 percent of all institutionalized mentally retarded individuals have Down's syndrome, making it the single most important cause of retardation. In terms of expense, the societal cost is at least $10,000 per year for institutional care for the average lifespan of 20 years, or about $200,000 for each Down's individual. In 1971 there were 44,000 Down's individuals in the United States. The lifetime cost of the care of these Down's syndrome patients now alive will thus be about $8.8 billion. However, the financial cost is minor compared with the emotional cost to parents.

Many women worry about reproducing later in life because of such chromosomal abnormalities as Down's syndrome. Much of the worry of pregnancies later in life can be reduced by amniocentesis, a procedure that permits karyotyping of the fetal chromosomes, as described in Box 6.1. In one study of amniocenteses performed on 3,012 pregnant women 35 years of age and older, 79 (2.6 percent) had fetuses with chromosomal abnormalities (Epstein and Golbus, 1977). The other 2,933 women (97.4 percent) no longer had to worry about the possibility that the pregnancy might result in a child with a major chromosomal anomaly.

Although Down's syndrome is the best-studied chromosomal abnormality, much continues to be learned both about biomedical and behavioral aspects of the disorder (Pueschel and Rynders, 1982).

Box 6.1

Prenatal Diagnosis

Chromosomal anomalies and certain single-gene problems can be detected before birth. Amniocentesis is the procedure by which fluid with cast-off fetal cells is obtained from the amniotic cavity. (See the figure opposite.) After these fetal cells are grown in culture, they can be karyotyped to detect chromosomal abnormalities and analyzed for enzyme deficiencies characteristic of single-gene disorders (Epstein and Golbus, 1977). Unfortunately, sufficient amniotic fluid for the test does not exist until after 13 weeks of pregnancy, and it then takes two weeks to grow the cells in culture. Thus, pregnancy will progress to four months before a diagnosis can be made. However, this is still early enough for an abortion if the parents so choose.

There were fears that the fetus might be injured by the hypodermic needle that extracts the amniotic fluid. But these early fears have been quelled by research indicating that fetuses are not injured, nor is there increased spontaneous abortion following amniocentesis. A four-year study of more than 1,000 amniocentesis cases and a control group of pregnant women who did not undergo amniocentesis indicates that the technique is safe (Culliton, 1975a).

The positive effects of amniocentesis cannot be denied. In one survey of 10,754 amniocenteses on women of all ages, a total of 209 chromosomal abnormalities were found. Many of these could have produced mental retardation. Also diagnosed were 199 X-linked diseases, 106 biochemical defects, and 86 neural-tube defects (Epstein and Golbus, 1977).

Fetal cells can also be obtained from the chorion as early as 9 to 12 weeks. The technique, called *chorionic villous sampling*, is still under development (Thompson & Thompson, 1986).

TURNER'S SYNDROME (XO)

The degree of mental defect in individuals with an extra small autosome in group G (trisomy-21) is more severe than that of individuals with X chromosome anomalies, assuming they have at least one X chromosome. This lesser deficit in individuals with X chromosome anomalies is apparently due to the inactivation of all but one of the X chromosomes in each cell, which was described as the Lyon hypothesis in the previous chapter. Thus, although individuals with trisomy-21 have a less deviant total amount of DNA than individuals with sex chromosome abnormalities, their genetic imbalance is actually greater. This is also the reason why so few autosomal trisomies involving the larger chromosomes have been found; they are presumably lethal. Even in the case of Turner's syndrome, 98 percent of 45,XO fetuses do not survive to birth. This demonstrates that the second X chromosome is not completely inert and that much remains to be learned about the issue of the "deactivation" of the second X chromosome in females.

Turner's syndrome is a particularly interesting exception to the rule that chromosomal abnormalities cause general retardation. Turner's syndrome

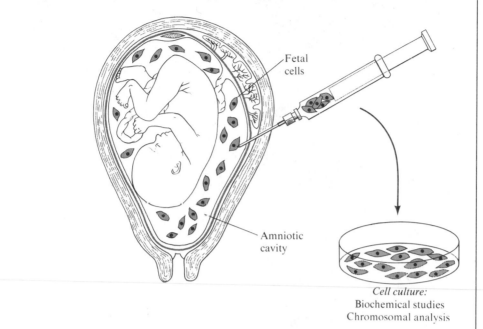

Fetal
cells

Amniotic
cavity

Cell culture:
Biochemical studies
Chromosomal analysis

Amniocentesis. (From "Prenatal diagnosis of genetic disease" by T. Friedmann.
Copyright © 1971 by Scientific American, Inc. All rights reserved.)

occurs only in females and nearly always involves sterility and limited secondary sexual development. (For example, fewer than 10 percent menstruate.)
There are also other physical stigmata, such as short stature and a webbed
neck. (See Figure 6.7.)

The incidence of Turner's syndrome is about 1 in 2,500 births, and it is
also frequently found among spontaneous abortions. About 60 percent of
women with Turner's syndrome have only 45 chromosomes, and their cells
have no Barr bodies (see Chapter 5), even though they are females. Their
karyotypes indicate that they have only one X and no Y chromosome, as
shown in Figure 6.8. Classical Turner's syndrome is caused by nondisjunction
but shows no increase with parental age. Another type of Turner's is caused
by the loss of an X chromosome early in embryonic development, so that
some cells have two X chromosomes and others have only one. These are
called XO/XX mosaics. Another cause of the syndrome is deletion of part of
one X chromosome.

Although it was once thought that Turner's females were below average
in IQ, we now know that this is not so. For example, there are no more
Turner's females in institutions for the mentally retarded than we would

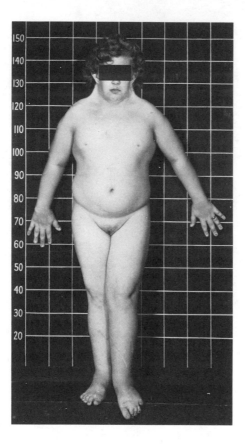

Figure 6.7
Patient with Turner's syndrome.
(Courtesy of John Money.)

expect on the basis of the frequency of these females in the population at large. In fact, most of the problems faced by Turner's females are cosmetic ones caused by their short stature and failure to develop sexually (for which they can now receive hormonal therapy) and problems caused by sterility.

Expression of Turner's syndrome is highly variable, and many individuals with this syndrome are diagnosed only because of infertility; Turner's accounts for about one-third of patients with primary amenorrhea. However, Turner's females are likely to have a highly specific cognitive defect. In 1962 it was first reported that Turner's females have low "perceptual organization" scores, although they have nearly normal overall IQ scores (Shaffer, 1962). Their average scores on verbal sections of IQ tests are about 20 points higher than their scores on nonverbal, performance tests. Subsequent research by John Money (1964, 1968) has shown that the most serious deficiency in Turner's females is in spatial ability and directional sense. For example, many have difficulty in copying a geometric design or following a road map. Studies during the past decade have attempted to discern specific deficits using neuropsychological (e.g., McGlone, 1985) and information-processing (e.g., Rovet and Netley, 1982) techniques.

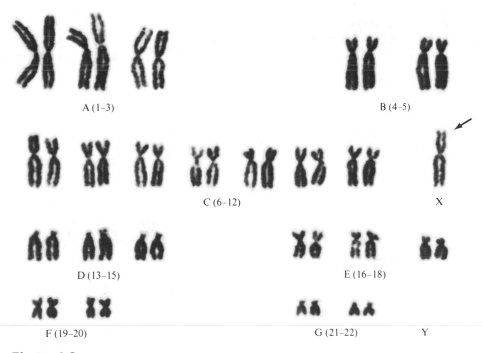

Figure 6.8
Karyotype of a Turner's female, 45, X. (From *Slide Guide for Human Genetics* by
R. A. Boolootian. Copyright © 1971 John Wiley & Sons, Inc. Reprinted by
permission of John Wiley & Sons, Inc.)

Turner's females also display X-linked recessive characteristics just as
males do. In Chapter 3 we discussed the hypothesis that spatial ability was
influenced by a major recessive allele on the X chromosome. This hypothesis
was developed to explain the greater spatial ability of males. We concluded
that the evidence to date is against this hypothesis, and we can now add to this
negative conclusion the data from the Turner's females. If the hypothesis were
correct, Turner's females should have higher spatial ability scores than other
females because they have only one X chromosome, just like normal males.
The fact that their spatial ability scores are lower negates this hypothesis.

FEMALES WITH EXTRA X CHROMOSOMES

One of the X chromosomes is partially inactivated in each somatic cell of
normal females, and additional X chromosomes in the nucleus are apparently
also inactivated. Those females with extra X chromosomes often show more
than one Barr body (see Figure 5.9). Although the partial inactivation greatly

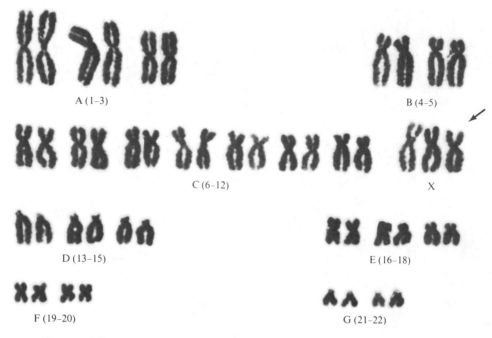

A (1–3)

B (4–5)

C (6–12)

X

D (13–15)

E (16–18)

F (19–20)

G (21–22)

Figure 6.9
Karyotype of an XXX female. (From *Slide Guide for Human Genetics* by R. A. Boolootian. Copyright © 1971 John Wiley & Sons, Inc. Reprinted by permission of John Wiley & Sons, Inc.)

decreases the imbalance due to the extra genetic material, some problems do occur and worsen with increasing numbers of additional X chromosomes.

Most females with an extra X chromosome have 47,XXX karyotypes, as in Figure 6.9. However, 48,XXXX karyotypes and 49,XXXXX karyotypes have been found. The incidence of trisomy-X is about 1 per 1,000 births. The characteristics of trisomy-X females are not distinctive. Earlier studies had sampled individuals from institutionalized populations, and thus tended to find retardation. Studies in the 1970s (Tennes et al., 1975; Gorlin, 1977), however, found that about one-fourth are essentially normal, one-fourth have mild developmental lags, one-fourth have some type of congenital problem, and one-fourth have possible cognitive or emotional problems. Overall, some tendency toward retardation is suggested by the finding that the incidence of trisomy-X is about 4 per 1,000 in institutions, whereas it is about 1 per 1,000 in the general population. Recent studies suggest that XXX females are at considerable risk for retardation — about three-quarters of 47,XXX girls evidenced learning problems (Stewart, Netley, and Park, 1982).

Problems tend to multiply as additional X chromosomes are found. For example, Figure 6.10 is a photograph of a rare case of pentasomy X (that is, 49,XXXXX). This individual has a host of problems, including severe retardation, uncoordinated eye movements, undeveloped uterus and breasts, and

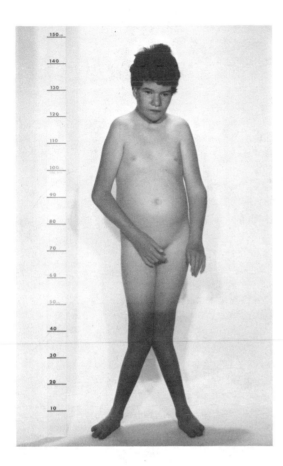

Figure 6.10
An XXXXX female. (From "The 49, XXXXX chromosome constitution: Similarities to the 49, XXXXY condition" by F. Sergovich et al., *Journal of Pediatrics*, 1971, 78, 285–290.)

many skeletal defects. Among about two dozen individuals who have been found to have at least four X chromosomes, mental retardation was a common feature.

KLINEFELTER'S SYNDROME

Individuals with Klinefelter's syndrome (see Figure 6.11) are phenotypic males with extra X chromosomes. They represent nearly 1 percent of males institutionalized for retardation, epilepsy, or mental illness, although their incidence in the general population is about 1 per 1,000 newborn males. Clinical features include the presence of abnormally small testes after puberty, low levels of the male hormone, testosterone, and sterility. Some of these individuals are mentally retarded (although three-quarters have IQs within the normal range). Language delay is a consistent finding across prospective studies, affecting at least half of 47,XXY boys (Robinson et al., 1979). Studies also tend to find that 47,XXY boys are less active, less assertive, and more

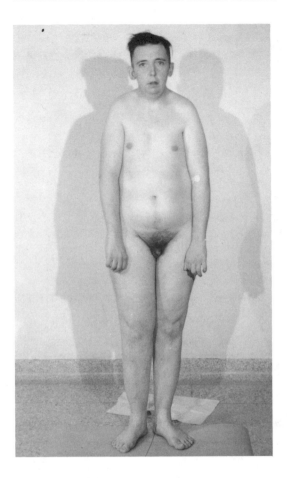

Figure 6.11
The XXY Klinefelter's syndrome. (Courtesy of R. J. Gorlin, M.D.)

susceptible to stress than controls (Stewart et al., 1986). In adulthood, 47,XXY men were also more submissive and anxious, tended to avoid social contact and sexual encounters, and felt less masculine than controls (Theilgaard, 1981). Like XYY males discussed in the next section, Klinefelter's males are somewhat taller than average.

Even though individuals with this condition are phenotypic males, they usually test positively for the presence of a Barr body. In about two-thirds of the cases, the karyotype is 47,XXY, as shown in Figure 6.12. However, 48,XXXY; 49,XXXXY; 48,XXYY; 49,XXXYY; and various other arrangements, including mosaicisms, have been described. As in other chromosomal abnormalities, the symptoms become more severe as more genetic material is added. For example, in over 90 cases of 49,XXXXY males that have been described, nearly all were severely retarded, and had severe sexual deformities, as well as other anatomical problems.

Like most chromosomal abnormalities, except Turner's syndrome, there is an increased risk of Klinefelter's syndrome among children of older

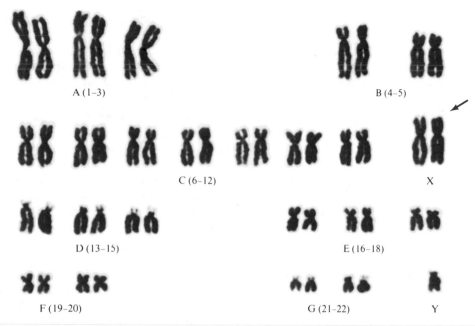

A (1–3)

B (4–5)

C (6–12)

X

D (13–15)

E (16–18)

F (19–20)

G (21–22)

Y

Figure 6.12
Karyotype of the XXY Klinefelter's syndrome. (From *Slide Guide for Human Genetics* by R. A. Boolootian. Copyright © 1971 John Wiley & Sons, Inc. Reprinted by permission of John Wiley & Sons, Inc.)

mothers, although this association is less marked than that of Down's syndrome. Klinefelter's syndrome is generally thought to be caused by nondisjunction during meiosis, resulting in a gamete that has an extra X chromosome. Fertilization of a normal X-bearing egg by an XY-bearing sperm, or fertilization of an XX-bearing egg by a normal Y-bearing sperm results in offspring with Klinefelter's syndrome. It is possible, however, that errors during early cell division in a normal zygote may also occasionally produce this syndrome.

Klinefelter's syndrome is not usually detected until after puberty, when some of the effects may be irreversible. Most of the problems are secondary to the low levels of hormones essential to proper development at puberty. Thus, there is a need for early identification so that hormonal therapy can begin soon enough to alleviate the condition, although infertility remains.

MALES WITH EXTRA Y CHROMOSOMES

The XYY chromosomal anomaly, first described in 1961, has received considerable publicity since 1965, when it was suggested that XYY males may be predisposed to commit violent acts of crime. It seems that fantasies were triggered by the notion of a "supermale" with exaggerated masculine characteristics.

An XYY karyotype is shown in Figure 6.13. Extra Y chromosomes are the consequence of nondisjunction during meiosis in the father. Nearly 1 percent of the sperm of normal males have two Y chromosomes (Sumner, Robinson, and Evans, 1971), but the incidence at birth of XYY males is closer to 1 in 1,000. This suggests considerable selection against sperm or zygotes with two Y chromosomes. A few cases of XYYY, and even XYYYY have been reported. Research on the Y chromosome abnormality is more difficult than investigation of X anomalies, because the simple Barr body test that reveals an inactivated X chromosome will not reveal Y chromosome anomalies. Until recently, the much more costly karyotype analysis was necessary to detect Y chromosomes. For this reason, prevalence of the XYY condition in the general population has been difficult to estimate, and much of the early research focused on institutionalized males.

In 1965 Jacobs and co-workers reported that the incidence of chromosomal anomalies among individuals institutionalized because of "dangerous, violent, or criminal propensities" was higher than that in the population at large. Of 197 institutionalized volunteers who were karyotyped, 12 were found to have a chromosomal anomaly of some kind. One was a 46,XY/

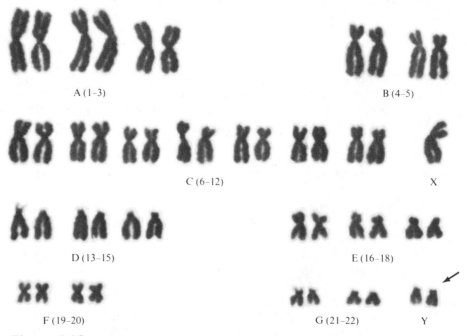

Figure 6.13
Karyotype of an XYY male. (From *Slide Guide for Human Genetics* by R. A. Boolootian. Copyright © 1971 John Wiley & Sons, Inc. Reprinted by permission of John Wiley & Sons, Inc.)

47,XXY mosaic, one was 48,XXYY, and seven were 47,XYY. Three had no sex chromosome anomalies but minor autosomal defects. The average height of the 47,XYY males was 73 inches, in contrast to an average height of 67 inches for the males of normal karyotypes in the institution. We now know that XYY boys are taller than 90 percent of their peers as early as six years of age.

Because of the importance of this initial discovery of a possible association between the presence of an extra Y chromosome and violent, aggressive behavior, a number of related studies have been undertaken. These surveys have usually been of tall prisoners confined to special security sections because of their violent behavior. Thus, the possibility of sampling bias is clear. Nevertheless, a fairly consistent pattern of results has been obtained. Data from 18 studies of 5,342 institutionalized males have been summarized (Shah, 1970). Of these, 103 (1.9 percent) possessed an extra Y chromosome. This outcome must be interpreted in the context of the prevalence of the condition in the general population. In a sample of 9,327 "normal" adult males, nine XYY karyotypes were identified (Price and Jacobs, 1970). This suggests a prevalence of about 1 in 1,000. Most reviews (e.g., Hook, 1973) concur in suggesting that the prevalence at birth in Caucasians is about 1 per 1,000, although it may be lower in other populations. Thus, even though it cannot be stated that the base rate in the general population has been firmly established, it is clear that the rate among institutionalized males is higher.

The possible association between antisocial behavior and XYY constitution has been widely debated. Many doubt that an important association exists (e.g., Noël et al., 1974). Most XYY males lead normal lives: fewer than 1 in 100 of XYY males are institutionalized (Kessler, 1975). Also, behaviors other than aggressiveness may be responsible for the increased rate of institutionalization of XYY males. For example, their greater-than-average height might make adjustment more difficult (Kessler and Moos, 1970). Poor coordination has also been frequently noted. Sexual development, however, appears to be normal.

Our discussion of other chromosomal anomalies suggests another behavior as a likely source of greater institutionalization of XYY males. Because nearly every chromosomal anomaly results in some cognitive deficiency, we might expect to find similar deficiencies associated with the XYY abnormality. In fact, the results of a study by Witkin and Mednick and a large group of co-investigators (Witkin et al., 1976) in Denmark led to that conclusion. Their study had several unique controls. First, they used a normal sample selected for height but not for institutionalization. Second, they evaluated the characteristics of males with normal and abnormal karyotypes. Third, they included a comparison group of XXY males.

All males born between 1944 and 1947 and still living in Denmark were scored for height from draft board records. The tallest 16 percent of these 28,884 men were selected for chromosomal analysis. Karyotypes were obtained for 4,139 men. Intellectual functioning was estimated by scores on an army selection test and by educational attainment tests routinely administered in Denmark. Criminal records of all subjects were detailed in terms of types of offenses.

Twelve XYYs and 16 XXYs were identified. The frequency of XYYs is higher (about 3 per 1,000) than the usual estimate of 1 per 1,000. However, these males were selected for height. Five of the XYYs (42 percent) had criminal records, as compared to three of the XXYs (19 percent) and 9 percent of all of the XY men. This indicates that the XYY males were, in fact, more likely to be incarcerated. But were they more violent? The answer is no. The nature of their crimes was no more violent than that of the control samples. Only 1 of the 12 XYY men had been convicted of a crime of violence against another person. In fact, the XYYs had been convicted of rather minor crimes and had received mild penalties.

Of the five XYYs with criminal records, four had army selection scores well below average and that of the fifth was somewhat below. It should be noted that criminality was also inversely related to intellectual functioning in the control sample of XY men. For both the army selection test and the educational index, men with criminal records had substantially lower scores. Also, the study found no support for the hypothesis that the greater incidence of criminality in XYY males is caused by adjustment problems due to their height. In the control sample, the noncriminal males were actually slightly taller than the criminal ones.

The authors conclude: "The elevated crime rate of XYY males is not related to aggression. It may be related to low intelligence" (Witkin et al., 1976, p. 547). They also caution: "The elevated crime rate found in our XYY group may therefore reflect a higher detection rate rather than simply a higher rate of commission of crimes" (p. 553). However, these conclusions must be tempered by the fact that, although the 47,XXY Klinefelter's males had average test scores almost identical to those of the XYY males, their rate of criminality was considerably lower (19 percent versus 42 percent). In addition, other studies—although less well controlled—have found greater antisocial behavior in XYYs than in XXYs and no intellectual deficit among XYYs.

In summary, no clear picture of behavioral disorder emerges from the worldwide literature on XYY males, except that, like XXY males, about half of XYY males exhibit delayed language development and have learning problems at school (Bender et al., 1984).

FRAGILE X AS A MARKER
FOR RETARDATION

An excess of mental retardation for males has long been noted, and in 1969 a structural abnormality was discovered in a family with what appeared to be an X-linked form of mental retardation (Lubs, 1969). In some of the X chromosomes of the retarded males the distal end of the long arm had a constriction so that the tip appeared to be connected to the rest of the chromosome by a thin region that frequently broke. However, this finding generated little immediate excitement because researchers in the 1970s were unable to find other cases. We now know that detection of fragile sites on the X chromosome depends on the type of culture medium, and the medium used in the 1960s was able to detect fragile sites, whereas the media of the 1970s were not. Recent research indicates that fragile sites do in fact lead to chromosomal breakage and translocations (Warren et al., 1987).

About 1 in 2,000 live male births have the fragile X site, although at least 2 percent of the male residents of schools for mentally retarded persons have the fragile X, making it the second most common chromosomal cause of mental retardation, exceeded only by Down's syndrome (Neri et al., 1988). Other fragile sites have been found (for example, on chromosomes 2, 10, 11, 16, and 20), although they do not appear to be associated with clinical abnormalities. The percentage of cells with fragile X varies widely across studies and even within individuals. For example, for some individuals, fragile X chromosomes are found in cells involved in connective tissue but not in blood cells. Neither the cause nor the consequences of this variability is understood. It has been suggested that the fragile X site is located in a repetitive region of DNA and that the variable length of repetitive DNA might be responsible for variable expression (Brown et al., 1986). Genetic markers for the fragile X site that have been found will help shed light on these and other issues (Arveiler et al., 1988). It has also been suggested that the number of cells showing fragile X is related to the degree of mental retardation (Silverman et al., 1983).

The fragile X abnormality is also highly variable in expression, although fragile X males are usually short in stature but have a relatively large head, ears, hands, testes, and feet (Nussbaum and Ledbetter, 1986). The most notable feature, however, is behavioral: Many individuals show mild retardation, although as many as one-quarter of fragile X males are of normal intelligence (Sherman et al., 1985). Individuals with the fragile X syndrome are usually among the least retarded members of institutions and many live on their own. Although fragile X is transmitted in many pedigrees as an X-linked recessive marker, cases of transmission through phenotypically normal males have been documented, and its transmission is considered to be more complicated than a single-gene recessive trait (Pembrey, Winter, and Davies, 1985). Also surprising is the fact that as many as one-third of heterozygote females show significant cognitive impairment. On average, only about 1 percent of the cells of heterozygote females exhibit the fragile X site; affected women generally have a larger proportion of fragile X chromosomes in their cells (Bronum-Nielsen et al., 1983). As mentioned in Chapter 5, the Lyon hypothesis proposes that one of the two X chromosomes in every cell of a female is inactivated early in development. The Lyon hypothesis could explain the varying degrees of mental retardation in females: retardation depends on the percentage of cells with an X chromosome with the fragile X site. Finally, it should be noted that an intense research effort is beginning to associate fragile X with autism as well as retardation (Brown et al., 1986).

It should be noted that early studies painted a grim picture of the likely outcomes for individuals with chromosomal abnormalities because these studies were undertaken with institutionalized individuals who have the most problems. In the 1960s geneticists began chromosomal screening studies of newborns. The three largest studies were conducted in Edinburgh, Scotland (Court-Brown, 1969), Denver, Colorado (Robinson and Puck, 1967), and Toronto, Canada (Bell and Corey, 1974). Together, these researchers screened chromosomes of 140,000 newborns and identified nearly 200 infants with sex chromosomal anomalies. In general, these newborn-screening studies show markedly less abnormality than the older studies of institutionalized individuals, and they also indicate that expression of deficits varies widely among children with the same karyotype (Netley, 1986). Another interesting possibility that is beginning to emerge from this research is an interaction between

Table 6.2

Summary of chromosomal anomalies

	Type of Anomaly	Incidence per Live Births	Symptoms
Autosomal anomalies:			
Edward's syndrome	Trisomy-18	1 in 5,000	Early death; many congenital problems
D-trisomy syndrome	Trisomy-13	1 in 6,000	Early death; many congenital problems
Cri du chat	Deletion of part of short arm of chromosome 4 or 5	1 in 10,000	High-pitched monotonous cry; severe retardation
Down's syndrome	Trisomy-21; 5 percent involve translocation	1 in 700	Congenital problems; retardation
Sex chromosomal anomalies:			
Turner's syndrome	XO or XX-XO mosaics	1 in 2,500	Some physical stigmata and hormonal problems; specific spatial deficit
Females with extra X chromosomes	XXX XXXX XXXXX	1 in 1,000	For trisomy X, no distinctive physical stigmata; perhaps some retardation
Kleinfelter's males	XXY XXXY XXXXY XXYY XXXYY	1 2 in 1,000	For XXY, sexual development problems; tall; perhaps some retardation
Males with extra Y chromosomes	XYY XYYY XYYYY	1 in 1,000	Tall; perhaps some retardation

genotype and environment: Children with sex chromosomal anomalies from stable families seem to show few differences from their chromosomally normal siblings, whereas less stable families appear to induce developmental problems in children with sex chromosomal anomalies (Bender, Linden, and Robinson, in press).

The discovery of these anomalies, particularly the finding that Down's syndrome is due to the presence of an extra chromosome, must be regarded as an extremely important breakthrough in the genetic analysis of behavior. Although several other chromosomal anomalies have since been described, it would seem that human chromosome analysis is still in its infancy.

The new molecular genetic techniques discussed in the following chapter are revolutionizing the study of cytogenetics. For example, it is now possible to identify the paternal, maternal, and even grandparental origin of each chromosome of an individual, which greatly facilitates linkage studies as well as genetic counseling. Even with the more refined banding patterns, the resolution of cytogenetics is measured in terms of millions of base pairs, whereas the resolution of molecular genetic techniques is one base pair.

SUMMARY

Almost all chromosomal abnormalities influence general cognitive ability and growth. Trisomy-21 is the most common autosomal problem, and the most frequent cause of mental retardation. Its strong relationship to maternal age suggests an important role for genetic counseling. Sex chromosomal abnormalities include Turner's syndrome, females with extra X chromosomes, males with extra X chromosomes, and males with extra Y chromosomes. These chromosomal anomalies are summarized in Table 6.2. Fragile X is a chromosome marker for an X-linked form of mental retardation that is the second most important cause of mental retardation.

CHAPTER·7

The New Genetics

During the past decade, advances in molecular genetics have led to the dawn of a new era for behavioral genetic research. The "new genetics" involves techniques that make it possible to study DNA variation of any species directly. As discussed at the end of this chapter, the power of molecular genetic techniques can be harnessed even for complex traits such as behavior. These techniques are already beginning to revolutionize behavioral genetic research in some areas, especially psychopathology. Without doubt, molecular genetic techniques will become a standard part of the tool kit of behavioral geneticists of the future. Students of behavioral genetics need to be aware of these techniques, even though we are only at the dawn of their application to behavior. The goal of this chapter is not to be a substitute for a course in molecular genetics nor to provide a laboratory manual to run these assays. Its goal is to introduce students of behavioral genetics to the future of the field. For this reason, we focus on those aspects of the new genetics most relevant to research in behavioral genetics.

Although knowledge gained by use of the new genetic techniques has not contradicted the central dogma of molecular genetics that was described in Chapter 4, it has greatly expanded our understanding of both genes and chromosomes, and the techniques have revolutionized the way genes and chromosomes are studied. Most of the research that led to the central dogma was based on single-celled organisms called *prokaryotes*, such as bacteria, which have no nucleus in their cells. During the past decade, the new techniques have made it possible to study the function of DNA in *eukaryotes*, organisms with cell nuclei, and this research indicates that DNA functioning is more complex in eukaryotes. The chapter begins with this expanded discussion of the mechanisms of heredity, then describes the new genetic techniques, and ends with a discussion of their potential relevance for behavioral genetics.

EUKARYOTIC COMPLEXITIES

The most notable difference between prokaryotic and eukaryotic DNA is that less than 10 percent of eukaryotic DNA is actually involved in protein synthesis. The other 90 percent in part involves segments, called *introns*, that are transcribed into RNA but are spliced out before the RNA leaves the nucleus, except in rare instances (Huang et al., 1988). To distinguish between RNA before and after splicing, the prespliced version is called *precursor* RNA and the sliced and spliced version is called *messenger* (mRNA). The segments of a gene that are spliced back together as mRNA are called *exons*.

Such so-called split genes were first discovered in 1977 (Chambon, 1981; Crick, 1979). We now know that nearly all eukaryotic genes have introns, often more introns than exons. Exon sizes typically involve about 150 to 300 nucleotide bases, but the sizes of introns vary widely, from 50 to 20,000 base pairs in length. For example, the human β-globin gene includes 92 base pairs in its first exon, 138 in the first intron, 223 in the second exon, 889 in the second intron, and 123 in the third exon. Other globin genes also have two introns, as does insulin; however, some genes, such as the collagen genes, have over 50 introns, varying in size from 100 to 2,500 base pairs. The function of introns is not known.

The existence of split genes means that a complex process is required to snip out the introns and then splice together the exons. Although specialized enzymes are probably involved (perhaps coded by the introns themselves), an exciting discovery made in 1981 is "autonomous splicing," in which RNA splices itself, a finding that will alter views of the capabilities of RNA (Cech, Zaug, and Grabowski, 1981). An important feature of splicing is that the same precursor RNA can be spliced in alternative ways, thus producing several gene products from the same gene. This alternative splicing can occur in different tissues in the same individual — for example, a gene on chromosome 11 has been found that is spliced to produce a certain protein (calcitonin) in thyroid but the same gene produces a different product in the hypothalamus.

Two other examples of eukaryotic complexities are repetitive DNA and transposable genes.

Repetitive sequences comprise as much as a third of the human genome. Repetitive sequences are of two types. Highly repetitive DNA consists of 100- to 300-base pair sequences repeated thousands of times; its function is not known. Middle repetitive DNA involves longer sequences (700 to 1,400 bases) repeated hundreds of times; it codes for ribosomal and transfer RNA and other housekeeping products.

Transposable genes have been called jumping genes. In the 1950s studies with corn suggested the presence of movable regulatory genes that inhibit the expression of genes when they move (McClintock, 1957). In the late 1960s the same phenomenon was observed in *Drosophila*. It turns out that these genes are transposable rather than movable: A copy of the gene is transposed, and the original gene is left unaltered. In bacteria, segments of DNA called *transposons* have the ability to make copies of themselves that are inserted elsewhere in the chromosomal DNA (Shapiro, 1983). In all cases, the transposon displaces the old gene. For this reason, the process is referred to as the cassette model, in which gene cassettes can be inserted in expression sites with strong promoters that ensure that the cassette will be played (Herskowitz et al., 1980).

Another type of gene rearrangement involves inversion, in which a DNA segment flips around. DNA rearrangements allow a relatively small number of gene segments to generate a large number of different products, as in the case of the rearrangement of immunoglobulin genes. Once an undifferentiated antibody-producing cell is rearranged, its daughter cells will continue to recognize and attack a specific antigen. It is thought that gene rearrangements may be importantly involved in gene regulation by altering gene expression (Borst and Greaves, 1987).

These complexities of eukaryotic DNA indicate that much remains to be learned about mechanisms of gene action. The study of eukaryotic DNA has been made possible by recombinant DNA technology, which is described in the following section.

RECOMBINANT DNA TECHNIQUES

As mentioned earlier, the essence of the new genetics lies with techniques that make it possible to study the DNA of any species, including the human species. These techniques are known as recombinant DNA techniques because they involve snipping out pieces of DNA from one species and recombining them with DNA from another species, typically bacteria. The major advantage of recombinant DNA technology is that eukaryotic DNA can be studied even if its gene product is unknown, or even if the DNA is not normally expressed.

Restriction Enzymes

The critical advance that led to recombinant DNA technology was the discovery of restriction endonucleases (enzymes) that cleave DNA at specific sites. These enzymes have become the scalpel for dissecting the genome of higher organisms. Restriction enzymes were found in certain bacteria that have the ability to destroy foreign DNA by severing it. Some restriction enzymes recognize a specific group of four or five bases; most cleave a particular group of six bases. For example, one commonly used restriction enzyme, EcoRI, recognizes the sequence GAATTC and severs the DNA molecule between the G and A bases. More than 400 restriction enzymes have been found that recognize a hundred different DNA sites, although only a dozen restriction enzymes are typically used to chop up DNA (Roberts, 1983).

Recombinant DNA and Cloning

Restriction enzymes made it possible to recombine any modest-sized DNA fragment—about 8,000 bases, referred to as 8 kilobases (kb)—with bacterial DNA. The restriction enzyme EcoRI mentioned above cuts one DNA strand GAATTC between the G and A bases, leaving a G base on that strand; the complementary DNA strand CTTAAG will be cut between the A and G, leaving CTTAA on that strand. The extra nucleotide bases on the second strand make the end "sticky," in the sense that the TTAA bases will tend to stick to its AATT complement. These sticky ends make it possible to join one fragment of DNA to another derived from a different source but cut with the same restriction enzyme. The first recombinant DNA was formed in 1973 (Cohen et al., 1973). A foreign bacterial DNA fragment cut by a particular restriction enzyme was inserted into the DNA of a different bacterium that was cut by the same restriction enzyme. The host DNA was a plasmid, a circular, double-stranded minichromosome that bacteria possess in addition to their main chromosome; plasmids can multiply, producing up to 200 copies per cell. Plasmids carry genes that code for enzymes that neutralize antibiotics. The importance of plasmids for recombinant DNA technology is that they are readily absorbed by bacterial cells and can thus serve as a vector to carry recombined DNA into bacterial cells, which can then express the foreign DNA. The recombined DNA can also be inherited. The steps in generating recombined plasmids are summarized in Figure 7.1.

Two other vectors have been developed that make it possible to insert larger foreign DNA fragments and that infect bacteria more efficiently than do plasmids. These vectors involve a virus that infects bacteria, called bacteriophage (literally, "bacteria-eater"). The most widely used bacteriophage is lambda phage, which accepts DNA inserts of about 15 kb. Another vector, called a cosmid, a hybrid between plasmids and lambda phage, permits packaging much longer (35 to 45 kb) DNA fragments. Vectors that permit packaging even longer DNA fragments are being developed to accommodate long eukaryotic genes with their flanking control regions.

Recombinant DNA plasmids or phages are allowed to enter bacteria that are grown on nutrient agar in petri dishes. The most widely used bacte-

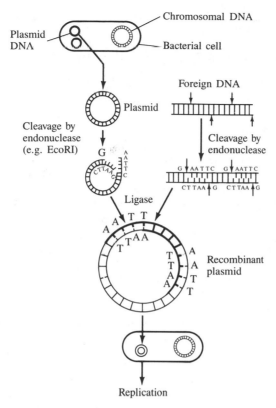

Figure 7.1

Generation of a recombinant plasmid. (From *An Introduction to Recombinant DNA* by A. E. H. Emery. Copyright © 1984 John Wiley & Sons, Inc. Reprinted by permission of John Wiley & Sons, Inc.)

rium, *Eschericia coli*, divides every 20 minutes and thus produces millions of copies in a few hours. When spread on a gel with antibiotics, each transformed cell (with plasmids that confer drug resistance) will appear as a discrete colony called a *clone*. Three steps are necessary to select clones with recombined plasmids of interest: (1) selecting clones with the inserted plasmid; (2) from these, selecting clones with plasmids with the foreign DNA; and (3) from these, selecting clones with plasmids with foreign DNA of interest. These steps are described in Box 7.1.

Recombinant DNA techniques have become valuable commercially. For example, the human insulin gene was cloned in 1979, and now bacteria provide a plentiful supply of the drug for diabetics. Interferon, human growth hormone, and other proteins have also been cloned. In addition, these techniques are useful in the production of safe vaccines in which recombinant

Box 7.1

Isolating Recombinant DNA Clones

Recombinant DNA techniques yield transformed bacterial cells that contain foreign DNA. The techniques used to isolate recombinant DNA clones that contain the foreign DNA of interest introduce additional tools of recombinant DNA that are relevant to behavioral genetics.

As mentioned in the text, three steps are needed to select clones with the target foreign DNA. Because many bacterial cells will not have been infected by the plasmid, the first step involves selecting bacterial cells with an inserted plasmid. Clones that have incorporated the plasmid can be selected by using bacteria that normally do not carry resistance to a particular antibiotic and a plasmid that does carry resistance to the antibiotic; clones that have incorporated the plasmid will grow in the presence of the antibiotic. (See Figure 7.2.) However, many of these plasmids will not have incorporated the foreign DNA. The second step, selection for clones with the recombined plasmid, can be accomplished by using plasmids with genes for resistance to several antibiotics and using a restriction enzyme that cuts the plasmid within one of the antibiotic genes. Thus, clones with the recombined plasmids will not grow in the presence of that antibiotic but will grow when other antibiotics are present. Colonies with the recombined plasmids can be picked off the petri dish and cultured separately.

These colonies will contain plasmids with different fragments of foreign DNA cut by the restriction enzyme. A colony with a particular gene can be isolated if a genetic probe is available, a third step in isolating recombinant DNA clones. As described later in this chapter, a genetic probe is a segment of single-stranded DNA or RNA that has been labeled; labeling typically involves a radioactive marker incorporated into DNA. Labeled probes can be used to search out and hybridize with complementary single-stranded DNA sequences in the presence of a large amount of noncomplementary DNA. Hybridization refers to the sticking together of two complementary single-stranded DNA molecules or of a single-stranded DNA and an RNA molecule to form a double-stranded molecule. The probe is hybridized, not with the colonies themselves, but with DNA bound to a nitrocellulose filter put on the master plate. The cells are lysed to release the DNA, the DNA is made single stranded (simply by heating) and fixed to the filter by baking, and the probe is washed on the filter. The colonies with which the radioactive probe hybridizes are detected by means of autoradiography, which involves exposing an x-ray film to the filter in the dark so that the "hot spots" light up. The locations of these "hot spots" are identified on the master plate and these colonies can be picked off and transferred to another petri dish so that all the colonies in the subsequent culture will contain copies of the desired DNA, a process called cloning. These procedures are illustrated in Figure 7.3. A more precise procedure for hybridizing a probe with DNA, called Southern blotting, is described in Box 7.2.

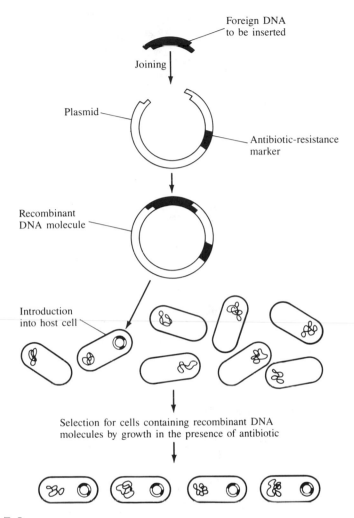

Figure 7.2
Cloning of recombinant DNA in plasmid. (From *Recombinant DNA: A short course* by J. D. Watson, J. Tooze, and D. T. Kurtz. W. H. Freeman and Company. Copyright © 1983).

DNA includes only those viral DNA sequences whose gene product is necessary to produce an immune response.

In addition to recombinant DNA techniques using bacterial cells, three approaches have been used to recombine foreign DNA with eukaryotic cells. Microinjection can be used to introduce foreign DNA directly into the nuclei of cells; sometimes the foreign DNA is integrated and expressed. Another approach recombines foreign DNA with a retrovirus, a virus that injects RNA rather than DNA (Varmus, 1988). Retroviruses invade the host by

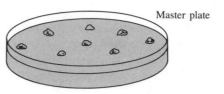

Master plate

Colonies of ampicillin-resistant bacteria
originating from single recombinant cells

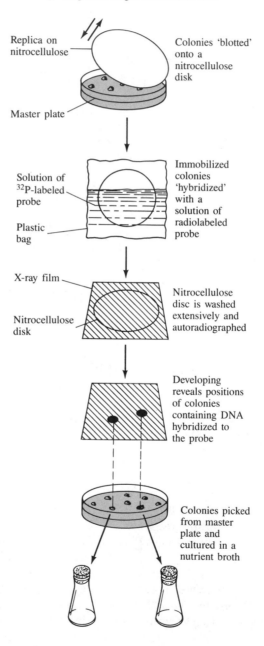

Replica on
nitrocellulose

Colonies 'blotted'
onto a
nitrocellulose
disk

Master plate

Solution of
^{32}P-labeled
probe

Plastic
bag

Immobilized
colonies
'hybridized'
with a
solution of
radiolabeled
probe

X-ray film

Nitrocellulose
disk

Nitrocellulose
disc is washed
extensively and
autoradiographed

Developing
reveals positions
of colonies
containing DNA
hybridized to
the probe

Colonies picked
from master
plate and
cultured in a
nutrient broth

Box 7.2

Southern Blotting

Southern blotting, named after the person who introduced the technique in 1975, is an alternative procedure for hybridization of DNA with a probe; it is more precise than the procedure described in Box 7.1. Southern blotting has additional applications in DNA technology, including some applications that are very relevant to behavioral genetics. DNA extracted from cells is exposed to a restriction enzyme and the fragments are subjected to electrophoresis (literally, "carried by electricity"), a procedure that subjects electrically charged molecules—in this case, negatively charged DNA—to an electric current for several hours. Only fragments smaller than 20 kb can be separated; however, a new technique called pulsed-field gel electrophoresis makes it possible to separate fragments as long as several thousand kilobases. Each DNA fragment migrates through the gel at a rate determined by its net electrical charge and molecular weight; smaller fragments move faster. The bands created by fragments of different length are calibrated by comparison to fragments of known length, which makes it possible to estimate the number of bases of each fragment. An example of the results of electrophoresis of low-weight restriction fragments is shown in Figure 7.4.

If, however, an entire genome is chopped up by a restriction enzyme, a million fragments of varying length are produced, which results in a smear that cannot be resolved by electrophoresis; Southern blotting solves this problem. The DNA on the gel is made single stranded and transferred to a nitrocellulose filter and then fixed by baking. The filter is placed in a solution of a radioactively labeled probe and hybridization is detected by autoradiography—that is, "hot spots" appear, bands that light up because the radioactive probe has hybridized to them. Southern blotting, summarized in Figure 7.5, can detect sequences as small as 20 bases long among all the DNA in a complex genome. A modification of the procedure—perversely called Northern blotting—has been developed to hybridize messenger RNA probes.

coding for reverse transcriptase, an enzyme that synthesizes a DNA copy of the RNA, which is incorporated into one of the host cell's chromosomes. Recombined retroviruses are particularly useful because they have strong promoters that ensure efficient expression of the DNA; they represent the main hope for gene transplants in humans (Baskin, 1984). The third approach involves the introduction of foreign genes into fertilized mouse eggs: Instead of implanting a foreign gene into nuclei of cells in culture, it is injected into the male pronuclei of fertilized mouse eggs, which are then placed in the oviduct of foster mothers (Gordon et al., 1980). In contrast, the

Figure 7.3

Detecting a genetic probe from cloned recombinant DNA. (From *Gene Cloning: The Mechanics of DNA Manipulation* by D. M. Glover. Chapman and Hall. Copyright © 1984.)

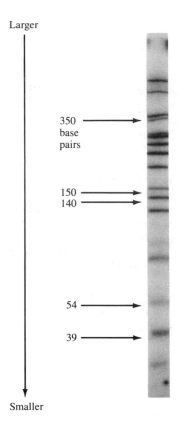

Larger

350 base pairs

150
140

54

39

Smaller

Figure 7.4
Separation of restriction fragments of DNA by electrophoresis. (From *Recombinant DNA: A Short Course* by J. D. Watson, J. Tooze, and D. T. Kurtz. W. H. Freeman and Company. Copyright © 1983.)

other approaches rely on cells in culture that have already differentiated and are thus not as useful for understanding developmental gene regulation.

Genetic Probes

Recombinant DNA techniques have also been valuable for obtaining and cloning genetic probes. Three types of probes are used in molecular genetics research. Probes of a few nucleotide bases (thus called *oligonucleotide probes*) can be synthesized and recombined with bacteria. Probes for genes themselves, which are often many thousands of base pairs in length, can be obtained from mRNA that has been isolated from cells that produce many copies of mRNA for a particular protein. The mRNA is transcribed into DNA using reverse transcriptase, which, as just mentioned, is an enzyme obtained from retroviruses that reverses the normal process of transcribing DNA into RNA. This procedure copies mRNA obtained from cells into so-called complementary DNA (cDNA), in contrast to genomic DNA on the chromosomes. These single-stranded cDNA copies of mRNA are allowed to form their complementary halves, and the double-stranded cDNA is inserted

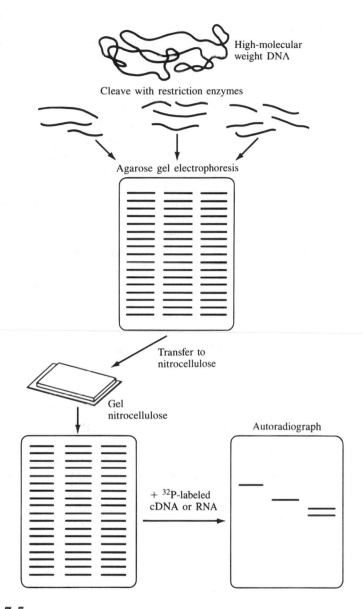

High-molecular weight DNA

Cleave with restriction enzymes

Agarose gel electrophoresis

Transfer to nitrocellulose

Gel nitrocellulose

Autoradiograph

+ ^{32}P-labeled cDNA or RNA

Figure 7.5
Gel transfer hybridization: Southern blotting. (From *Recombinant DNA: A Short Course* by J. D. Watson, J. Tooze, and D. T. Kurtz. W. H. Freeman and Company. Copyright © 1983.)

into bacterial plasmid and then transferred into bacterial cells. Recombinant DNA cells that yield the appropriate protein are cloned to produce a plentiful supply of a DNA probe for that protein. An important by-product of this work was the discovery about eukaryotic DNA mentioned earlier: When these cDNA copies of mRNA were compared to genomic DNA, genomic DNA was found to contain much "extra" DNA not represented in the cDNA copies of mRNA. The third type of probe uses fragments of genomic DNA chopped up by restriction enzymes and cloned using the recombinant DNA techniques described earlier. This is now the major approach used to map the human genome, a topic discussed later. These techniques involve the use of Southern blotting, which is described in Box 7.2.

Another use of these techniques is to obtain a "shotgun" library of DNA fragments that can be used as probes. For example, your genomic DNA (from blood or any tissue—genomic DNA is the same in all cells) can be digested by a restriction enzyme that cuts the DNA into fragments of reasonable length. Millions of different fragments can be inserted into plasmids, which are transferred to bacteria that can then be grown to produce many copies of this library of your DNA. Using the techniques described above, this library could be screened for any genetic probe. The first human gene library was constructed in 1978, a procedure that can now be accomplished commercially for a few thousand dollars. Just as we now routinely blood type people, in the future it may be possible to "gene type" people.

Restriction Map and Chromosome Walking

Another use of restriction enzymes is to map DNA. Each restriction enzyme produces a different set of DNA fragments; the fragments can be sorted by molecular weight using electrophoresis. A *restriction map* can be deduced from comparisons of overlapping fragments cut by different restriction enzymes to show where each restriction enzyme severs a sequence of DNA to produce fragments. An example of a restriction map for just three restriction enzymes of the human β-globin gene on chromosome 11 is shown in Figure 7.6. Restriction maps make it possible to excise a specific sequence of DNA and also to order correctly much larger pieces of DNA than can be cloned in one piece, a procedure known as "chromosome walking."

In relation to behavioral genetics, the importance of these techniques is that loss or gain of a restriction site signals a genetic difference that results in fragments of different lengths. These genetic differences are called *restriction fragment-length polymorphisms.*

Restriction Fragment-Length Polymorphisms and Variable Number Tandem Repeats

If total genomic DNA is cut with a restriction enzyme, a large number of DNA fragments will be produced—about a million fragments on average. As described in Box 7.2, these fragments can be separated on a gel by electrophoresis, and Southern blotting and autoradiography can be used to determine

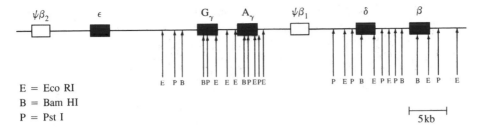

E = Eco RI
B = Bam HI
P = Pst I

5kb

Figure 7.6

Restriction sites in the region of the human β-globin gene on chromosome 11. The black boxes indicate functional genes, the open boxes indicate pseudogenes. (From *An Introduction to Recombinant DNA* by A. E. H. Emery. Copyright © 1984 John Wiley & Sons, Inc. Reprinted by permission of John Wiley & Sons, Inc.)

whether a particular radioactive DNA probe finds its match in the DNA fragments. Restriction fragment-length polymorphisms (RFLPs) occur, for instance, when a restriction enzyme does not cut a particular DNA sequence because the sequence is missing, thus leaving a longer fragment instead of two shorter ones.

For example, a DNA probe for a gene within the β-globin cluster of genes will light up fragments at molecular weights representing 2.7, 3.5, 7.2, and 8.0 kb after digestion with a particular restriction enzyme. Four fragments are obtained because the restriction enzyme happens to cut the gene in three places and the gene probe is able to hybridize with each fragment even if the fragment has only a small portion of the gene. Individuals with sickle-cell anemia, however, do not show the bands at 2.7 and 7.2 kb because the sickle-cell β-globin gene is missing a nucleotide base involved in the site that the restriction enzyme normally recognizes, severing a 9.9 kb fragment into the two smaller fragments of 2.7 and 7.2 kb. Carriers for sickle-cell will show hybridization at all five bands because the normal allele produces the bands at 2.7 and 7.2 kb and the sickle-cell allele results in the 9.9 kb fragment.

The first human DNA polymorphism was found in 1978 (Kan and Dozy, 1978); the first "anonymous" (function unknown) RFLP was found in 1980 (Wyman and White, 1980). There are now more than 1,000 RFLPs, although fewer than a fifth are highly variable from person to person (called *polymorphic*).

Not all DNA fragment-length polymorphisms are due to nucleotide base differences that differ in recognition sites for restriction enzymes. A restriction fragment may be longer for some individuals because the fragment produced by a particular restriction enzyme contains more repetitive DNA sequences than in other individuals, even though the same restriction sites are present. These genetic differences are called variable number tandem repeats (VNTRs); several hundred have been identified. A new technique called DNA fingerprinting uses DNA probes for the tandem repeat sequences that hybridize with many DNA fragments containing these polymorphic DNA sequences (Jeffreys et al., 1985). Each hybridization serves as a genetic marker, and the pattern of hybridization across the various fragments is a powerful tool for determining genetic relatedness, which has practical utility in determining paternity and for assessing zygosity of twins.

Both VNTRs and RFLPs are inherited as simple Mendelian traits and are anonymous DNA markers in the sense that their genetic function is not known. Their value lies in the fact that they serve as markers for DNA that is not expressed and can be found throughout the genome. Figure 7.7 shows patterns of DNA fragments hybridized by a particular DNA probe with four haplotypes (types of fragments rather than alleles) in a three-generation family (White et al., 1985). The parents of the eight children are heterozygous (A1A4 and A2A3, respectively) and the genetic markers are distributed nearly evenly in the children: Two offspring are A1A2, two are A1A3, three are A2A4, and one is A3A4.

The practical significance of RFLPs and VNTRs lies in their use as markers of genetic differences among individuals. These genetic markers can be used for genetic counseling. For example, genetic markers are available for sickle-cell anemia, thalassemias, and phenylketonuria, even when the disease occurs as a result of new mutations, as sometimes happens with Lesch-Nyhan syndrome and Duchenne muscular dystrophy (Caskey, 1987). RFLPs and VNTRs can also be used in linkage studies to determine whether a single-gene disease can be linked to a particular gene on a particular chromosome, as

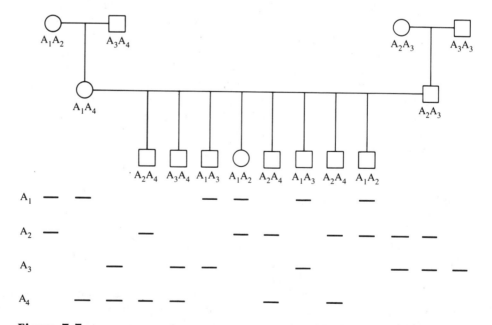

Figure 7.7
RFLPs are inherited as simple Mendelian traits. The pattern of fragments detected by the probe is displayed under their position in the pedigree. Females are represented by circles and males by squares. (Based on R. White et al., 1985, p. 101.)

discussed in the following section. Most importantly for behavioral genetics, genetic markers can be used in genetic analyses of complex behavioral traits, as indicated in the final section of this chapter.

LINKAGE ANALYSIS

Recombinant DNA techniques have been a boon to the study of linkage, introduced in the preceding chapters on chromosomes. RFLPs and VNTRs provide a battery of polymorphic genetic markers that can be used to find a linkage between a single-gene disease and a genetic marker, thus linking the trait to a single gene on a particular chromosome. Using several markers on that chromosome, we can begin to determine the precise location of the trait gene on the chromosome. An even more basic point is that linkage to a specific genetic marker demonstrates the influence of a single gene, which has greatly expanded our knowledge of single-gene influences, a topic introduced in Chapter 3.

As indicated in Chapter 5, linkage analysis uses large pedigrees to determine whether loci are transmitted on the same chromosome. The method rests on recombination: If two genes are close together, little recombination occurs and the recombination fraction is close to zero. For unlinked loci, half of the offspring are recombinants, and the recombination fraction is 0.5; recombination fractions less than 0.5 suggest linkage. Rather than simply showing linkage to a single marker, the use of closely linked multiple markers is much more powerful in identifying linkage. Most importantly for behavioral genetics, an RFLP battery of closely spaced markers can find linkages even for complexly determined traits and can consider genes in combination as they explain transmission in different families (Lander and Botstein, 1986).

In recent years, two direct mapping methods involving recombinant DNA techniques have greatly assisted mapping efforts. *In situ hybridization* (Harper, Ulrich, and Saunders, 1981) involves hybridization of radiolabeled DNA segments to the chromosomes of individual cells undergoing division (called *metaphase spreads*). The site of the gene on the chromosome can be seen by means of autoradiography. In the second method, called *high-resolution chromosome sorting*, chromosomes are stained with a fluorescent dye and then sorted by a fluorescence-activated cell sorter; the chromosomes are then analyzed by Southern blots to determine hybridization with a radiolabeled DNA segment (Lebo et al., 1984). Typically, when a gene is cloned, human-rodent somatic hybrid cells are used to determine the chromosome on which the gene resides; somatic cell hybridization was described in Chapter 5. In situ hybridization and high-resolution chromosome sorting are used to determine the gene's regional localization.

RFLPs have been used to demonstrate linkage for Huntington's disease (Gusella et al., 1983), polycystic kidney disease (Reeder et al., 1985), cystic fibrosis (White et al., 1985), Duchenne muscular dystrophy (Davies et al.,

1983), and half a dozen other diseases (Martin, 1987). The linkage of the gene for Huntington's disease with a new RFLP marker on chromosome 4 was a particularly important event because it was the first time that a disease whose gene product was unknown had been linked — it is not known how this gene has its devastating effect. A portion of the pedigree of over 550 family members of the Venezuelan family in which this linkage was discovered is shown in Figure 7.8. It can be seen that in this family the C allele for a particular RFLP is on the same chromosome as the gene for Huntington's disease. The easiest case to see is the three children in generation V, in which the affected father had the *A* allele on one chromosome and the *C* allele on the other. The mother had the *A* and *B* alleles. The two sons in the fifth generation who received the chromosome with the *C* marker from the father were affected; the one son who received the chromosome with the *A* marker from the father was not affected. The only case in the pedigree not consistent with linkage to the chromosome 4 *C* marker is the first individual listed in generation IV. This affected daughter has the *A* marker on both chromosomes. Although the affected father's markers are not known, the daughter must have received the *A* marker from her father. Because this affected daughter did not receive chromosome 4 with the *C* marker from the father, she is counted as a recombination. That is, if there is linkage with the *C* marker within this family, she received that part of her father's chromosome with the gene for Huntington's disease normally linked with the *C* marker but now recombined with the father's other chromosome with the *A* marker. This woman showed the only recombination in the pedigree of 550 individuals, suggesting that Huntington's disease is linked with this RFLP marker on chromosome 4. By linking Huntington's disease with an RFLP marker on chromosome 4, it is hoped that the gene itself can be cloned, and then its gene product can be studied.

Several linkages have been reported for the affective disorders, although failures to replicate are also common (reviewed by Gershon, in press). One of the earliest linkages to be discovered for any disorder with primarily behavioral manifestations is the familial variety of Alzheimer's disease, which has been linked to markers on chromosome 21 in four large pedigrees (St. George-Hyslop et al., 1987); however, another study excluded linkage to these markers in different pedigrees (Schellenberg et al., 1988). Moreover, most cases of Alzheimer's disease are not hereditary. Linkage involving manic-depressive mental illness and schizophrenia is discussed later. At the 1986 Human Gene Mapping meeting, chromosomal mapping of a total of 376

Figure 7.8

A portion of the Venezuelan Huntington's disease pedigree that shows linkage of the HD gene on a chromosome bearing the C allele for an RFLP. A single individual shows a recombination (marked with an arrow) in which HD occurred in the absence of the C allele. (From "DNA markers for nervous system disease" by J. F. Gusella et al., *Science*. Copyright © 1984. Used with permission of the American Association for the Advancement of Science.)

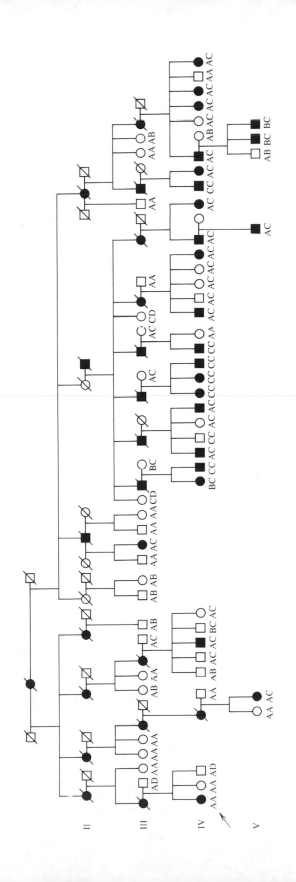

autosomal and 124 X-chromosome loci was confirmed; 222 other assignments were deemed provisional. These successes are recent and much remains to be done: Fewer than 2 percent of known human genes have been mapped (McKusick, 1986b).

Even when linkage is found between a trait and a genetic marker, it is difficult to isolate the gene responsible for the trait. A candidate gene for cystic fibrosis was found by demonstrating a direct association between the disease and a genetic marker (Estivill et al., 1987). However, other examples in which the gene was isolated after detecting linkage are rare. For example, the linkage between Huntington's disease and a marker on chromosome 4 established in 1983 showed one recombination (see Figure 7.8) that suggested a map distance of at least 2 cM (Gusella et al., 1983). Each recombination event marks a point closer to the gene than the original marker; a much tighter linkage of less than 1.5 cM was found in 1987 (Gilliam et al., 1987). However, the gene responsible for Huntington's disease has not yet been cloned. The reason is that a linkage of 1 cM (1 percent recombination) involves a DNA distance of about 1 million base pairs, and there may be a thousand genes in such a region. Even when more markers are available, detecting recombinations that occur fewer than 1 percent of the time will be difficult, especially for rare disorders. The goal, however, is to find a sequence of DNA that shows a direct association with the disease in individuals; genetic markers for Huntington's disease have not yet been found that show such an association. Other techniques for isolating and cloning such genes are being developed (Gusella et al., 1984).

The Human Gene Map and DNA Sequencing

In addition to mapping specific genes of interest, a goal of research is to obtain a complete map of the human genome (DeLisi, 1988). The idea behind a genetic map is to blanket each of the 23 chromosomes with highly polymorphic genetic markers, evenly spaced and as close as possible. Hundreds of such markers are needed to guarantee that linkage for a new single-gene trait can be identified. We are very near that goal: In 1987 a gene map involving 403 RFLPs was published, promising a 97 percent probability that a new single-gene trait can be linked (Donis-Keller et al., 1987). Chromosomes 14 and 19 have yielded the fewest markers. The goals of current research are to fill in the holes in the gene map, to reduce the distance between markers, and to find markers that are more polymorphic—genetic markers that only rarely show differences between individuals are of less use in linkage studies.

A related goal is to identify the sequence of the 3 billion base pairs of human DNA. The first DNA sequencing took about a year for a team of researchers to sequence 5,000 bases; techniques introduced in 1977 permitted DNA sequencing of 15,000 base pairs per year by a single individual. New automated DNA sequencing machines that attach different-colored fluorescent dyes to each of the four bases can now sequence as many bases in two days. Lengthy genes—for example, the 170 kb of the Epstein-Barr virus—have been sequenced, and the complete sequencing of one of the smaller human chromosomes is expected in the next few years.

It is not unreasonable to expect that the entire human genome may be sequenced by the turn of the century—120 of the new DNA-sequencing machines could complete the task in ten years, and faster machines are under development. The magnitude of the task is awesome: Printing the sequence of the haploid nuclear genome of a single person will require the equivalent of 13 volumes of *Encyclopedia Brittanica* (McKusick, 1986b). The enormous variations among individuals greatly complicates the task; although genetic variation is the essence of behavioral genetics, genetic variation has scarcely been considered in discussions of *the* human gene map other than in terms of genetic markers such as RFLPs. About one in 1,000 bases differ for unrelated humans; this means that about 3 million of our 3 billion bases will vary between any two individuals picked at random. It should be mentioned, moreover, that knowing the sequence of the 3 billion nucleotide bases in human DNA is not the end of the story: Understanding the function of genes is much more difficult than determining their nucleotide base sequence.

MOLECULAR GENETICS AND BEHAVIORAL GENETICS

The power of molecular genetic techniques can be harnessed even for complex traits, such as behavior. The use of these techniques to understand behavioral variation will without a doubt represent a major new direction for behavioral genetic research:

> The challenge now is to use the new genetics to climb from the DNA level at the bottom up to the analysis of variation, whose genetics and corresponding gene products have hardly been defined. Obvious examples include cystic fibrosis and Huntington's chorea. The still greater challenge is to deal with diseases and other variations that are not simple Mendelian traits but are multifactorial, a term that basically implies that there are inherited components, but these are not well defined and may be due to a mixture of the effects of several genes and the environment. Heart disease, mental disease, cancer, and complex normal variation, including behavior as well as general physical attributes, all fall in this category. The molecular challenge, which has as its goal the characterization of the whole human genome, is to use this knowledge to unravel the genetic contributions to complex human traits. (Bodmer, 1986, p. 1)

Quantitative genetic techniques, which have been the mainstay of behavioral genetic research, do not assess genetic variation directly. This has been and will continue to be a source of great strength because quantitative methods address the "bottom line" of genetic influences on quantitative trait variability. That is, these methods assess the total impact of genetic variability of any kind, regardless of its molecular source. However, the advent of RFLPs has yielded many genetic markers that can be screened for their contribution, individually and in combination, to the variance of quantitative traits, even in the case of traits for which perhaps scores of genes each contribute small portions of variances in the population and for which environmental factors are important. Such multiple loci that affect quantitative traits have been called *quantitative trait loci* (Gelderman, 1975). Behavioral traits are affected by quantitative trait loci, each with small effects on average in the population. As more genetic markers become available, it will be possible to account for most genetic variability with measured genetic markers. The use of molecular techniques to clarify the inheritance of complex traits is not new (Kloepfer, 1946), but the idea gains tremendous power from the large number of RFLPs that are available. The point is that it is possible to measure DNA variation directly for individuals rather than relying on comparisons among genetically related individuals. The significance of this measured genotype approach is that it permits the systematic identification of associations as well as within-family linkages between specific genetic markers and quantitative traits. This will lead to more powerful means for investigating genetic action such as epistasis, pleiotropy, and genetic correlations among variables.

Linkage Analysis for Complex Traits

An important direction for behavioral genetic research is to use such markers of genetic differences to study complexly determined traits. Traditional linkage analysis assumes that the mode of inheritance is known and that a single gene is involved; however, most behavioral traits involve multiple gene effects and environmental influence. Moreover, behavioral disorders are often genetically heterogeneous—that is, different genes can lead to the same disorder. These complex modes of inheritance are common in genetically well-studied organisms like bacteria, yeast, nematodes, and fruit flies, and evidence is accumulating that humans are no exception. Although traditional linkage methods have been unable to map such traits, a complete RFLP linkage map with highly polymorphic markers evenly spaced at about 20 cM or less throughout the genome—a map that will soon be available—will make it possible to detect some single-gene effects even for complexly determined traits (Lander and Botstein, 1986).

As an illustration, the first solid evidence for a single-gene effect in psychopathology has recently been reported in the Older Order Amish in Pennsylvania, a population of over 12,000 people who descended from about 50 couples who emigrated from Germany during the first half of the eighteenth century. A severe mental illness—manic-depressive psychosis, which is marked by extreme mood swings—was shown to be linked to a RFLP marker on the short arm of chromosome 11 in one family pedigree of 81 individuals, 19 of whom were affected (Egeland et al., 1987). Expression of the gene appears limited because over a third of individuals with the gene in these pedigrees did not show the disorder. Moreover, two other studies of Icelandic and non-Amish North American families have shown that manic-depressive illness is *not* linked to chromosome 11 in these populations, suggesting that chromosome 11 linkage might be unique to the particular Amish family that was examined (Hodgkinson et al., 1987; Detera-Wadleigh et al., 1987). Indeed, two years later, the linkage to chromosome 11 had yet to be found in another pedigree (Lander, 1988). Other studies have suggested that manic-depressive illness may be linked to the X chromosome (Baron et al., 1987).

A linkage for schizophrenia on chromosome 5 has been identified in two Icelandic families with a high incidence of schizophrenia (Sherrington et al., 1988). However, another linkage study of a large Swedish pedigree ruled out the possibility of linkage to chromosome 5, suggesting that the chromosome 5 genetic marker may be limited to the Icelandic families (Kennedy et al., 1988). It is likely that these initial reports of linkages are limited to a few pedigrees with a high frequency of these rare disorders. Selection for such families makes it likely that a major gene mutation will be found, but it also makes it unlikely that such linkages will generalize to other families.

In addition to isolating single-gene effects that occur in certain pedigrees, these techniques are likely to be useful in breaking down the heterogeneity of behavioral traits. That is, different causal factors can lead to similar behavioral problems, such as schizophrenia, as well as to many physical disorders, such as cancer. The goal is to find a major gene effect that accounts for a particular subtype of such heterogeneous disorders.

A major limitation of traditional linkage analysis for behavior is that it can only be applied to qualitative, either–or traits, such as the presence or absence of disease. Most behavioral characteristics, however, are quantitatively distributed; one approach is to assess directly the association between genetic markers and quantitatively distributed traits.

Association

Linkage analysis attempts to show that a marker and a trait gene are on the same chromosome. Potentially more useful for analysis of quantitatively distributed traits is association, in which genetic variation assessed by RFLPs is related directly to behavioral variation. Although an RFLP may be due to variation at a single nucleotide base, it can also serve as a genetic marker for DNA that is sufficiently close (recombination fraction of about 0.5 percent or less), so that the RFLP is virtually never separated from the DNA by recombination. The aim is to find RFLPs that account for some of the genetic variability that affects a trait, even a highly polygenic behavioral trait for which environmental as well as genetic variance is important.

The first association between genetic markers and quantitative traits was found over 60 years ago (Sax, 1923), and many associations have subsequently been reported (Thompson and Thoday, 1979). Recent plant research has been particularly impressive in demonstrating associations with continuously varying phenotypes (e.g., Edwards, Stuber, and Wendel, 1987; Tanksley, Medina-Filho, and Rick, 1982). For example, Edwards and colleagues studied associations between 20 genetic markers and 82 quantitative traits in maize plants. Their findings may be summarized as follows: (1) Significant associations were found for each of the 82 quantitative traits; (2) the maximum variance of any quantitative trait explained by a single marker was 16 percent; (3) over half of the significant associations accounted for less than 1 percent of the trait variance; (4) only 5 percent of the marker loci accounted for more than 5 percent of the variance; and (5) in concert, the genetic markers predicted between 8 and 37 percent of the variance of a subset of 25 relatively independent traits. In other words, quantitative traits appear to be highly polygenic. In addition, this research suggests that dominance is important for many traits.

Associations between genetic markers and human behavior have been explored primarily using the 80 markers in blood, saliva, or urine. A consistent association has been shown between a blood marker (HLA A9) and paranoid schizophrenia (McGuffin and Sturt, 1986). Despite the high statistical significance of the association, it accounts for only about 1 percent of the incidence of paranoid schizophrenia. Because the marker accounts for only a small portion of variance, linkage studies have not yet found evidence of close linkage between the marker and schizophrenia. This work illustrates another important benefit of the measured genotype approach: the analysis of heterogeneity. Associations between the genetic marker and schizophrenia were not found until the associations focused only on the paranoid subtype of schizophrenia.

Studies using traditional blood groups (e.g., ABO) have found few associations with quantitative traits. For example, a recent study found only slightly more than a chance number of significant associations between 18 blood markers and cognitive ability (Ashton, 1986). It would seem unlikely that such functionally focused markers as blood proteins will show widely diffuse effects on behavior. Moreover, in humans, studies of proteins are mostly limited to circulating red blood cells, whereas we are most interested in proteins in the central nervous system. RFLPs assess genomic DNA, which is not limited to genes or gene products expressed in peripheral physiological systems.

With a battery of RFLPs, it will be possible to determine the extent to which these genetic markers, individually and in combination, are able to account for genetic variation in behavior. In the future, as more RFLPs become available, this approach is likely to build a major bridge between molecular genetics and quantitative genetics. It will be a long time before we can identify all quantitative trait loci responsible for genetic influence on behavior. It will take even longer to understand the physiological functions of genes linked to quantitative trait markers. In the nearer future, the measured genotype approach is likely to sharpen our phenotypes by breaking down heterogeneity of complex phenotypes, perhaps leading to differential diagnosis of strengths and risks and thus to more targeted interventions. For example, it is possible that markers that account for only a small proportion of variance on average in the population are of overwhelming importance for certain individuals. The value of this knowledge is that it will enable us to make predictions about genetic strengths and risks *for an individual,* rather than relying on familial resemblance to provide such predictions. Finally, this approach offers the prospect of identifying parts of causal subsystems that can enrich biochemical and physiological studies of complex phenotypes, "providing discrete windows into complex metabolic and biochemical systems" (White and Caskey, 1988, p. 1487). In short, successful identification of associations between quantitative trait loci and behavior will open new vistas for research on the genetics of behavior.

SUMMARY

The techniques of recombinant DNA that emerged from the new genetics of the past decade will eventually revolutionize the study of behavioral genetics. These techniques have made it possible to study DNA of higher organisms, which has led to recognition of the greater complexity of eukaryotic DNA, including split genes, repetitive sequences, and transposable genes. Recombinant DNA techniques make it possible to assess DNA variation among individuals directly. These techniques can be used to isolate DNA variation responsible for behavioral variation, the overarching goal of behavioral genetics. Restriction enzymes can be used to cut DNA of higher organisms and insert the DNA into other organisms such as bacteria. This recombined DNA can be cloned so that a particular fragment of DNA from a higher organism can be obtained as a probe or the gene's product can be obtained in large quantities; the latter has been the main commercial use of recombinant DNA technology. Southern blotting hybridizes a radioactively labeled probe with DNA, a procedure that can be used to detect restriction fragment-length polymorphisms (RFLPs) and variable number tandem repeats (VNTRs). These genetic markers have greatly assisted linkage analyses and will form a bridge between molecular genetics and behavioral genetics.

CHAPTER·8

Population Genetics

$\mathbf{M}$any of the early approaches to the study of evolution were more art than science. However, the theories and methods of population genetics have now provided evolutionary biology with a quantitative basis. Its unique contribution is to assess allelic and genotypic frequencies in groups of breeding organisms and to study the forces that change these frequencies. In this chapter we shall consider characters influenced by single genes. In the next chapter, our focus will shift to polygenic characters—that is, those influenced by many genes. Although population genetics has traditionally employed single-gene univariate approaches, a new direction is this field is the application of polygenic, multivariate approaches (e.g. Arnold and Wade, 1984; Lande, 1979). For a more detailed discussion of population genetics, several excellent textbooks are available (e.g., Crow, 1986).

ALLELIC AND GENOTYPIC FREQUENCIES

When mice from a purebreeding albino strain are crossed to a black strain, all F_1 offspring are black. Thus, albinism is completely recessive to black coat color. In the F_2 generation, as expected, a classical Mendelian ratio of 3 black to 1 albino is observed. (See Chapter 3.) Now, if several mice from the F_2

generation are mated without regard to coat color, the ratio observed in the F_3 generation is again 3:1. In fact, as long as the sample size is large and mating occurs at random, this ratio will recur generation after generation, i.e., in the F_2, F_3, F_4, . . . F_n. How is this possible?

First, recall that 3:1 is a *phenotypic* ratio. The ratio of genotypes is actually $1A_1A_1:2A_1A_2:1A_2A_2$. Therefore, the *genotypic frequencies* are $0.25A_1A_1$, $0.50A_1A_2$, and $0.25A_2A_2$. What are the allelic frequencies? All of the alleles carried by A_1A_1 animals and half of those carried by A_1A_2 animals are A_1. Therefore, the frequency of the A_1 allele in an F_2 generation is 0.50 [i.e., $0.25 + \frac{1}{2}(0.50) = 0.50$]. Because there are only two alleles at this locus, the frequency of the other allele, A_2, is also 0.50, and the proportional frequencies of the two alleles add up to 1.0. In summary, in the F_2 population, the frequency of each allele is 0.50.

Now, what would be the frequencies of these two alleles and the three genotypes in the next (F_3) generation created by randomly mating F_2 individuals with one another? Common sense tells us that, if Mendel was right in suggesting that genes are discrete units, the alleles ought to show up in the F_3 population just as they did in the F_2. We can check this by studying the results of random crosses between F_2 individuals. Because the frequency of each allele is 0.50 in the F_2, eggs and sperm have an equal chance of carrying an A_1 allele or an A_2 allele. Table 8.1 describes the genotypic results of such a cross. For example, the frequency of A_1A_1 genotypes will be 0.25, which is the probability that an A_1 sperm (frequency $= 0.50$) will fertilize an A_1 egg (frequency $= 0.50$). We multiply the frequencies because A_1 sperm are just as likely to fertilize A_2 eggs as A_1 eggs (which is another way of saying that the F_2 individuals are mated at random). Thus, as in the F_2 population, the genotypic segregation ratio for the F_3 population is 1:2:1. Of course, the allelic frequencies ($0.50A_1$ and $0.50A_2$) do not change.

Table 8.1
Genotypic frequencies in the F_3 population

		Allelic Frequencies in Sperm Produced by F_2 Males	
		$0.50A_1$	$0.50A_2$
Allelic Frequencies in Eggs Produced by F_2 Females	$0.50A_1$	$0.25A_1A_1$	$0.25A_1A_2$
	$0.50A_2$	$0.25A_1A_2$	$0.25A_2A_2$

HARDY – WEINBERG – CASTLE EQUILIBRIUM

We have shown that genotypic and allelic frequencies remain stable generation after generation. This equilibrium is the cornerstone of population genetics. Population geneticists study departures from the equilibrium resulting from such factors as selection, migration, mutation, random genetic drift, and nonrandom mating.

Figure 8.1 summarizes the results of random mating in more general terms. The symbol p is often used to refer to the frequency of one allele, while q symbolizes the frequency of the other. If there are only two alleles for a particular locus, $p + q = 1$. When gametes with p A_1 and q A_2 frequencies randomly unite, the probability of producing offspring homozygous for the A_1 allele is p^2. The probability of homozygotes for the A_2 allele is q^2. The probability of heterozygotes is $pq + qp$, or $2pq$. All p^2 of the A_1A_1 genotypes have the A_1 allele, and half of the alleles for the A_1A_2 genotypes [$\frac{1}{2}(2pq)$] are A_1. Using a bit of algebra, we can see that the frequency of A_1 is still p after one generation of random mating:

$$p^2 + \tfrac{1}{2}(2pq) = p^2 + pq = p(p + q) = p$$

Because $p + q = 1$, this expression reduces to p, meaning that the value of p has not changed after one generation of random mating. Therefore, with continued random mating, the genotypes will also remain stable with frequencies of $p^2 + 2pq + q^2$.

In an F_2 generation, the frequencies of the two alleles are equal. However, an equilibrium may also occur when allelic frequencies are different. Suppose that the initial frequency of one allele in a population was 0.80 and the frequency of the other was 0.20. Table 8.2 shows the resulting genotypic frequencies. After one generation of random mating, $p = 0.64 + \frac{1}{2}(0.32) = 0.80$, and $q = 1 - 0.80$, or $0.04 + \frac{1}{2}(0.32)$; either way, $q = 0.20$. The point is that the frequencies of the alleles have not changed after one generation of random mating.

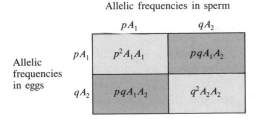

Figure 8.1
Genotypic frequencies after one generation of random mating.

Table 8.2
Genotypic frequencies after one generation of
random mating when allelic frequencies are
not equal

	Allelic Frequencies in Sperm	
	$0.80A_1$	$0.20A_2$
Allelic Frequencies in Eggs $0.80A_1$	$0.64A_1A_1$	$0.16A_1A_2$
$0.20A_2$	$0.16A_1A_2$	$0.04A_2A_2$

This law of equilibrium for genetic variability was apparently so obvious that no one really wanted to take credit for discovering it. In 1908 an English mathematician, G. H. Hardy, and a German obstetrician, W. Weinberg, independently published papers describing the equilibrium. Hardy began his note in *Science* almost apologetically: "I am reluctant to intrude in a discussion concerning matters of which I have no expert knowledge, and I should have expected the very simple point which I wish to make to have been familiar to biologists" (1908, p. 49). More recently, it has been pointed out that W. E. Castle, an American geneticist, utilized and even extended this relationship in a paper published in 1903. For this reason, it has been suggested that we place in our textbooks a belated recognition of "Castle's law" (Keeler, 1968). Although, with hindsight, the law of equilibrium may seem obvious, it is the key concept of population genetics, and has many uses.

Uses of the Hardy–Weinberg–Castle Equilibrium

The Hardy–Weinberg–Castle relationship can be used to determine whether a population is in equilibrium, as well as to estimate allelic and genotypic frequencies. If a population is in equilibrium, the genotypic frequencies should correspond to $p^2 + 2pq + q^2$ for a single-locus, two-allele character. The easiest example is a co-dominant system in which both alleles are expressed. The MN antigens carried on the surface of red blood cells provide an example of a co-dominant system. Individuals with only the M antigen are MM homozygotes; individuals with only the N antigen are NN homozygotes; and individuals with both antigens are MN heterozygotes. In the United States, the frequencies of the MM, MN, and NN genotypes are 0.30, 0.50, and 0.20, respectively. What is the frequency of the M allele that produces the M antigen? It includes all of the alleles for the MM genotypes and half of the alleles for the MN genotypes. Therefore, $p = 0.30 + \frac{1}{2}(0.50) = 0.55$. If $p = 0.55$, $q = 1 - 0.55$, or 0.45.

Now we can ask whether the MN blood system is in genetic equilibrium — that is, whether systems of mating, mutation, migration, or selection cause significant departures from the expected genotypic frequencies of $p^2 + 2pq + q^2$. If p is 0.55 and q is 0.45, then $p^2 = 0.30$, $2pq = 0.50$, and $q^2 = 0.20$. Thus, this character is in equilibrium in this population because the expected genotypic frequencies are the same as those that are observed (MM = 0.30, MN = 0.50, NN = 0.20). Satisfy yourself that the system would not have been in equilibrium if the frequencies for the MM, MN, and NN genotypes were, for example, 0.35, 0.40, and 0.25, respectively.

If heterozygotes can be distinguished from homozygotes, as in the MN example, then allelic frequencies can be determined exactly for representative samples. When there is complete dominance, allelic frequencies can be estimated if the population is in equilibrium. Let us assume that a population is in equilibrium for a particular single-locus, two-allele trait. Then, the frequency of the homozygous dominant genotype (A_1A_1) is p^2, the frequency of the homozygous recessive genotype (A_2A_2) is q^2, and the frequency of the heterozygous genotype (A_1A_2) is $2pq$. The heterozygotes are called *carriers* because they carry only one recessive allele and thus do not display the trait. We can estimate the frequency of the recessive allele and the number of carriers in a population if we know the number of individuals displaying the trait. For example, suppose that 16 percent of the population shows a recessively determined trait (as is true for the recessive blood factor called rh-negative). Because the allele is recessive, these individuals are of the A_2A_2 genotype and have a frequency in the population of q^2. Because the frequency of the recessive allele is q, all we have to do is to take the square root of q^2 to find the frequency of q. The square root of 0.16 is 0.40. Thus, although only 16 percent of the population has the homozygous recessive genotype, the frequency of the recessive allele is 40 percent.

Given this information, one can determine the number of carriers for this recessive allele. If q is 40 percent, the frequency of the other allele is 60 percent. Given an equilibrium, the frequency of carriers will be $2pq$, which is $2(0.60)(0.40) = 0.48$. Thus, 48 percent of the population is made up of carriers for the recessive allele. Remember, however, that these estimates are correct only if the population is in equilibrium. Using the earlier example of genotypic frequencies that are not in equilibrium ($0.35A_1A_1$, $0.40A_1A_2$, and $0.25A_2A_2$), we would mistakenly estimate q as 0.50 instead of 0.45.

Tasting PTC

As discussed in Chapter 3, about 70 percent of the Caucasians in the United States experience a very bitter taste when a solution of phenylthiocarbamide (PTC) is applied to the tongue, whereas about 30 percent find it virtually tasteless at the same concentration. If there is random mating for tasting PTC, which seems likely, then the Hardy–Weinberg–Castle equilibrium can be used to test hypotheses concerning the mode of transmission. Typically, we assume the simplest genetic hypothesis, test it, and then discard it only if there is a significant departure from the expected result. Thus, we shall retain the simplest model until we are compelled by the data to consider more complex hypotheses. The simplest genetic model to account for the family data in Table 8.3 involves two alleles at one autosomal locus, with one allele (the one for tasting PTC) completely dominant over the other.

If the population described above is in equilibrium, the frequency of the homozygous recessive genotype is q^2 (i.e., 29.8 percent). The frequency of q is thus the square root of 0.298, which is 0.546. The frequency (p) of the dominant allele for tasting PTC is $1 - 0.546 = 0.454$. Tasters can be homozygous or heterozygous for the dominant allele. The frequency of the homozygous tasters is p^2, which is 0.206, and the frequency of the heterozygous tasters is $2pq$, or 0.496. Thus, 70.2 percent of the individuals are tasters (that is, $0.206 + 0.496 = 0.702 = 70.2$ percent). Of these tasters, 29.3 percent are homozygous for the PTC-tasting allele ($0.206 \div 0.702 = 0.293 = 29.3$ percent). The rest of the tasters, 70.7 percent, are heterozygous ($0.496 \div 0.702 = 0.707 = 70.7$ percent).

We can check this hypothesis by comparing the numbers of expected offspring from certain mating combinations with the actual data for PTC tasting presented in Table 8.3. One aspect of the table conforms to the hypothesis that nontasters of PTC are homozygous for a recessive allele at one autosomal locus: Matings between two nontasters almost always produce nontaster offspring. The five taster children out of 223 could be due to variable gene expression, misclassification, or illegitimacy.

However, the rest of the data are not as straightforward, since the tasters could either be homozygous or heterozygous, which affects the expectations for the offspring. For example, if all of the tasters in the taster/nontaster matings were homozygous, then all of their offspring would be tasters. If all

Table 8.3

Data on the inheritance of ability to taste phenylthiocarbamide

Mating	Number of families	Offspring		Fraction of nontasters among offspring
		Tasters	Nontasters	
Taster × taster	425	929	130	0.123
Taster × nontaster	289	483	278	0.366
Nontaster × nontaster	86	5	218	0.978

SOURCE: From Stern, *Principles of Human Genetics*, 3rd Ed. W. H. Freeman and Company. Copyright © 1973; after Snyder.

the tasters were heterozygous, then half of their offspring should be nontasters: they have a 50 percent chance of receiving the dominant allele from the heterozygous taster parents. However, the taster parents can actually be either homozygous or heterozygous. Thus, as expected, Table 8.3 indicates that the fraction of nontasters from this type of mating is neither 0 nor 50 percent, but rather 36.6 percent.

In order to determine how well our model fits the data for taster/nontaster matings, we need to ask what percent of the tasters are homozygous. As indicated above, 29.3 percent of the tasters should be homozygous according to our model. Thus, of the 761 offspring from taster/nontaster matings, we would expect that 29.3 percent, or 223, result from matings between homozygous tasters and nontasters. All of these 223 children will be tasters. The other 538 children resulted from matings between heterozygous tasters and nontasters. For these matings, half of the offspring, 269, should be tasters. The other 269 should be nontasters. These calculations correspond well to the observed data for taster/nontaster matings. We have predicted that 492 (that is, 223 + 269) will be tasters and 269 will be nontasters; the observed data for these matings are 483 taster offspring and 278 nontaster offspring. If we performed similar calculations for the taster/taster matings, we would also find that our predictions correspond to the data. Thus, the data presented in Table 8.3 closely conform to that expected on the basis of a single-locus, two-allele model, where the allele for tasting PTC is completely dominant.

Although the Hardy–Weinberg–Castle equilibrium is useful for estimating allelic and genotypic frequencies, its most important use in population genetics is as a standard, like standard temperature and pressure in chemistry and physics. Forces that change allelic and genotypic frequencies are measured in relation to this standard.

FORCES THAT CHANGE ALLELIC FREQUENCY

In this section, we shall examine the changes in allelic frequency caused by migration, random drift, mutation, and selection. Because systems of mating (assortive mating and inbreeding) affect genotypic frequencies, not allelic frequencies, they will be considered in a following section.

Migration

If two populations have the same allelic frequencies at a particular locus, the frequencies will not be altered if individuals from one population randomly migrate to the other. However, if the allelic frequencies at a locus differ, then the frequencies will change due to random migration. Assume that the frequency of some allele is q in a certain population. We will designate the frequency of that allele in this original population as q_o. If individuals from another population immigrate into this population, the resulting frequency will depend on the frequency of the allele in the immigrant population (q_m) and the rate of immigration. In fact, the frequency of the allele after one generation of immigration (q_1) will be the frequency of q_m weighted by the proportion of immigrants plus q_o weighted by the proportion of natives. If m is the proportion of immigrants and $1 - m$ is the rest of the population (the natives), we can express the effect of migration algebraically:

$$q_1 = mq_m + (1 - m) q_o$$

This expression can be simplified by multiplying and then factoring out m:

$$q_1 = mq_m + q_o - mq_o = m (q_m - q_o) + q_o$$

The change in allelic frequency (Δq) is simply the difference between the frequency after immigration (q_1) and the original frequency (q_o). If we take the above expression for q_1 and subtract q_o, we are left with

$$\Delta q = q_1 - q_o = m(q_m - q_o)$$

Thus, the change in allelic frequency is a function of the rate of immigration (m) and the difference between the frequencies in the immigrant and native populations.

For example, assume a rate of migration of 10 percent, where the frequency of some allele is 0.20 in the natives and 0.30 in the immigrants. The change in allelic frequency will be 0.01:

$$\Delta q = 0.10(0.30 - 0.20) = 0.01$$

Although this change in frequency may appear small, sustained immigration of this magnitude over many generations would have a substantial effect.

The opposite side of the coin is emigration. Allelic frequencies will change if the frequency in the emigrant population is different from that of the remaining population. Like immigration, emigration will affect allelic frequencies in the population of individuals who remain (q_r) as a function of the rate of emigration (n) and the difference between frequencies in the emigrant (q_n) and remaining populations (q_r):

$$\Delta q = n(q_r - q_n)$$

Random Genetic Drift

Immigration and emigration are systematic migrational influences that affect allelic frequencies. Chance is also an important factor. If a population is small, the random sampling of genotypes may lead, purely by chance, to changes in allelic frequency. This type of unsystematic change in allelic frequency from one generation to the next is called *random drift* (Wallace, 1968). In a large population, random drift will not be an important source of allelic frequency change.

Mutation

Mutation is the ultimate source of genetic variability. However, our major concern here is mutation as a force affecting the frequency of a specific allele. Because the spontaneous mutation rate is low, the random event of mutation is not an important source of change for a specific allele, except when considered on an evolutionary time scale.

We can be more specific about expected changes due to mutation. Assume that there are two alleles at an autosomal locus and that A_1 mutates to A_2 with frequency u per generation. We also need to consider the possibility that A_2 can mutate back to A_1 (rate v). The change in frequency of the A_2 allele is the extent to which A_1 mutates to A_2 minus the extent to which A_2 mutates back to A_1:

$$\Delta q = up - vq$$

No one knows the exact spontaneous mutation rate (u), and it can be different for various alleles. But let us assume that A_1 mutates to A_2 once in a million DNA replications. If u is 10^{-6} (i.e., 0.000001) and v is 10^{-7}, mutation will not be a major source of gene frequency change. Suppose that p is 0.90 and q is 0.10, then $\Delta q = (0.000001)(0.90) - (0.0000001)(0.10) = 0.00000089$ per generation.

Selection

Selection occurs when there are differences in reproductive rates among individuals in a population, and it can be a powerful force in changing allelic frequencies. Of course, individuals, not genes, are the targets of selection. With complete dominance of A_1 over A_2, individuals with an A_1A_2 genotype could not be distinguished from individuals with an A_1A_1 genotype. If neither of these genotypes reproduced at all, the frequency of the A_1 allele would be zero after just one generation of such severe selection. There are so few lethal dominant alleles because individuals with such alleles are quickly selected out of the population. However, recessive alleles are a different story. If recessive homozygotes (A_2A_2) did not reproduce for one generation, the A_2 allele would not be eliminated from the population, since A_1A_2 heterozygotes would continue to reproduce.

Selection can also operate against both the A_1A_1 homozygote and the A_2A_2 homozygote, thus favoring the A_1A_2 heterozygote. This heterozygote advantage leads to a *balanced polymorphism*. When A_1A_2 heterozygotes mate, they continue to produce both A_1A_1 and A_2A_2 homozygotes. Balanced polymorphisms ensure genotypic variability within the population. They will be considered in greater detail after we discuss other types of selection.

We can estimate the change in allelic frequencies caused by selection by starting with the Hardy–Weinberg–Castle equilibrium. The first row of Table 8.4 lists the frequencies of the three genotypes for a single-locus, two-allele character. The frequencies are $p^2 + 2pq + q^2$. Now we shall consider the effect of selection against one or more of these genotypes. However, rather than focusing on selection against a certain genotype, we will concentrate on the relative fitness of the genotype. This is the reproductive rate of that genotype compared with the genotype having the highest reproductive rate. Thus, if A_1A_1 and A_1A_2 genotypes average 20 offspring, while A_2A_2 genotypes produce only 5, the relative fitnesses of these three genotypes would be 1, 1, and ¼ (i.e., ⁵⁄₂₀ = ¼), respectively. Selection against the A_2A_2 is ¾ (i.e., 1 − ¼). We shall label this selection coefficient s, and the relative fitness as $1 − s$. When $s = 0$, the relative fitness of all genotypes is 1, and individuals of all genotypes will contribute equally to the next generation. However, when the relative fitness of a certain genotype is less than 1, individuals of that genotype will contribute relatively less to the next generation.

The relative contribution of each genotype to the next generation is the genotypic frequency weighted by the relative fitness value. Table 8.4 describes three examples of selection. The first case involves complete dominance of the A_1 allele and selection against A_2A_2. The frequencies of the three genotypes after selection are $p^2 + 2pq + (1 − s)q^2$. The frequency of the A_2 allele after one generation of selection will be determined by the A_2A_2 individuals who reproduce $[(1 − s)q^2]$ and half of the alleles from the A_1A_2 heterozygotes (½ × 2pq = pq). Because we express allelic frequencies as proportions, this frequency must be divided by the genotypic frequencies after selection $[p^2 +$

Table 8.4

Relative fitness of genotypes for three different cases of selection

	Genotype		
Item	A_1A_1	A_1A_2	A_2A_2
Frequency	p^2 +	$2pq$ +	q^2 = 1
Relative fitness:			
A_1 completely dominant, selection against A_2A_2	1	1	$1 − s$
A_1 completely dominant, selection against A_1A_1 and A_1A_2	$1 − s$	$1 − s$	1
Overdominance, selection against A_1A_1 and A_2A_2	$1 − s_1$	1	$1 − s_2$

$2pq + (1 - s)q^2]$. Thus, after one generation of selection, the A_2 frequency will be

$$q_1 = \frac{(1 - s)q^2 + pq}{p^2 + 2pq + (1 - s)q^2}$$

The term $[(1 - s)q^2]$ is equivalent to $q^2 - sq^2$. If we make this substitution in the denominator, it becomes $p^2 + 2pq + q^2 - sq^2$. Because $p^2 + 2pq + q^2 = 1$, the denominator simplifies to $1 - sq^2$.

$$q_1 = \frac{(1 - s)q^2 + pq}{1 - sq^2}$$

For example, if $q = 0.10$ (frequency of the A_2 allele) and the relative fitness of the A_2A_2 genotype is zero, the frequency of q after one generation will be 0.09:

$$q_1 = \frac{(1 - 1)0.01 + 0.9(0.10)}{1 - (1)(0.01)} = \frac{0.09}{0.99} = 0.09$$

The change in the frequency of the A_2 allele is simply $q_1 - q$. For this example, the change in frequency is -0.01.

However, the effects of selection must be considered for intervals longer than one generation. After n generations of complete selection against A_2A_2, if q_o is the frequency in the original generation, the frequency of A_2 will be

$$q_n = \frac{q_o}{1 + nq_o}$$

As an example, consider some deleterious condition determined by an autosomal recessive allele (A_2). How much could the frequency of this recessive allele be lowered if A_2A_2 individuals did not reproduce for a number of generations? Because alleles with detrimental effects tend to have a relatively low frequency, let us assume that $q_o = 0.02$. If no A_2A_2 individuals reproduced for 50 generations, the frequency of this undesirable allele would become

$$q_{50} = \frac{0.02}{1 + 50(0.02)} = 0.01$$

In other words, after some 1,500 years of intense selection against A_2A_2 (assuming a generation interval of about 30 years, as is the case for human populations), the frequency of this allele would only change from 0.02 to 0.01. This demonstrates the relative ineffectiveness of this form of selection if the frequency of a recessive allele is initially low.

The second example in Table 8.4 involves selection against the A_1A_1 homozygote and A_1A_2 heterozygote. Obviously, the frequency of the A_2 allele will increase, because the relative fitness of the A_2A_2 genotype is greater than that of the other two genotypes. In fact, if selection is complete ($s = 1$) against both A_1A_1 and A_1A_2, only individuals with the A_2A_2 genotype will reproduce. Thus, after one generation of such selection, no more A_1 alleles will be produced, and the frequency of the A_2 allele will become 1.00.

However, selection is not likely to be complete. The more general case can be determined similarly to the first example of selection against the A_2A_2 genotype. The frequency of the A_2 allele will come from the A_2A_2 genotypes (q^2), and half of the alleles from the A_1A_2 heterozygous individuals who reproduce [$\frac{1}{2}(1 - s)2pq = (1 - s)pq$]. In order to express this as a proportion of the total genotypic frequency after selection, we will divide this by the total genotypic frequencies weighted by their relative fitnesses [$(1 - s)p^2 + (1 - s)2pq + q^2$], as in the previous example.

Thus, after one generation of selection, the frequency of the A_2 allele will be

$$q_1 = \frac{q^2 + (1 - s)pq}{(1 - s)p^2 + (1 - s)2pq + q^2}$$

We can simplify the denominator by multiplying through and reducing, remembering that $p^2 + 2pq + q^2 = 1$. This leaves the denominator as $1 - s(p^2 + 2pq)$. This can be further simplified because $p^2 + 2pq = 1 - q^2$. Thus,

$$q_1 = \frac{q^2 + (1 - s)pq}{1 - s(1 - q^2)}$$

This represents the more general case, in which selection may not be complete against the dominant homozygote and the heterozygote. However, to get a feeling for this equation, consider the case in which selection is complete ($s = 1.0$) against both A_1A_1 and A_1A_2. As expected, the new frequency of the A_2 allele is 1.00 because the genotypes with the A_1 allele do not reproduce:

$$q_1 = \frac{q^2 + (1 - 1)pq}{1 - 1(1 - q^2)} = \frac{q^2}{1 - 1 + q^2} = 1.00$$

In a similar manner, one generation of complete selection against the A_2A_2 genotype and the carriers (A_1A_2) would result in the elimination of the recessive A_2 allele. As in the first example, the frequency of a recessive allele is slowly changed when only the recessive homozygote is selected against. When the frequency of A_2 is low, most of the A_2 alleles will remain undetected in heterozygous carriers. However, if carriers could be detected and if they did not reproduce, it would be possible to eliminate the detrimental allele in one generation.

The third type of selection in Table 8.4 results in a *balanced polymorphism*, such that selection maintains different alleles rather than favoring just one. Heterozygotes can be distinguished from homozygotes in the case of overdominance. (See Figure 3.2.) Suppose that selection operated against both homozygous genotypes. The A_1A_2 genotypes would reproduce relatively more than the two homozygous genotypes, but they would continue to produce both homozygotes along with heterozygotes. Thus, genetic variability would be maintained.

The frequency of the A_2 allele can be determined as before. However, in this case, we need to allow for the possibility that the relative fitnesses of the two homozygotes differ. For this reason, Table 8.4 indicates s_1 for the A_1A_1 genotype and s_2 for the A_2A_2 genotype. Once again, the frequency of the A_2 allele after one generation of balanced selection will be determined by the surviving A_2A_2 genotypes $[(1 - s_2)q^2]$ and half of the alleles from the heterozygotes $[(\frac{1}{2}(2pq)) = pq)]$. The denominator is the sum of the genotypic frequencies weighted by their relative fitnesses. Thus,

$$q_1 = \frac{(1 - s_2)q^2 + pq}{(1 - s_1)p^2 + 2\,pq + (1 - s_2)q^2}$$

Given that $p^2 + 2pq + q^2 = 1$, the denominator can again be simplified by multiplying through and reducing:

$$q_1 = \frac{(1 - s_2)q^2 + pq}{1 - s_1p^2 - s_2q^2}$$

If $p = q = \frac{1}{2}$ and selection is equal against both homozygotes, the frequencies of p and q will remain equal. For example, substitute in the above equation: $s_1 = 1$, $s_2 = 1$, and $p = q = \frac{1}{2}$.

If the relative fitnesses of the two homozygotes are not equal, the allelic frequencies will change. However, after many generations of such selection, an equilibrium will be reached in which the allelic frequencies no longer change. At that point, the frequency of the A_2 allele is simply a function of the two selection coefficients:

$$q_1 = \frac{s_1}{s_1 + s_2}$$

For example, if the relative fitness of the A_1A_1 genotype $(1 - s_1)$ is 0.25 and that of the A_2A_2 genotype $(1 - s_2)$ is 0.50, the frequency of q will stabilize at 0.60. (Check this by substituting $s_1 = 0.75$ and $s_2 = 0.50$ in the above equation.) Heterozygote advantage is one of several types of selection resulting in a balanced polymorphism or selectional balance. Because of the importance of balanced polymorphisms in maintaining genetic variability, let us consider this topic in more detail.

STABILIZING SELECTION AND BALANCED POLYMORPHISMS

In the past, selection has often been regarded simply as a force that molded individuals to a particular environment. If this were the case, then we would expect to find very little genetic variability within a local group. To the contrary, however, it appears that at least a third of all loci are polymorphic. Although everyone now agrees that there is considerable genetic variability, there are differing opinions concerning the importance of selection in maintaining variability. *Neutralists* argue that most genetic variability has no selective value. They suggest that it is maintained simply by an equilibrium of backward and forward mutations. *Selectionists*, however, argue that the variability is maintained by selection. Although both positions are correct in some instances, in recent years several interesting examples of balanced polymorphisms have been discovered.

Francis Galton, who was responsible for so many firsts, was one of the first to consider the evolutionary consequences of stabilizing selection. During his travels in Africa, Galton noticed the strong herding behavior of his pack-oxen. He considered the survival value of this behavior in a paper published in 1871, and concluded that stabilizing selection, rather than directional selection, was the best evolutionary bet. A moderate amount of gregariousness appeared optimal for both grazing and protection. Too little gregariousness would leave the oxen in a vulnerable position for predators, and too much gregariousness would be inefficient for grazing.

Even with artificial selection, stabilizing selection operates to keep the organism in balance. For example, as hens are selected for extreme egg production, their fertility and general viability usually diminish, thus counterbalancing the effects of extreme artificial selection. Stabilizing selection is not nearly as dramatic as directional selection, but it is very common in nature (Wilson, 1975).

Four specific sources of balanced polymorphisms have been studied in recent years: heterozygote advantage, environmental diversity, frequency-dependent selection, and frequency-dependent sexual selection.

Heterozygote Advantage

Sickle-cell anemia in humans is a specific example of a balanced polymorphism of the sort described as the third type of selection in Table 8.4. Although few individuals afflicted with this most serious disease (recessive homozygotes) survive to reproduce, the allele is nonetheless maintained in relatively high frequency in some African populations and among Afro-Americans. This high frequency of an essentially lethal recessive allele is apparently due to the high relative fitness of heterozygotes. Carriers seem to be more resistant than normal homozygotes to a form of malaria prevalent in certain parts of Africa. Although sickle-cell anemia is one excellent example of heterozygote advantage, other examples of this source of balanced polymorphisms are speculative.

Environmental Diversity

If environments encountered by a species are diverse, selection pressures can differ and thus foster genetic variability, as noted by Darwin. For example, different genotypes can adapt to different ecological niches. Moreover, rare genotypes are likely to find niches that are not completely full. Box 8.1 describes an example in which shell markings of snails differ in woodlands and grasslands.

Frequency-Dependent Selection

Selection that favors rare alleles produces genetic variability (Crow, 1986). For example, it is possible that rare genotypes are able to exploit environmental diversity, that is, to utilize resources that are not used by other members of the species.

Predator–prey relationships may also help to maintain genetic diversity. For example, minnows will prey on the more common type of water bug, leaving the rarer forms at a reproductive advantage. Predatory birds and mammals also tend to attack more common types of prey.

Frequency-Dependent Sexual Selection

Another type of balanced polymorphism results from mating preference. In *Drosophila*, it has been shown that rarer males are relatively more likely to reproduce. As the rare type of male reproduces more and thus becomes more common, the reproductive edge vanishes. This rare-male advantage may be a general process for maintaining genetic variability in a population. The greater relative reproductive success of rare males was independently discovered by Claudine Petit and Lee Ehrman. Petit (1951) first discovered this phenomenon in a study of the mating success of two strains of *Drosophila*. One strain was the wild type; the other was bar-eyed, a mutant, sex-linked condition that changes eye morphology. The two strains were permitted to breed freely for a number of generations. During the course of the experiment, randomly chosen females were occasionally separated from the population and allowed to lay their eggs in individual vials. By examining the offspring of these females, it was possible to determine whether the male with which the female had mated was wild type or bar-eyed.

The reproductive success of males of a certain type is measured by the number of females mated by that type of male divided by the total number of

Box 8.1

Maintaining Genetic Variability in Snails

Shell markings of a single species of land snail display great variation. (See the figure on the facing page.) The pattern of stripes and the color of the shell are controlled by different genes. Many combinations of stripes and colors are thus possible. Such genetic variability has been around for a long time; fossil snails tens of thousands of years old have similar varieties of shells.

In this case, genetic variability is maintained by selection. When snails are found in woodlands, their shells are likely to be without bands. However, snails in grasslands are likely to have banded shells. The fact that shell banding is correlated with habitat suggests that selection is at work.

Direct evidence for selection comes from an examination of the shells of snails captured by thrushes, who smash the snails on stones to break them open (Clarke, 1975). The most conspicuous snails in a particular habitat (banded snails in woodlands, unbanded snails in grasslands) are most often preyed on by the thrushes. Given that this species of land snail occupies both woodlands and grasslands, genetic variability for shell banding will continue as the result of such selection.

Variation in the shell markings of a single species of land snail, *Cepaea nemoralis.*
(From "The causes of biological diversity" by B. Clarke. Copyright © 1975 by Scientific American, Inc. All rights reserved.)

those particular males. The coefficient of mating success (*K*) is useful in describing the results of experiments on frequency-dependent sexual selection. *K* is the ratio of the reproductive success of one type of male to the reproductive success of the other type. In Petit's experiment,

$$K = \frac{\text{no. of females mated by mutants/no. of mutant males}}{\text{no. of females mated by wild types/no. of wild-type males}}$$

Thus, *K* is the ratio of the number of females mated per mutant male to that of females mated per wild-type male. If the two types of males have equal reproductive success, *K* will equal 1.00. If *K* is less than 1.00, the mutant males are at a disadvantage. Conversely, if *K* is greater than 1.00, the mutant

males mate more females than would be expected simply on the basis of their frequency.

The coefficient of mating success observed by Petit during the course of her experiment is graphed in Figure 8.2. The frequency of bar-eyed males fell from 93 percent of the total male population early in the experiment to 6 percent, due to their relatively low mating success. Although K was less than 1.00 throughout the experiment, the reproductive success of bar-eyed males was frequency dependent, i.e., their mating success increased as they became rarer.

The results of more recent work by Ehrman have been even more striking in suggesting frequency-dependent sexual selection. Rather than relying on progeny testing as an index of mating success, Ehrman has employed

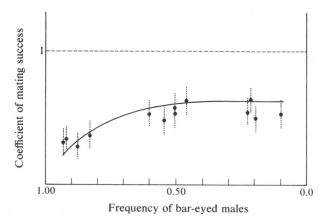

Figure 8.2
Sexual selection in *Drosophila melanogaster*. Reproductive success of bar-eyed males. See text for explanation. (From Petit and Ehrman, 1969.)

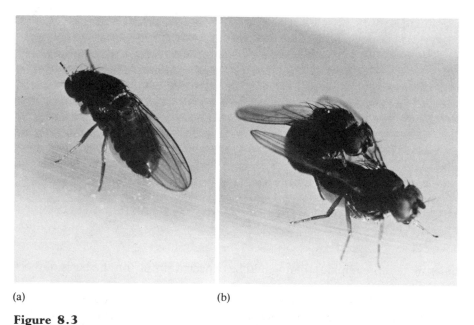

(a)   (b)

Figure 8.3
(a) The female *Drosophila pseudoobscura*, her abdomen distended with eggs, preens with both forelegs, apparently unaffected by (b) the mating couple sharing the observation chamber with her. Both females are orange-eyed mutants. The male has the dark red, nonmutant eye color of the wild-type *Drosophila*. The wings of the copulating female provide a base of support for her mate; she may even fly while carrying him. (Photograph by A. Heder. Reprinted by permission from *American Scientist*, journal of Sigma Xi, The Scientific Research Society of North America.)

direct behavioral observation in her studies of the rare-male advantage in *Drosophila*. (See Figure 8.3.) Females and males are placed into a mating chamber in which they are observed for several hours. Males and females from each of two strains are placed together, resulting in four possible mating combinations. Ehrman has found that the rare male is at a reproductive advantage in a number of different test situations. For example, it has the advantage when the two strains possess different chromosome arrangements, are mutant versus wild-type, and are positively versus negatively geotactic (that is, move down rather than up). The rare-male advantage has now been demonstrated in seven species of *Drosophila*, as well as in a species of beetle and a species of wasp (Ehrman and Parsons, 1981).

Frequency-dependent sexual selection—at least in *Drosophila*—is limited to males. Rare females have no mating advantage. Although female *Drosophila* appear to be passive during courtship, they clearly exercise discrimination. Males, on the other hand, are indiscriminately active and attempt to mate with anything resembling another *Drosophila*—including other males, females of other *Drosophila* species, dead or etherized flies, and even inanimate objects. An examination of the cues that may be involved when a female chooses one male over another points to olfactory cues as the critical factor (Ehrman and Probber, 1978). A recent summary of research involving mate selection in *Drosophila* has been edited by Ehrman and Seiger (1987).

FORCES THAT CHANGE GENOTYPIC FREQUENCIES

Migration, mutation, and selection change both allelic and genotypic frequencies. Certain systems of mating, however, change only genotypic frequencies. We shall consider inbreeding and assortative mating, which differ from random mating.

Inbreeding

Inbreeding is a nonrandom system of mating between genetically related individuals. If inbreeding occurs, offspring are more likely than average to have the same alleles at any locus. In this sense, inbreeding is not specific to any particular character. We shall see in the next section that the other major system of nonrandom mating, assortative mating, is character specific.

Sewall Wright (1921) defined the *coefficient of inbreeding* as the correlation between uniting gametes. Another way of looking at the coefficient of inbreeding involves the probability that both alleles at a locus carried by an individual are identical by descent, i.e., are replicates of those carried by a common ancestor. An easier way of thinking about inbreeding is to consider the coefficient of inbreeding as the percentage decrease in heterozygosity. Inbreeding will change genotypic frequencies by reducing the frequency of heterozygotes, but in the absence of selection, it will not change allelic frequencies. Consider a self-fertilizing type of plant such as Mendel's garden peas, and a single locus with A_1A_1, A_1A_2, and A_2A_2 genotypes. If the plants fertilize themselves, the homozygotes will produce only homozygotes. However, half of the offspring of the heterozygotes will be homozygotes, as illustrated in Figure 8.4. Each generation of self-fertilization reduces heterozygosity by half. Coefficients of inbreeding for self-fertilization are indicated in Figure 8.4. Heterozygosity is reduced by one-quarter in matings between full siblings, by one-eighth in matings between half-siblings, and by one-sixteenth in matings between cousins.

Because inbreeding operates equally across all loci to decrease heterozygosity, it can significantly increase homozygosity in a population. The increase in homozygosity has important implications. Recessive alleles are more likely to be expressed. Because most harmful genetic traits are attributable to recessive alleles, offspring of matings between genetically related individuals are more likely to exhibit recessive genetic problems. For this reason, tradition and law have prohibited matings between closely related individuals in most populations. However, low levels of inbreeding occur in a few populations either by choice (for example, in Japan) or because a small population size limits options (for example, among the American Amish).

The fact that inbreeding reduces heterozygosity can be used to create strains of animals that are essentially genetically identical. For example, each generation of brother–sister matings reduces heterozygosity in the succeeding generation by one–fourth. After 20 generations of such inbreeding, at least 98 percent of all loci are fixed—i.e., all individuals have the same alleles at 98

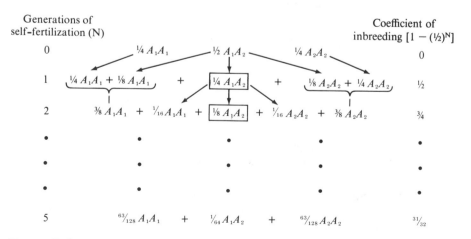

Figure 8.4
Reduction of heterozygosity in self-fertilizing plants. Dots represent a continuing series. See text for explanation.

percent of their loci. Such homozygosity means that regardless of which chromosomes are passed on from parent to offspring, the same alleles will be on each chromosome. This method has been used to create inbred strains of mice or rats. In mice, the term "inbred strain" is reserved for strains that are products of at least 20 generations of full-sibling matings. In fact, most commercially available pedigreed inbred mice have been mated brother-to-sister for 50 to 100 or more generations. Inbred strains of mice are used in much behavioral genetic research and will be described in more detail in Chapter 10.

Many attempts to create inbred strains fail because harmful recessive alleles become incorporated into the strain by chance, and the reproductive ability of the strain drops. Even those inbred strains that survive usually have some problems. The term *inbreeding depression* has been used to describe the general malaise of inbred individuals caused by the increase in homozygosity. The other side of the coin is *hybrid vigor*, or *heterosis*, which is the increase in viability and performance when inbred strains are crossed to produce an F_1 generation. Crossing inbred strains reintroduces heterozygosity at all loci for which the two strains differ, thus masking the effects of deleterious recessive alleles (Wright, 1977).

In addition to changing genotypic frequencies, inbreeding can alter the average phenotype of a population. This occurs when there is dominance. Complete dominance means that the heterozygote (A_1A_2) will have the same phenotypic value as the dominant homozygote (A_1A_1). If inbreeding occurs, the alleles of these heterozygotes will gradually be distributed into homozygotes, as shown in Figure 8.4. Thus, the frequency of a phenotype influenced by a dominant allele will diminish over the generations because some of the heterozygotes will produce recessive homozygotes (A_2A_2). As a result, the average phenotypic value in the population will be lowered. However, if there were no dominance, the heterozygote would have a value intermediate to the two homozygotes. Because inbreeding causes the alleles of heterozygous individuals to be distributed evenly among the two homozygous types, the resulting average value in a population will not change when the alleles operate in an additive (nondominant) mode.

We can use this fact to determine whether a particular character is influenced by a dominant allele. We have already noted that inbreeding depression frequently occurs with inbred strains of mice. Inbreeding depression is a change in the average value of some trait in a population. Although we are focusing on single-gene characteristics in this chapter, there is some evidence that even a complexly determined trait (such as IQ) is somewhat affected by inbreeding depression, due to the expression of dominant alleles. In other words, inbreeding tends to result in lower IQ scores (Vandenberg, 1971). For example, the risk of mental retardation is more than 3.5 times as high among children of marriages between first cousins as among unrelated controls (Böök, 1957). In addition, children of such cousin marriages generally perform worse on subtests of the Wechsler intelligence test than do children of unrelated spouses (Cohen et al., 1963; Schull and Neel, 1965). Some of these results are summarized in Figure 8.5.

Another study (Bashi, 1977) included a representative sample of Arabs living in Israel, a group in which the frequency of marriages among relatives is about 34 percent. Because such consanguineous marriages are encouraged, even marriages between "double-first cousins" are fairly common (about 4 percent). Double-first cousins are children of siblings who are married to another pair of siblings. Raven's Progressive Matrices, a test of general reasoning, was administered to large samples of children of both first cousins and double-first cousins, as well as to fourth- and sixth-grade children of unrelated marriages. The results in Table 8.5 indicate a slight depression for children of cousin marriages, and a greater depression for children of double-first-cousin marriages. This demonstration of inbreeding depression again suggests that dominant alleles at some loci affect IQ scores. A recent study of Indian Muslim teenagers whose parents are first cousins yielded similar results (Agrawal, Sinha, and Jensen, 1984).

Inbreeding can change the variability as well as the mean of a population. However, even though the effects of inbreeding are severe in individual cases, inbreeding probably does not have an appreciable effect on the means or variances of human traits in the population as a whole. In present-day human populations, the inbreeding coefficient is almost always less than 0.04, even for very small breeding isolates. Thus, changes in population means and variances resulting from inbreeding should be negligible.

Assortative Mating

Old adages are sometimes contradictory. Do "birds of a feather flock together," or do "opposites attract"? Studies of assortative mating, or phenotypic similarity between mates, seek answers to this question. It turns out that

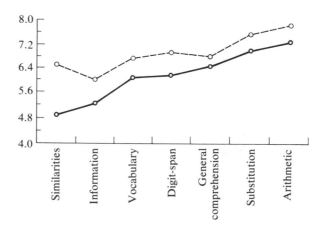

Figure 8.5

Scores on seven subtests of the WAIS achieved by 38 children of first cousins, plotted on the solid line. Those of 47 matched controls are plotted on the broken line. (After "School attainment in an immigrant village" by R. Cohen et al. In E. Goldschmidt, ed., *The Genetics of Migrant and Isolate Populations.* Copyright © 1963. Used with permission of Foundation for Child Development.)

Table 8.5
Effect of consanguinity on Raven's progressive matrices

	Grade 4		Grade 6	
Degree of consanguinity	Number	Mean	Number	Mean
Children of unrelated marriages	1,054	8.8	1,054	13.1
Children of first-cousin marriages	503	8.6	467	12.3
Children of double-first-cousin marriages	71	7.9	54	10.6

SOURCE: After Bashi, 1977.

assortative mating is almost always in a positive direction. "Birds of a feather" do "flock together," in the sense that individuals who mate tend to be similar in certain characteristics. In contrast to inbreeding, assortative mating is much more common and it is character-specific. Thus, individuals sort themselves into mating couples on the basis of certain phenotypic characteristics.

Like inbreeding, assortative mating affects only genotypic frequencies, not frequencies of alleles. If we think about the influence of a single locus on a trait for which positive assortative mating occurs, assortative mating, like inbreeding, reduces heterozygosity. Homozygotes tend to mate with homozygotes, and some heterozygous individuals in each generation have homozygous offspring. However, for characters influenced by genes at many loci, assortative mating will not greatly reduce heterozygosity. On the other hand, assortative mating for such characters may substantially increase genotypic variability. For example, differences in height are mostly the result of genetic differences. If random mating were to prevail for height, tall women would be just as likely to mate with short men as tall men. Offspring of the matings of tall women and short men would be of moderate stature. If, however, there is positive assortative mating for height (as we know there is), children with tall mothers are also likely to have tall fathers, and the offspring themselves are likely to be taller than average. Positive assortative mating thus increases the variance, in the sense that the offspring differ more from the average than they would if mating were random.

Assortative mating for some traits is the rule rather than the exception for most species (Ehrman and Parsons, 1981). In human populations, assortative mating is common. The highest correlation between husband and wife—about 0.75—is for age. Approximately one-third of the 290 correlations for physical characters summarized by James Spuhler (1968) were in the range 0.10 to 0.20. For example, the correlation for height is about 0.25 and for weight it is about 0.20. Few correlations are negative. Thus, it appears that in general there is some positive assortative mating for physical characters; however, the correlations are relatively low.

Among behavioral characters, most personality-rating correlations between mates are found to be in the 0.10 to 0.20 range, comparable to values observed for the physical characters (Vandenberg, 1972). Correlations for cognitive measures, most notably IQ, have been thought to be much higher. However, IQ and age, for which there is much assortative mating, are related to some extent. It has been shown that correlations for specific cognitive abilities, as well as for overall cognitive ability, are also in the 0.10 to 0.20 range, when scores are adjusted for age. Of the specific cognitive abilities (such as memory, spatial ability, verbal ability, and perceptual speed), verbal ability seems to show the most assortative mating (Johnson et al., 1976). Although one might expect that psychopathology would involve the attraction of opposites in the sense that a vulnerable person might seek a solid rock to steady their psyche, some evidence suggests that positive assortative mating occurs even for psychopathology (Rosenthal, 1975a).

Assortative mating is one of the oldest theoretical problems in behavioral genetics, but its mechanisms are still not well understood (Jensen, 1978). The basic issue is that assortative mating is phenotypic, and it is difficult to determine the extent to which the source of assortative mating is phenotypic, genetic, or environmental. Genetic or environmental resemblance is imposed if matings take place within groups that are differentiated from each other genetically or environmentally (Crow, 1986). If assortative mating is entirely induced environmentally, no genetic consequences of assortative mating will occur. However, if, as is likely, assortative mating leads to some genetic resemblance between spouses, it will have several genetic consequences. For example, even though spouse correlations are not large, assortative mating can greatly increase genotypic variability in a population because its effects accumulate generation after generation. It has been estimated that assortative mating for IQ may be responsible for over half the frequency of individuals with IQs above 130 and for four out of five individuals with IQs over 145, as compared to the IQ distribution if mating were random (Jensen, 1973). Assortative mating also affects estimates of genetic and environmental influence, as discussed in the following chapters. A relatively new behavioral genetic method that focuses on adult twins and the families that they rear is particularly powerful in studying the processes that underlie assortative mating, because the spouses of identical twins have, genetically speaking, married the same person. This so-called families-of-twins method is described in Chapter 12. Assortative mating is also becoming an important topic in its own right (Buss, 1984a, 1984b, 1985).

GENETIC VARIABILITY

Population genetics provides another perspective for understanding genetic variability. The point of the Hardy–Weinberg–Castle equilibrium is that genetic variability will be maintained generation after generation in the absence of forces that change the frequency of alleles. The forces that change allelic frequencies can also enhance genetic variability. New alleles can be introduced into a population through migration and mutation. Selection can also maintain genetic variation in a population. Selection can be considered a dynamic flux that maintains genetic variability. Balanced polymorphisms provide a reservoir of genetic variation for future selection, in the face of changing environmental circumstances. Thus, they may have considerable evolutionary significance.

Equilibrium genotypic frequencies can also be changed by certain systems of mating. Even here it seems that the deck is stacked in the direction of genetic variability. While inbreeding can reduce the frequency of heterozygotes, its rarity in the human species makes its effects quite negligible. Assortative mating, on the other hand, increases genotypic variability for many polygenic characters.

SOCIOBIOLOGY

A major impetus in recent years for the study of genetic variability in populations as it relates to behavioral evolution is sociobiology. *Sociobiology: The New Synthesis* is the title of a book by E. O. Wilson (1975) that capped a new wave of research on an old problem. The old problem that sociobiologists have addressed is "altruism," self-sacrificing behavior. How can such behavior be explained in Darwinian terms? Darwin himself worried about this problem:

> I will not here enter on these several cases, but will confine myself to one special difficulty, which at first appeared to me insuperable, and actually fatal to the whole theory. I allude to the neuters or sterile females in insect-communities; for these neuters often differ widely in instinct and in structure from both the males and fertile females, and yet, from being sterile, they cannot propagate their kind. (Darwin, 1859, pp. 203–204)

Darwin's answer to the problem involved group selection:

> . . . if such insects had been social, and it had been profitable to the community that a number should have been annually born capable of work, but incapable of procreation, I can see no especial difficulty in this having been affected through natural selection. (Darwin, 1859, p. 204)

In other words, if individual members of a group sacrifice themselves for the good of the group, then that group is more likely to survive. The group selection hypothesis was brought into prominence by a book in which V. C. Wynne-Edwards (1962) used group selection to explain why many animals seem to reduce their reproduction altruistically.

However, group selection is difficult to reconcile with individual selection. In an altruistic group, one might expect selfish individuals to have a selective edge over the altruists. If the altruists are selected against because some do not reproduce, and if the selfish individuals are favored in selection, then the offspring of the selfish individuals might be expected to take over.

In *Sociobiology*, Wilson suggests a different answer to the old problem of altruism: "The answer is kinship: if the genes causing the altruism are shared by two organisms because of common descent, and if the altruistic act by one organism increases the joint contribution of these genes to the next generation, the propensity to altruism will spread through the gene pool" (1975, pp. 3–4). Sociobiologists argue that the notion of individual fitness should be extended to inclusive fitness. *Inclusive fitness* is the fitness of an individual plus that part of the fitness of kin that is genetically shared by the individual (Hamilton, 1964). J.B.S. Haldane reportedly anticipated this view when he jokingly announced that he would lay down his life for two full siblings or eight first cousins. Either two siblings (each sharing approximately one-half of his genes) or eight cousins (each sharing one-eighth) would be his genetic equivalent. Fisher (1930) long ago suggested an example of kin selection that involves distastefulness of some butterfly larvae. A bird will learn that certain larvae taste bad but the lesson costs the larva its life. However, sibling eggs are laid in a cluster and if more than two siblings are saved, inclusive fitness is served by the sacrifice of one larva.

Inclusive fitness and kinship selection suggest that the unit of selection is not the group or the individual, but the gene. This is the premise of Dawkins's (1976) book, *The Selfish Gene*. The title of the book implies that genetic selfishness can explain seemingly altruistic acts of an individual if the net result of the altruistic act helps more of that individual's genes survive and helps transmit them to future generations. This is the point of considering inclusive fitness rather than individual fitness. Wilson (1975) summarizes the sociobiological view of altruism: "A genetically based act of altruism, selfishness, or spite will evolve if the average inclusive fitness of individuals within networks displaying it is greater than the inclusive fitness of individuals in otherwise comparable networks that do not display it" (p. 118).

Between-Species Examples of Sociobiology

Kinship selection theory makes the prediction that altruism should occur more frequently as genetic similarity increases. Nearly all of the tests of this hypothesis have come from comparisons between species. That is, why do some species show a certain kind of altruism while others do not? For example, in one of the papers forming the foundation for sociobiology, W. D. Hamilton (1964) used kinship selection theory to explain why some species, such as honey bees, ants, and wasps, have a caste of female workers who do not reproduce. In evolutionary terms, this is the ultimate act of altruism. Nonreproductive female workers share three-quarters of their genes with their sisters (future queens as well as other female workers) because of a special genetic arrangement in which the females, like most organisms, have two sets of genes (diploid), but the males have only one set (haploid). If these female workers reproduced, they would share only half of their genes with their offspring. Thus, their inclusive fitness is better served by working for the rest of the community than by rearing their own offspring (Smith, 1978).

There are other interspecies illustrations of these principles. For example, why do zebras, but not wildebeests, defend their calves? Zebras live in family groups, while wildebeest herds are mixed family groupings. The answer may be that adult zebras are genetically related to the calves they defend, whereas adult wildebeests may be risking their lives for an unrelated calf (West-Eberhard, 1975). Occasionally, sociobiological research has focused on populations within a species—for example, why paternal behavior of marmots in isolated family units differs from paternal behavior in populous colonies (Barash, 1975). Nonetheless, the perspective is typological—assuming invariant behavior for each group or species studied. (See Chapter 1.)

With regard to the human species, it has been argued that "sociobiology deals with biological universals that may underlie human social behavior" (Barash, 1977, p. 278). This is also a typological approach. One example of sociobiological speculation concerning human behavior is the prediction of conflicts between parents and their offspring. If you don't think too deeply about it, you might expect sociobiology to predict harmony between parents and offspring because they are, after all, kin. However, sociobiology reinterprets the relationship between parents and their offspring in terms of the genetic selfishness of inclusive fitness, which leads to conflict (Trivers, 1974). For parents, maximal inclusive fitness comes from having many offspring, which reduces their relative genetic investment in each offspring. For the offspring, however, it is a different story. The inclusive fitness of each offspring may be better served by attempting to maximize its parents' investment in itself rather than in its siblings, thus creating conflict between parents and offspring, as well as among siblings.

Sociobiology also provides an explanation for greater maternal than paternal care of offspring in the vast majority of mammalian species, including humans. Unless a species is completely monogamous (as are eagles, for example), males have less invested in the young. Males may have offspring by several females, but each female must devote energy to the pregnancy and, in mammals, continue to provide sustenance after birth. Although parental care generally increases with parental investment, the ultimate issue is inclusive fitness. Because of the investment females have made in their young, their fitness is better served by increased care of their present offspring. In many cases, however, the males' investment is little more than copulation, and they may maximize their inclusive fitness by having more offspring by different females. A related reason for greater maternal than paternal care may be that females can always be sure that they share half of their genes with their young. Males, however, cannot be so sure. According to sociobiology, the greater altruistic attention of mothers to their offspring is no less selfish from a genetic point of view than that of fathers.

Sociobiology has been criticized by the behavioral science community, in part because of the eagerness of writers to apply animal evolutionary research to important human issues such as rape, homosexuality, and ethics (e.g., Wilson, 1978a,b), and because of the use of colorful anthropomorphic similies such as adultery, suicide, and rape in relation, for example, to the behavior of birds (e.g., Barash, 1977). Another problem is that sociobiology has too often consisted of telling adaptive stories, Darwinian versions of Kipling's *Just So Stories*. An example of the problem with post hoc explanations concerns a sociobiological explanation of homosexuality. Homosexuality, as all social behavior, is assumed by sociobiologists to be governed by natural selection. In order to explain why homosexuality has not disappeared —assuming that homosexuals reproduce less than heterosexuals— sociobiologists propose that homosexuals help their siblings rear their children, thus compensating for their own lack of reproduction in terms of inclusive fitness (e.g., Wilson, 1978a,b). This is an interesting hypothesis, but it needs to be tested before it can be counted as an explanation (Lewontin, Rose, and Kamin, 1984): Do homosexuals in fact leave fewer offspring? (One could counter, for example, that many homosexuals are also bisexual and may be more sexually active in general.) Is there genetic influence on homosexuality? (Little is known as yet about this.) Do homosexuals in fact increase the reproductive rates of their sisters and brothers? (Nothing is known about this.) The need to test sociobiological hypotheses is now generally recognized and has become the focus of current sociobiological research. On the positive side, it must be recognized that sociobiology offers novel and interesting hypotheses that stem from the simple principle of kin selection and inclusive fitness.

Despite continued criticism, interest in sociobiology is growing among behavioral scientists, primarily because of its novel evolutionary explanation of differences between species in altruistic behavior in terms of inclusive fitness (e.g., Kitcher, 1985). Most social problems, however, involve differences among individuals within a species rather than average differences between species. For example, rather than asking why our species experiences conflict between parents and offspring, it is important to ask about differences within our species: For example, why do some siblings get along so well and others so poorly? Why does parent–offspring conflict occur more frequently and intensely in some families than in others? Why do some parents neglect or even abuse their children?

Sociobiology and Genetic Differences Within Species

Kinship selection requires that genetic variability exists within a species. If there were no genetic variability within a species, then all members of that species would be identical genetically and kinship would not matter. However, sociobiologists often assume that genetic variability is important, even

though this is the main question over which behavioral geneticists toil empirically. Moreover, kinship selection is limited in terms of explaining individual
differences such as those mentioned above because members of all families
share heredity to an equal degree. That is, because all parents and their
biological offspring are genetically related 50 percent, genetic relatedness
cannot in general explain why parent–child conflict is greater is some families
than in others. Nonetheless, as explained later, the theory of inclusive fitness
leads to some interesting predictions about behavioral differences within
species.

One study (Sherman, 1977) considered alarm calls in response to predators within a species of ground squirrels. (See Figure 8.6.) During more than
3,000 hours of observation, researchers recorded 102 predator encounters
that led to the deaths of nine squirrels. Because squirrels who call out in
warning are more likely to be stalked or chased by predators, alarm calls
qualify as altruism. Kinship selection predicts that alarm calls will be more
prevalent among individuals with more relatives. This prediction is supported
by the observation that females with kin gave more alarm calls than females
without kin. Females who were pregnant, lactating, or living with postwean-
ing young gave alarm calls 14 of the 19 times that they were present when a
predator appeared. Females with no known kin gave alarm calls when a

Figure 8.6
Ground squirrel in position for alarm call. (Photograph by George D. Lepp;
courtesy of Paul W. Sherman.)

predator appeared on only 2 of 14 occasions. The data for the males did not support kinship selection theory. For example, frequency of copulation by males was not related to the frequency of their alarm calls. However, it is difficult to interpret the data for the males because their total number of alarm calls was low and because ground squirrel society is decidedly matrilineal.

In this study, variation was observed in alarm calls within a species. The kinship selection theory prediction that such altruism will be more common among individuals who have more relatives was confirmed for females. However, this does not prove that altruism has a genetic base, because relatives share environments as well as genes. For example, it is possible that females who rear young learn to protect them as they were themselves protected. If kinship selection theory is to be accurately applied, it is necessary to determine the extent to which observed variability is genetic in origin.

Inclusive fitness has also begun to be tested within the human species. As indicated earlier, because first-degree relatives are genetically related to the same extent, kinship itself would not seem to be able to address issues of differences within a species. However, some predictions can be made: For example, less altruism should be seen for step-parents and step-siblings because these relatives are not kin; less altruism should be seen for fathers than for mothers because paternity is uncertain (compared to maternity, which is guaranteed) and because mothers invest more into their offspring; and older parents and parents with fewer children will show greater altruism because they have more invested in each child. Some support for the first hypothesis comes from animal research that has shown that full siblings interact more frequently with each other than do half siblings (that is, having one parent in common), who in turn interact more frequently than do nonrelated peers (Holmes and Sherman, 1983; Suomi, 1982). In another study that considered bereavement as an index of investment, a study of the grief of parents following the death of their child supported several sociobiological hypotheses (Littlefield and Rushton, 1986).

Sociobiology and Behavioral Genetics

Sociobiology and behavioral genetics are both concerned, in part, with the inheritance of social behavior. However, behavioral genetics differs from sociobiology in several important ways. First, behavioral genetics includes many genetic perspectives on behavior: Mendelian single-gene approaches, molecular genetics, quantitative genetics, population genetics, and evolutionary genetics. Sociobiology is concerned primarily with the last two, although an evolutionary approach to behavior can incorporate the other perspectives (Broadhurst, 1979). Second, behavioral genetics is not limited to studying social behavior. For example, in studies of human beings, individual behavior such as specific cognitive abilities and psychopathology have been the focus of many behavioral genetic investigations. Third, behavioral genetics focuses on differences within species, while sociobiology emphasizes differences between species. Although both perspectives are useful, the distinction between them is important. The causes of average differences between species may be unrelated to the causes of individual differences within species. For example, similarities between humans and chimps may be due to their genetic similarity, but either genetic or environmental factors may be primarily responsible for differences within either species. Similarly, differences within either species could be largely genetic in origin, yet average differences in behavior between the species could be caused by the influence of culture on human behavior. Finally, behavioral genetics has developed methods to assess the extent to which observed behavioral variation can be ascribed to genetic or environmental influences. It is important to note that most human and nonhuman behaviors studied by behavioral geneticists are found to be influenced by both heredity and environment. In contrast, sociobiological discussions of human behavior tend to rely on average differences and similarities between the human species and other contemporaneous species, average differences and similarities among contemporaneous human cultures, and speculations concerning the adaptive value of behaviors in hunter-gatherer societies. E. O. Wilson's (1978b) book, *On Human Nature*, exemplifies a sociobiological approach to human behavior.

For example, here is how a behavioral geneticist would approach the issue of parental care in humans. Rather than taking a typological approach by asking how paternal care differs on the average from maternal care, a behavioral geneticist would study the variability in parental care. Parents vary from smothering protectiveness of their children to neglect, rejection, and even abuse. After describing such variations in human parental care, the next question concerns the extent to which genetic and environmental influences are involved in causing such variation. Environmental causes seem likely at the outset. For example, a common hypothesis is that child abusers were themselves abused as children and thus learned to react to the frustrations of rearing a child in this destructive manner. However, it is also possible that genetic factors are involved. For example, child abuse may run in families because of an inherited propensity to be emotionally labile, or short-tempered. However, such armchair speculation will not answer the question. In families in which parents rear their own children, both genes and environment are shared, so we cannot separate their effects. Behavioral geneticists, therefore, use special designs, such as the adoption study, in which either genes are shared or environments are shared, but not both. In this way, a behavioral geneticist can study differences in parental care and determine the extent to which that variation is genetic or environmental in origin. These methods derive from quantitative genetic theory, which is the subject of the following chapter.

In summary, sociobiology and behavioral genetics have developed quite independently and differ in important ways in the questions asked, their theories, and their methods. Nonetheless, greater rapprochement between the two disciplines would help both fields (Fuller, 1983a). Sociobiology will be assisted by the rigorous analysis of genetic and environmental variance within populations provided by behavioral genetic studies. Behavioral genetics will benefit by being more evolutionary in its approach, and sociobiology provides an important approach to behavioral evolution. More behavioral genetic research is needed on evolutionarily important social behaviors such as sexual, agonistic, and parenting behavior. The need for more evolutionary research in behavioral genetics extends beyond sociobiological analyses of social behavior (Fuller, 1983b): Behavioral genetic studies of feeding behavior (Gray, 1981) and foraging behavior (Sokolowski, Hansell, and Rotin, 1983) are examples of this research direction. Most research on behavioral evolution has involved invertebrates (Huettel, 1986), although interest in human behavioral evolution is keen (e.g., Buss, 1984c). Another important direction for research on behavioral evolution is the application of the tools of the "new genetics," described in the previous chapter (Nei, 1987).

SUMMARY

Population genetics considers allelic and genotypic frequencies in groups of breeding organisms. It provides the basis for understanding how genetic variability is maintained in populations. This idea is formalized in the concept of the Hardy–Weinberg–Castle equilibrium, which shows that frequencies of alleles and genotypes remain stable generation after generation in the absence of forces of change. We can use this concept to estimate frequencies of alleles, since genotypic frequencies in a population in equilibrium should correspond to $p^2 + 2\,pq + q^2$ for a single-locus, two-allele character. Population genetics is also concerned with the forces that change frequencies, such as migration, mutation, and selection. Balanced polymorphisms can be caused by heterozygote advantage and frequency-dependent selection. Inbreeding and assortative mating change genotypic frequencies without affecting frequencies of alleles. Sociobiology employs kinship selection and inclusive fitness as evolutionary explanations for genetic differences between populations.

CHAPTER·9

Quantitative Genetic Theory

U ntil now, we have focused on single-gene influences. Characters influenced by only one gene are often called Mendelian because they show the classical segregation ratios described by Mendel. Although there are many examples of the effects of single genes on behavior (see Chapter 3), most of these interrupt the organism's normal course of development. For example, many of the single-gene influences on human behavior cause mental retardation. These examples demonstrate the power of a single gene to throw the organism out of kilter. However, the normal range of behavior variation is more likely to be orchestrated by a system of many genes.

Because behavioral genetics considers polygenic as well as single-gene influences on behavior, we need to study the theory underlying quantitative inheritance. Since quantitative genetic theory is somewhat abstract, some people erroneously believe that a character is really influenced by genes only if the character shows classical Mendelian inheritance. In this chapter we will show that one can generalize from single-gene, Mendelian theory to the quantitative effects of multiple genes.

Quantitative genetics is more abstract in that it considers variance in a population, rather than specific genotypes. If we just have two or three types of individuals in a population, as is the case for most Mendelian characters, we can simply count the different types. Figure 9.1a illustrates the distribution in a population of a character determined by a single gene with two alleles.

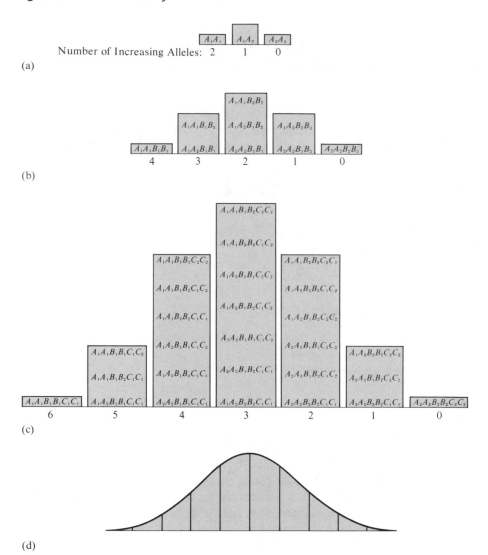

Figure 9.1
Single-gene and polygenic distributions for characters with additive gene effects. (a) Distribution of genotypes for a single locus with two alleles. (b) Distribution of genotypes for two loci, each with two alleles. (c) Distribution of genotypes for three loci, each with two alleles. (d) Continuous variation.

There are three distinct types of individuals. However, there are many characters that show continuous variability similar to the normal bell-shaped curve illustrated in Figure 9.1d. A normal distribution would be approximated if you tossed a handful of 20 coins hundreds of times and each time recorded the number of heads and tails. The average number of heads per toss would be 10, and the other numbers would be evenly distributed around 10, with 0 and

20 being extremely rare. The size of the pea seed is continuously distributed, as Galton discovered when he conducted his early studies of quantitative inheritance. In fact, continuous variation is the rule rather than the exception for behavioral characters. The genetic foundation for such variability has been a recurrent theme in the previous chapters.

Figure 9.1b and c suggest how the qualitative distribution of a single-gene character becomes a quantitative distribution as more loci become involved. For example, when a trait is influenced by two alleles at each of three loci (*A, B, C*), there are 27 different genotypes. Even if we assume that the alleles at the different loci equally affect the trait and that there is no environmental variation, there are still seven different phenotypes, as indicated in Figure 9.1c.

The point is that, even with just three loci and two alleles at each locus, the genotypes begin to approach a normal distribution in the population. When we consider environmental sources of variability and the fact that the effects of alleles at different loci may not be equal (as we assumed in our oversimplified example), it is easy to see that the effects of even a few genes will lead to an approximately normal distribution. Moreover, the complex characters that interest behavioral scientists may be influenced by hundreds of genes. Thus, we should not be surprised to find continuous variation at the phenotypic level.

Because variance is the core concept of quantitative genetic theory, a brief digression is in order.

BRIEF OVERVIEW OF STATISTICS

Statistics Describing Distributions

Figure 9.2 describes the results of testing a small sample of two inbred strains of mice for activity in an open field. The open-field apparatus (see Figure 3.5) is a brightly lit enclosure in which an animal's activity is measured. The activity scores in Table 9.1 were obtained by placing mice one at a time in the open field for 5 minutes. The number of squares entered during this observation period was recorded as each subject's score. How can we describe these two distributions? We first calculate an average score (or some measure of *central tendency*) and then describe the variability of the scores. The average score is not very useful in itself, because it adequately represents the scores only if there is little variability. The distributions in Figure 9.2 show substantial variability, as do distributions for most behavioral characteristics. The *average*, or *mean*, is simply the sum of scores divided by the number of scores:

$$\overline{X} = \frac{\Sigma X}{N}$$

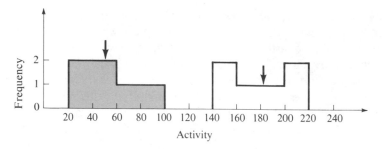

Figure 9.2
Frequency histograms of the activity scores of two inbred strains of mice: A (shaded) and C57BL. The means are indicated by arrows.

where $\overline{X}$ refers to mean, ΣX is the sum of scores, and N is the number of scores. The sum of the scores obtained from the six A subjects is 306. Thus,

$$\overline{X}_A = \frac{306}{6} = 51$$

The mean score of C57BL subjects in the sample is:

$$\overline{X}_C = \frac{1092}{6} = 182$$

These means are indicated by the arrows in Figure 9.2.

As the name implies, *variance* is a measure of variability or dispersion. The more spread out the distribution, the greater the variance. Variance is described relative to the mean of the sample. The difference between each subject's score and the mean is computed (i.e., $X - \overline{X}$). Some of these

Table 9.1
Activity scores
of two inbred
strains of mice

A	C57BL
29	155
29	157
44	161
58	199
63	202
83	218

deviations are above the mean and are thus positive numbers; those below the mean are negative numbers. We would like to obtain an average deviation from the mean in order to describe the variability in the distribution. However, if we simply summed the deviations from the mean, the positive deviations would balance the negative deviations and the sum would always be zero. The solution is to square the deviations from the mean and then calculate an average squared deviation. This is the definition of variance. However, the sum of the squared deviations from the mean is divided by N − 1 for technical reasons (in order to obtain an unbiased estimate of the variance). In short, the variance of a sample (V) is:

$$V = \frac{\Sigma(X - \overline{X})^2}{N - 1}$$

To illustrate the calculation of V, the data of Table 9.1 are presented again in Table 9.2, along with corresponding deviations from means and squared deviations. As you can see, the variance of activity scores in the C57BL sample is somewhat larger than that in the A sample.

Since variance is the average of the *squared* deviations from the mean, the values obtained are expressed in squared units, rather than in the actual units of measure. In spite of this, as you will see later in this chapter, variance has many important applications in genetics. Nevertheless, a measure of variability expressed in actual units, rather than squared units, is useful. Such a measure is provided by the square root of the variance, the so-called *standard deviation*. If our sample has been drawn at random from a population with a normal distribution of a trait (see Figure 9.3), the sample standard deviation (s) provides a useful estimate of dispersion of the trait within that population.

Approximately two-thirds of the population (68 percent) fall within one standard deviation above and below the mean, and about 96 percent of the observations fall within two standard deviations. Thus, in a large population of mice of the A strain with a mean of 51 and a standard deviation of 21.14, approximately two-thirds of their activity scores would fall within the range of 51 ± 21.14, i.e., between 29.86 and 72.14. The precision of such estimates increases along with the sample size.

Statistics Describing the Relationship Between Two Variables

When two variables are measured for each subject, or when the same variable is measured on pairs of subjects (for example, pairs of twins or parents and their offspring), we can analyze the relationship between the two measures. The question is usually phrased in terms of *covariance*, which literally means "shared variance." It tells us the extent to which the measures

Table 9.2
Examples of variance estimation from activity scores
of two inbred strains of mice

A		
X_i	$X_i - \overline{X}$	$(X_i - \overline{X})^2$
29	−22	484
29	−22	484
44	−7	49
58	+ 7	49
63	+12	144
83	+32	1024
$\Sigma X_i = 306$	$\Sigma(X_i - \overline{X}) = 0$	$\Sigma(X_i - \overline{X})^2 = 2234$
$\overline{X}_A = 51$		$V_A = \dfrac{2234}{5} = 446.8$
		$s_A = \sqrt{V_A} = \sqrt{446.8} = 21.14$

C57BL		
X_i	$X_i - \overline{X}$	$(X_i - \overline{X})^2$
155	−27	729
157	−25	625
161	−21	441
199	+17	289
202	+20	400
218	+36	1296
$\Sigma X_i = 1092$	$\Sigma(X_i - \overline{X}) = 0$	$\Sigma(X_i - \overline{X})^2 = 3780$
$\overline{X}_C = 182$		$V_C = \dfrac{3780}{5} = 756.0$
		$S_C = \sqrt{V_C} = \sqrt{756.0} = 27.50$

relate to one another. If there is substantial covariance between two variables (X and Y), then a subject above the mean on X will also likely be above the mean on Y. Like variance, the covariance statistic is based on deviations from the mean of each variable. It is computed by multiplying each subject's deviation from the mean of X by the subject's deviation from the mean of Y. Cross products of these deviations are summed across subjects and divided by the size of the sample (actually, $N - 1$). In short, the sample covariance between two variables (Cov_{XY}) is

$$Cov_{XY} = \frac{\Sigma[(X - \overline{X})(Y - \overline{Y})]}{N - 1}$$

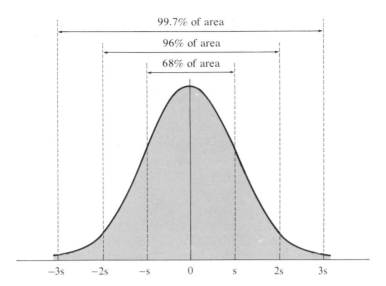

Figure 9.3
The normal distribution curve.

Consider the hypothetical data presented in Table 9.3 and plotted in Figure 9.4. Note that Y tends to increase as X increases. The variance of X is 2.5, and the variance of Y is 10. The covariance between X and Y is 3.

Covariances are easier to interpret if we divide them by an appropriate variance or a product of standard deviations. The two major statistics are *correlation* and *regression*. Sometimes one of these methods is more appropriate for certain quantitative genetic analyses, and sometimes the other is more suitable. A correlation coefficient standardizes the covariance by dividing it

Table 9.3
Sample calculation of a correlation coefficient, r_{XY}

X	$X - \overline{X}$	$(X - \overline{X})^2$	Y	$Y - \overline{Y}$	$(Y - \overline{Y})^2$	$(X - \overline{X})(Y - \overline{Y})$
1	−2	4	2	−4	16	+8
2	−1	1	8	+2	4	−2
3	0	0	6	0	0	0
4	+1	1	4	−2	4	−2
5	+2	4	10	+4	16	+8
Σ: 15	0	10	30	0	40	12

$$V_X = \frac{10}{4} = 2.5 \qquad V_Y = \frac{40}{4} = 10 \qquad Cov_{XY} = \frac{12}{4} = 3$$

$$r_{XY} = \frac{3}{\sqrt{(2.5)(10)}} = 0.6$$

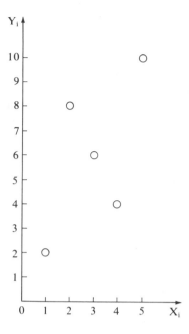

Figure 9.4
Plot of hypothetical data presented in
Table 9.3.

by the product of the standard deviations of X and Y. This is known as standardization because it results in equal units of X and Y. The regression coefficient divides the covariance by the variance of just one of the variables. For example, if we are predicting offspring scores (Y) from parental scores (X), the regression Y on X(b_{YX}) is the covariance divided by the variance of X. Thus, the regression coefficient is not standardized. It is expressed in terms of observed units of measure. It expresses the number of units, on the average, that Y changes corresponding to each unit change of X.

In summary, the formulas for a correlation coefficient (r_{XY}) and a regression coefficient (b_{YX}) are

$$r_{XY} = \frac{\text{Cov}_{XY}}{\sqrt{(V_X)(V_Y)}} \text{ and } b_{YX} = \frac{\text{Cov}_{XY}}{V_X}$$

Note that if the standard deviations of X and Y are equal, then the correlation coefficient and the regression coefficient are the same. If $\sqrt{V_X} = \sqrt{V_Y}$, then $\sqrt{(V_X)(V_Y)} = V_X$, so that both the denominator and the numerator are identical for the correlation and the regression.

Table 9.3 illustrates the computation of a correlation. The covariance (3) divided by the product of the standard deviations is 0.6. A correlation of zero (or near zero) indicates that the two variables are independent: scores on one variable tell us nothing about scores on the other. A high positive or negative correlation (close to +1 or −1) indicates a close relationship. Because correlations are standardized, they are easily related to variances. Squaring the correlations yields the percent of variance in one variable related to the variance of the other. The correlation of 0.6 in Table 9.3 indicates that 36 percent of the variance in Y is related to the variance of X (and vice versa). The variance of Y is 10.0. We can express this in terms of variance rather than percent by stating that $0.36 \times 10 = 3.6$ is the variance of Y related to the variance of X. This means that the rest of the variance of Y, 6.4, is not related to the variance of X.

We have been using the phrase "related to" rather than "caused by" because correlations do not in themselves prove the existence of a causal relationship. In genetics, however, there are clear causal associations between genotype and phenotype. When a causal relationship between two variables (X and Y) has been established, the correlation coefficient can be used to estimate the variance in Y caused by the variation in X. In the previous example, this would mean that, if X is held constant, 64 percent of the variance in Y will remain.

The regression of Y on X for the same data (Table 9.3) is 1.2:

$$b_{YX} = \frac{Cov_{XY}}{V_X} = \frac{3.0}{2.5} = 1.2$$

Thus, on the average, for every unit of change in X, Y changes 1.2 units. This regression coefficient can be used to show how the variance of Y may be partitioned into two parts—one due to variation in X, and one that is independent of X. In overview, we will use an equation to predict scores on Y, given scores on X. Then we will obtain the variance of the Y scores as they were predicted by X scores. The deviation of the actual Y scores from the predicted Y scores can be squared and averaged to produce the variance of Y, independent of X.

The regression coefficient describes the change in Y predicted by a unit change in X. Such a prediction may seem unnecessary, given that we already have information regarding both variables. However, from the sample regression, we may estimate Y for other members of the population for whom we have information only regarding variable X. More importantly, for our present purpose, the regression can be used to draw a straight line through the observed points, as in Figure 9.5. This line is called a "least squares" regression line because the sum of the squared deviations from the predicted points is at a minimum. This prediction equation is

$$\hat{Y} = \overline{Y} + b_{YX}(X - \overline{X})$$

where $\hat{Y}$ is the predicted value of Y, given information on X. Thus, the predicted value of Y is derived from the deviation of the X score from its mean, weighted by the regression coefficient. From the data of Table 9.3,

$$\hat{Y} = 6 + 1.2(X - 3) = 6 + 1.2X - 3.6 = 2.4 + 1.2X$$

Using this equation, we can calculate the expected value of Y corresponding to each observed value of X in Table 9.3. These observed and expected values

Table 9.4
Observed values of
X and Y and
expected values of Y

X	Y	$\hat{Y}$
1	2	3.6
2	8	4.8
3	6	6.0
4	4	7.2
5	10	8.4

are presented in Table 9.4 and graphed in Figure 9.5. For example, the X score of 2 predicts a Y score of 4.8, because $2.4 + 1.2\,(2) = 4.8$. This predicted value has been entered as a point on the straight line in Figure 9.5.

The variance of these predicted scores of Y is the variance of Y due to variation in X. As calculated in column 4 of Table 9.5, the variance of the predicted Y scores is 3.6, the same answer obtained using the correlation coefficient ($r^2 V_Y$). Of course, the variance of Y *not* predicted by X is the rest of the variance of Y (that is, $10 - 3.6 = 6.4$). However, we can directly calculate the variance of Y *not* predicted by X by obtaining the deviation of each Y value from its predicted value and then deriving the variance of these deviations as in column 6 of Table 9.5.

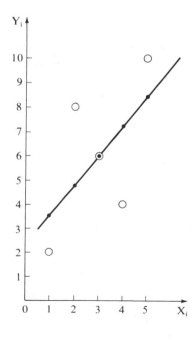

Figure 9.5
Plot of observed values (open circles) and expected values (small black dots) of Y, corresponding to observed values of X. Expected values are obtained from the regression equation, $\hat{Y} = 2.4 + 1.2X$.

Table 9.5
Calculation of the variance in Y due to both regression and deviations from regression

	Y	$\hat{Y}$	$\hat{Y} - \bar{\hat{Y}}$	$(\hat{Y} - \bar{\hat{Y}})^2$	$Y - \hat{Y}$	$(Y - \hat{Y})^2$
	2	3.6	−2.4	5.76	−1.6	2.56
	8	4.8	−1.2	1.44	+3.2	10.24
	6	6.0	0.0	0.00	0.0	0.00
	4	7.2	+1.2	1.44	−3.2	10.24
	10	8.4	+2.4	5.76	+1.6	2.56
Σ:	30	30.0	0.0	14.40	0.0	25.60

$$V_Y = 10 \qquad V_{\hat{Y}} = \frac{14.4}{4} = 3.6 \qquad V_{Y-\hat{Y}} = \frac{25.6}{4} = 6.4$$

HISTORICAL NOTE

In 1877, the first regression line was drawn by Galton, the father of behavioral genetics, to describe quantitative inheritance. As an example, he chose the size of the seed in the pea plant. He knew that parental plants with large seeds were likely to have offspring with larger than average seeds. He plotted parent and offspring seed sizes and drew the regression line, reproduced as in Figure 9.6. Galton noticed that the slope of the line described the following relationship: As parental size increases one unit, the offspring size increases one-third unit. Thus, the regression of offspring on parent was 0.33, which is the

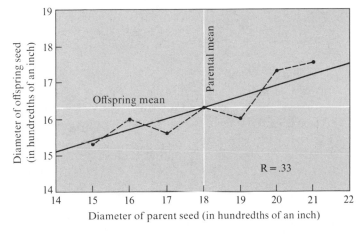

Figure 9.6
First regression line. Drawn by Galton in 1877 to describe the quantitative relationship between pea seed size in parents and offspring. (Courtesy of the Galton Laboratory.)

covariance divided by the variance of the parents. This similarity could be ascribed to inheritance because all plants were raised in similar environments.

Because the size of the pea seed varies continuously, the hereditary mechanism did not appear to operate like Mendelian elements. Such results led to a controversy between the so-called Mendelians and biometricians during the early twentieth century. After the rediscovery of Mendel's laws, some researchers began to equate heredity with Mendelian segregation ratios. It was difficult for Mendelians to reconcile continuous variation with the type of qualitative, discrete difference, mediated by single genes, with which they had worked. The biometricians, on the other hand, vigorously pursued Galton's approach to problems of the inheritance of continuously varying characteristics. However, they supported the blending hypothesis of inheritance, and generally regarded Mendelian inheritance as an unimportant exception to the general rule. With justification, they pointed to the importance of smoothly continuous, quantitative characteristics, such as height, weight, and intelligence. The biometricians thought that the characteristics investigated by Mendelians, resulting in qualitative differences and usually in abnormalities, could not account for such continuous distributions.

The groundwork for the resolution of this conflict was actually provided by Mendel himself, with his suggestion that a certain characteristic might be due to two or three genes. General acceptance of this idea, however, was not forthcoming until the work with plants by H. Nilsson-Ehle (1908) and E. M. East and collaborators (East and Hayes, 1911; Emerson and East, 1913). These researchers showed that, if it was assumed that a number of gene pairs, rather than just one pair, each exerted a small, cumulative effect on the same characters, and if the effects of environment were taken into consideration, the final outcome would appear to be a continuous distribution of the character instead of discrete categories such as those featured in the typical Mendelian researches. This was quite different from the blending hypothesis. In this multiple-factor hypothesis, it was not presumed that the hereditary determiners vary continuously in nature from one individual to another, creating a continuous distribution in the population. Rather, the genes were acknowledged to occur in discrete alternate states (typically two, sometimes more). But when a number of such discrete units bear on the same character, the final outcome approximates a continuous distribution, as discussed in the beginning of this chapter. Elaborate statistical development of this notion was provided by R. A. Fisher (1918) and by Sewall Wright (1921). Their work presented convincing demonstrations that the biometrical results, in fact, follow logically from a multiple-factor extension of Mendel's theory.

THE SINGLE-GENE MODEL

Although quantitative genetics was developed for application to characters influenced by genes at many loci, the underlying model is based on segregation at only a single locus. Once we have described gene action at a single locus, we can generalize to the polygenic case.

Genotypic Value

Genotypic values are expressed as deviations from the mid-homozygote point, as indicated in Figure 9.7. The homozygote with the higher value will be referred to as A_1A_1. The genotypic value for A_1A_1 will be $+a$. The genotypic value of other homozygote, A_2A_2, is $-a$. The values $+a$ and $-a$ are equidistant from the mid-homozygote point. However, the genotypic value of the heterozygote, A_1A_2 (symbolized by d), is dependent on the gene action at the locus. If there is no dominance, d will equal zero and will fall at the mid-homozygote point. If A_1 is partially dominant to A_2, d will be closer to a, as in the example in Figure 9.7. If dominance is complete, that is, if the observed value for A_1A_2 equals that of A_1A_1, then d $= +a$.

Additive Genetic Value

The additive effect of genes is merely the extent to which they "add up" or sum according to gene dosage. More specifically, the additive genetic value is the genotypic value expected from gene dosage, as illustrated in Figure 9.8. Gene dosage is the number of a particular allele (say, the A_1 allele) present in a genotype. As gene dosage increases by one (for example, from the A_2A_2 genotype to A_1A_2), the expected genotypic value increases by a constant unit. If the frequencies of the two alleles were not equal, we would need to weight each allele according to their respective frequencies in the population. But this will not affect our example. If there is no dominance, these expected genotypic values will be the same as the actual genotypic values. However, dominance can cause the actual genotypic values to deviate from expected values.

Another way of thinking about additive genetic values is to consider that every allele in the genotype has some average effect. In this sense, the additive genetic value is the sum of these average effects of alleles across the genotype. Additive genetic value is a fundamental component of genetic influence, because it represents the extent to which genotypes "breed true" from parents to offspring. If a parent has one "dose" of a particular allele, the offspring of that parent each have a 50 percent chance of receiving that allele. If the offspring receive that allele, its effect will be added in to the same extent it was added into the total genotype of the parent. It does not matter how many alleles are involved at a locus (or as we shall see in the next section, how many

Figure 9.7

Assigned genotypic values. (After *Introduction to Quantitative Genetics* by D. S. Falconer. Copyright © D. S. Falconer, 1960, p. 113, Longman Group, Ltd., London and New York.)

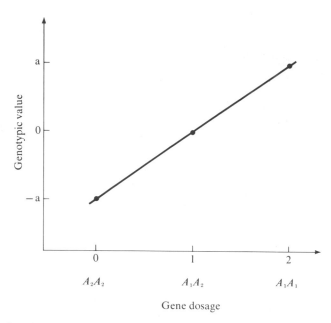

Figure 9.8
Additive genetic values predicted by gene dosage, when d = 0. Because there is no dominance, the genotypic values are the same as the additive genetic values.

loci are involved). Additive genetic values are simply the extent to which the effects of the alleles add up according to gene dosage.

Dominance Deviation

If there is dominance, genetic values do not simply add up according to gene dosage as in Figure 9.8. Dominance deviations are the difference between the expected (or additive) genotypic value and the actual genotypic value. Dominance allows for the fact that alleles at a given locus can interact with one another, rather than simply adding up in a linear fashion. For example, if there is complete dominance, genotypic values will fall on points as plotted in Figure 9.9. In Figure 9.8 the genotypic values were the same as the additive genetic values, and they fell on a straight line. In Figure 9.9, however, the genotypic values are not on a straight line. For this reason, we use a regression equation to fit the best straight line through these points. Regression of genotypic value on gene dosage yields the genotypic values predicted by gene dosage. This, of course, is our definition of additive genetic values. Thus, the crosses on the regression line in Figure 9.9 are additive genetic values. If there is dominance, this prediction of genotypic values from gene dosage will be slightly off. Dominance, as represented by the Ds in Figure 9.9, is thus the deviation of the genotypic value from the regression line, which represents the predicted genotypic values based on gene dosage.

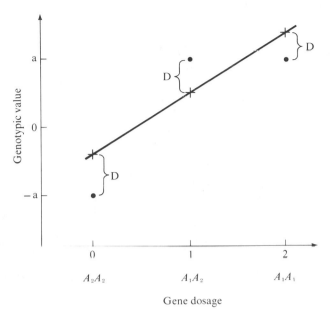

Figure 9.9
Genotypic values (black dots) when dominance is complete. Regression line predicts additive genetic values (crosses) based on gene dosage. Dominance deviations (D) are the difference between the additive genetic values and the actual genotypic values.

Dominance is important because it represents genetic influence that does not "breed true." If dominance occurs, a parent's genotypic value is due to some particular combination of alleles at a locus. Offspring cannot receive both of those alleles from the parent. Therefore, they will be genetically different from the parent to some extent if alleles do not add up in their effect. In summary, we have partitioned the genotypic value into two parts—one predicted by gene dosage, and one that is not. Additive genetic values are the extent to which genotypic values add up or sum according to gene dosage; dominance is the extent to which they do not add up.

THE POLYGENIC MODEL

Not only can we consider the additive and nonadditive effects of alleles at a single locus, we can also sum these effects across loci. This is the essence of the polygenic extension of the single-gene model. Just as additive genetic values are the summation of the average effects of two alleles at a single locus, they may also be summed across the many loci that may influence a particular phenotypic character. Similarly, dominance deviations from additive genetic values may also be summed for all the loci influencing a character. Thus, it is relatively easy to generalize the single-gene model to a polygenic one with many loci, each with its own additive and nonadditive effects. However, we need to introduce one more concept: *epistatic interaction*.

Epistatic Interaction Deviation

Dominance is the nonadditive interaction of alleles at a single locus. When we consider several loci, we need to consider the possibility that a particular allele interacts not only with the allele at the same locus on the homologous chromosome, but also with alleles at other loci. This type of interaction is called *epistasis*. In other words, dominance is *intralocus* inter-action, and epistasis is *interlocus* interaction. For example, consider two loci (*A* and *B*) that affect a phenotypic character. Both the additive genetic values and the dominance deviations are summed across the two loci. However, a particular combination of a certain allele at locus *A* and another allele at locus *B* may influence the phenotype in ways not explainable by the additive and dominance effects. *Epistasis* refers to this sort of effect.

In summary, we may partition genetic effects into three components: additive, dominance, and epistatic. At a single locus, the genotypic value includes additive and dominance effects. When we consider the effects across two or more loci, the additive and dominance effects are summed but may not yield the joint genotypic value, due to epistatic interaction among alleles at different loci. In symbolic terms,

$$G = A + D + I$$

where G is the genotypic value due to all loci, A is the sum of the additive genetic values across all loci, D is the sum of the dominance deviations across all loci, and I symbolizes the deviations due to epistatic interactions. Epistatic interactions may be of several types. They may involve interactions between additive genetic values at different loci, between dominance deviations at different loci, between additive genetic values at one locus and dominance deviations at another locus, and so on.

Variance

Up to this point, we have considered only genetic influences on a phenotype. Although that is a workable approach for traits such as those that Mendel studied in pea plants in a controlled environment, environmental influences are so important for other traits that analyses must consider both the genetic and environmental factors. The basic model of quantitative ge-netic theory simply says that the phenotype of an individual is due to a genotypic value (including A, D, and I), and an environmental effect due to all nongenetic causes. However, science seldom studies a single individual.

Our focus is on phenotypic differences in a population and on the genetic and environmental differences that create those differences. So, instead of thinking of P as an individual's phenotypic value, we will consider it as the individual's deviation from the population mean.

Thus, quantitative genetic theory begins with a model in which observed (phenotypic) deviations from the mean for some character in a population are a function of environmental (E) and genetic (G) deviations, which combine in an additive (linear) manner. However, this model may also include a nonadditive, or interaction, term (G × E) to deal with possible nonadditive combinations of genetic and environmental effects, just as dominance and epistasis allow for the possibility of nonadditive effects for single and multiple loci. Symbolically,

$$P = G + E + (G \times E)$$

The symbol G × E does not necessarily refer to multiplication of G and E. It designates the contribution of some nonadditive function of G and E to the phenotype, independent of the main effects of G and E. That is, an environmental factor may have a greater effect on some genotypes than on others, and a genotype may be expressed differently in some environments than others.

Each of the components is expressed as a deviation from the mean, but we want to express them in terms of variance. As described earlier in this chapter, variance is the sum of individuals' squared deviations from the mean, divided by the number of individuals. Let us take the G deviation and express it as variance. The variance of G simply involves squaring the genetic deviations, summing the squared deviations, and dividing by the sample size. Let us also obtain the variance of each of the components of $G = A + D + I$. The variance of G (V_G) can be expressed as the covariance of G with itself. (The covariance of a variable with itself is the same as its variance.) Therefore,

$$
\begin{aligned}
V_G &= \mathrm{Cov}(G)(G) \\
&= \mathrm{Cov}(A + D + I)(A + D + I) \\
&= V_A + V_D + V_I + 2\mathrm{Cov}(A)(D) + 2\mathrm{Cov}(A)(I) + 2\mathrm{Cov}(D)(I)
\end{aligned}
$$

Because A, D, and I are not correlated, we are left with

$$V_G = V_A + V_D + V_I$$

In other words, genetic variance is due to additive genetic variance, dominance variance, and variance resulting from epistatic interactions. Additive genetic values are equivalent to genotypic values expected from gene dosage. Thus, additive genetic variance may be thought of as genetic variance due to variation in gene dosage. In the same way, dominance and epistatic variance (or nonadditive genetic variance) is the genetic variance that is not predicted by gene dosage. It should be noted that, even if dominance is complete (that is, d = +a), genetic variance may still have a substantial component due to additive genetic variance.

In a similar manner, we can determine the variance for the general model P = G + E + (G × E). The symbol G × E is defined as being uncorrelated with either G or E; however, G and E may themselves be correlated. The variance of the phenotypic deviations (V_P) is a function of the squared deviations for the other components, as follows:

$$
\begin{aligned}
V_P &= \mathrm{Cov(P)(P)} \\
&= \mathrm{Cov[G + E + (G \times E)][G + E + (G \times E)]} \\
&= V_G + V_E + 2\mathrm{Cov(G)(E)} + V_{G\times E}
\end{aligned}
$$

In other words, observed variance in a population includes components due to genetic variance (V_G) and those due to environmental variance (V_E). Phenotypic variance also contains components added by the correlation between genetic and environmental effects [2Cov(GE)], as well as by the interaction between G and E. Although error of measurement is also likely in the variance of a phenotype, we will ignore it for now.

An Example of the Polygenic Model

An example illustrating this model may be helpful. The example is hypothetical because we cannot often measure genotypic values, and we do not know the environmental values. We measure phenotypes. Behavioral genetics employs methods to estimate genetic and environmental variance from observed phenotypic values, as discussed in the next section. For now, we will use a hypothetical example to clarify the underlying model.

Suppose that we knew the genetic, environmental, and phenotypic deviations from the mean for a number of individuals, as indicated in Table 9.6. Because these values are expressed as deviations from the mean, the mean in all cases is zero. In this example, the genetic variance is 2.0, the environmental variance is 2.0, and the phenotypic variance is 4.0. Thus, $V_P = V_G + V_E$. There is no variance added by the covariance between G and E because there is no covariance between G and E in this example. (Satisfy yourself that this is true by multiplying the deviations of G by the deviations of E and then summing the cross products.)

Now let us suppose that genes and environment are perfectly correlated, as in Table 9.7. The genetic and environmental variances remain the same (2.0), but the phenotypic variance is now 8.0 instead of 4.0. The added variance is due to the correlation between genetic and environmental deviations: $V_P = V_G + V_E + 2\mathrm{Cov(G)(E)} = 2 + 2 + 4 = 8$. It should be noted that even if we somehow removed variance due to the correlation between G and E, V_G and V_E would remain unchanged. In fact, correlation between G and E will contribute substantially to V_P only when both V_G and V_E are substantial (Jensen, 1974). Our example illustrates a positive correlation between G and

Table 9.6
Hypothetical genetic, environmental, and phenotypic deviations from the mean for five individuals

Individual	G	+	E	=	P
1	-2		$+1$		-1
2	-1		-2		-3
3	0		0		0
4	$+1$		$+2$		$+3$
5	$+2$		-1		$+1$
	$V_G = 2.0$		$V_E = 2.0$		$V_P = 4.0$

NOTE: To keep the example as simple as possible, we will consider these individuals as constituting a population rather than a sample, thus ignoring problems of sampling. As a result, variances are obtained by dividing by N, rather than by N − 1.
SOURCE: After Plomin, DeFries, and Loehlin, 1977.

E, in that large deviations in G correspond to large deviations in the same direction in E. Negative correlation between G and E would decrease rather than increase V_P.

Let us now add the $G \times E$ interaction to the example, retaining the positive correlation between G and E. We said that $G \times E$ refers to any nonadditive effect of G and E. In our example, however, we will assume that the nonadditive function is, in fact, G multiplied by E. (See Table 9.8.) The variance of the $G \times E$ values around their mean of 2.0 is 2.8. Genetic

Table 9.7
Hypothetical genetic, environmental, and phenotypic deviations from the mean for five individuals when genetic and environmental deviations are perfectly correlated

Individual	G	+	E	=	P
1	-2		-2		-4
2	-1		-1		-2
3	0		0		0
4	$+1$		$+1$		$+2$
5	$+2$		$+2$		$+4$
	$V_G = 2.0$		$V_E = 2.0$		$V_P = 8.0$

SOURCE: After Plomin, DeFries, and Loehlin, 1977.

Table 9.8
Hypothetical genetic and environmental deviations from the mean
and phenotypic values for five individuals when genetic and
environmental deviations are perfectly correlated and when there is
an interaction between G and E

Individual	G	+	E	+	G × E	=	P
1	−2		−2		+4		0
2	−1		−1		+1		−1
3	0		0		0		0
4	+1		+1		+1		+3
5	+2		+2		+4		+8
	$V_G = 2.0$		$V_E = 2.0$		$V_{G \times E} = 2.8$		$V_P = 10.8$

SOURCE: After Plomin, DeFries, and Loehlin, 1977.

variance, environmental variance, and variance due to the correlation between G and E [2Cov(G)(E)] remain 2.0, 2.0, and 4.0, respectively. Adding
the $V_{G \times E}$ term yields 10.8, which is the phenotypic variance.

Although we cannot often measure genetic variance, environmental
variance, or genotype–environment interaction directly, this hypothetical
example indicates that all four components can contribute to phenotypic
variance for a character. Because we cannot measure these components directly, we estimate them indirectly from the resemblance of relatives.

COVARIANCE OF RELATIVES

If we could measure genetic and environmental effects for individual subjects,
we could directly estimate V_G and V_E in populations. Instead, our analyses
proceed indirectly, estimating the various genetic and environmental components of variance from relationships that differ in genetic or environmental
relatedness. For example, full siblings who have both parents in common are
twice as similar genetically as half-siblings with only one parent is common. If
genes influence a particular behavior, then the double genetic similarity of full
siblings should make them more similar for that behavior than half-siblings.
Quantitative behavioral genetic methods involve comparisons of several such
relationships, in which genetic similarity is varied randomly while environmental similarity is held constant, or vice versa. The purpose of this section is
to provide the theoretical background for behavioral genetics studies of familial resemblance.

Covariance

Earlier in this chapter, we discussed covariance, correlation, and regression. Covariance between X and Y is the sum of the cross products of the deviations from the mean of X and the corresponding deviations from the mean of Y, divided by $N - 1$. Correlation and regression express covariance as a proportion of variance. For now, we will focus on covariance. Previously, we considered the covariance between two variables, X and Y, for many individuals—that is, the extent to which individuals' scores on X covaried with scores on Y. Now, we will consider covariance between relatives rather than between variables. For example, instead of considering the covariance between the two traits, X and Y, for many individuals measured on both traits, we will consider the covariance between twins or between parents and their offspring for a single variable. If members of a family are more similar than individuals picked at random from the population (i.e., if their deviations from the mean are in the same direction), there is covariance.

Both genetic and environmental hypotheses predict similarities between relatives. Relatives share genes to some extent and thus should be similar if genes affect the particular behavior under study. Environmental hypotheses also predict that members of the same family should be similar because they are subject to much the same environmental influences. For example, if certain child-rearing practices in human families are thought to be important influences on the development of personality, then children in the same family subjected to similar child-rearing practices should be similar in those aspects of personality. Later, we shall see how the knowledge that certain family relationships are not as similar genetically or environmentally as others provides the basis for untangling genetic and environmental influences. However, the point here is that both genetic and environmental hypotheses predict covariance among relatives living together.

There is zero covariance between pairs of unrelated individuals picked at random. Because such individuals share neither genes nor environment, their scores do not covary. Other relationships, however, share both genes and environment. We can describe covariance between relatives as $Cov(P_1)(P_2)$, where P_1 is the phenotype of one relative and P_2 is the phenotype of the other. In the previous section, we noted that $P = G + E$, and we can substitute that for $Cov(P_1)(P_2)$:

$$Cov(P_1)(P_2) = Cov(G_1 + E_1)(G_2 + E_2)$$

Remember that quantitative genetics always addresses differences rather than universals. That is, when we say that genetically unrelated individuals do not share genes, we mean that they are uncorrelated for polymorphic loci. Much DNA is identical for all humans, and much DNA is even identical for all mammals. Similarly, when we say that individuals living in uncorrelated environments do not share environment, we mean that their experiences do not make them similar. Many environmental factors, from oxygen to caregivers, can be viewed as important environmental constants shared by all members of our species. To the extent that such genetic and environmental factors are constants, they do not contribute to differences among individuals. However, even factors such as these could be viewed as contributing to individual differences. In terms of oxygen, do children living at high altitudes differ developmentally from those at sea level? A major concern of developmental psychologists is the extent to which different styles of caregiving affect development. However, our point is that quantitative genetics only addresses genetic and environmental factors that make a difference, not the genes and environments shared by everyone.

Shared and Independent Influences

Not all genetic, nor environmental, influences for a particular behavior make family members similar to one another. Identical twins are, of course, identical genetically and thus share all genetic influences. However, for other family relationships there are both shared genetic influences and those that are not shared (due to segregation). Genetic theory predicts differences between genetically related individuals other than identical twins. In contrast, environmental theories rarely predict differences for members of the same family.

Although V_A, V_D, and V_I contribute in various ways to different familial relationships, for the moment we shall consider only genetic variance that the relatives have in common. Parents and offspring are first-degree relatives, as are full siblings. Consider a single locus with two alleles. An offspring has a fifty-fifty chance of inheriting one particular allele rather than the other from the parent. For this reason, first-degree relatives are 50 percent similar genetically; in other words, half of the genetic variance is shared between them. The other half of the genetic variance does not covary between them, so it makes them different from one another. Such reshuffling of genes is the consequence of meiosis and the source of genetic variability. Thus, we can divide the genetic contribution to the phenotype of an individual into two parts—that part the individual shares, or has in *common*, with the relative (G_c); and the part that is not shared with the relative (G_w). Influences not shared by family members have traditionally been labeled with a *w* to indicate differences *within* families.

Similarly, some environmental influences are shared by relatives, while other aspects of the environment make family members different from one another. Some parents are physically punitive toward their children. If punitiveness affects some aspect of personality development (such as aggression), then it will make the children in the family more similar to each other in aggressiveness. There are very few known examples of systematic environmental factors that make family members different from one another, although such influences are evidently of crucial importance in behavioral development, as indicated in the following chapters (Plomin and Daniels, 1987).

For example, the order of birth may well cause behavioral differences among full siblings. If earlier-born children are different from later-born children in the same family, some environmental factor (perhaps prenatal influences, child-rearing practices, or interactions with siblings) operating within the family makes them different from one another. It is clearly not a genetic factor; regardless of birth order, siblings are equally similar genetically to each other and to their parents. Other environmental influences of this type include differential experiences within the family, such as differential parental treatment of children, as well as those factors that are independent of the family relationship, such as interactions in school and with peers, not shared by family members. Thus, we can also divide the environmental contribution to the phenotype into influences shared with the relative (E_c) and those independent of the relative (E_w). In the previous equation, by definition, only G_c and E_c can contribute to the phenotypic covariance between relatives:

$$Cov(P_1)(P_2) = Cov(G_c + E_c)(G_c + E_c)$$

The covariance of G_c with G_c is equivalent to the variance of G_c (that is, V_{G_c}). As we indicated earlier, a variable completely covaries with itself, meaning that the covariance of a variable with itself is the same as its variance. In the same way, $Cov(E_c)(E_c) = V_E$.

Now we can express the phenotypic covariance between relatives in terms of components of variance:

$$Cov(P_1)(P_2) = V_{G_c} + V_{E_c}$$

In other words, for a particular character, the covariance between relatives includes the genetic variance and the environmental variance resulting from shared genetic and environmental influences.

Genotype – Environment Correlation and Interaction

The model we have used up to this point is oversimplified. Earlier, we mentioned the correlation and interaction between genetic and environmental factors. These components of variance also enter the picture when we consider the covariance among relatives. $Cov(G_c + E_c)(G_c + E_c)$ also includes the covariance between G_c and E_c, in that it is equivalent to $V_{G_c} + V_{E_c} + 2Cov(G_c)(E_c)$. Covariance between genetic and environmental deviations can add to phenotypic variance. It can also add to the covariance between relatives. In addition, when we substituted $G + E$ for P, we did not consider the $G \times E$ interaction. The $G \times E$ interaction shared by relatives will also contribute to their phenotypic covariance.

Genetic Covariance Among Relatives

Our general model for the covariance of relatives is also too simple because it treats only shared genetic variance, rather than distinguishing between V_A, V_D, and V_I. These components of genetic variance contribute variously to different types of family relationships. (See Table 9.9.) Parents and their offspring share one-half of their additive genetic variance, as discussed in the previous section. (For this reason, additive genetic variance provides a measure of the extent to which characters "breed true.") However, parents and offspring do not share genetic variance due to dominance. Remember that dominance is the result of nonadditive combinations of alleles at loci. Offspring cannot obtain a chromosome *pair* from one parent. Thus, although dominance may contribute to the phenotypes of parent and offspring, this genetic factor will not be shared by them.

Another factor that contributes to genetic covariance among relatives is assortative mating (discussed in Chapter 8). Assortative mating adds to the genetic similarity between parents and their offspring, as well as between siblings (Jensen, 1978). For example, if assortative mating exists, a correlation between mothers and their children will include not only the genetic similarity between the mothers and their children, but also some part of the genetic similarity between the children and their fathers.

Siblings, like parents and their offspring, share half of the additive genetic variance that influences a character. However, siblings also share one-fourth of the dominance variance, since full siblings can be expected to receive the same alleles from both parents one-fourth of the time and thus have the same dominance deviation.

Fraternal twins are just siblings who happen to be born at the same time. Two eggs are fertilized by different sperm. For this reason they are sometimes referred to as dizygotic (two-zygote) twins. Like other siblings, dizygotic (DZ) twins can be the same sex or of opposite sexes, and they share half of the additive genetic variance and one-fourth of the variance due to dominance. Twins are born about once in every 83 births, and two-thirds of these are fraternal twins. The other third of twin births are identical twins. They are called monozygotic (MZ) twins because they begin life as a single zygote that

Table 9.9

Contribution of additive genetic (V_A), dominance (V_D), and common environmental (V_{E_c}) influences to the phenotypic covariance of relatives

Phenotypic covariance between:	V_A		V_D		V_{E_c}
Parents and offspring (PO)	½	+	0	+	$V_{E_{c(PO)}}$
Half-siblings (HS)	¼	+	0	+	$V_{E_{c(HS)}}$
Full siblings (FS)	½	+	¼	+	$V_{E_{c(FS)}}$
Fraternal twins (DZ)	½	+	¼	+	$V_{E_{c(DZ)}}$
Identical twins (MZ)	1	+	1	+	$V_{E_{c(MZ)}}$

splits sometime during the first few weeks of life. Because they are genetically identical, identical twins are always of the same sex. They share all genetic variance — V_A, V_D, and V_I.

Finally, half-siblings who share only one parent thus share only one-fourth of the additive genetic variance (half as much as full siblings). However, unlike full siblings, half-siblings do not share any dominance variance. Because half-siblings have only one parent in common, they cannot inherit the same chromosome pairs and thus cannot share in allelic interactions at a given locus.

Sometimes we need to consider the covariance of behavioral measures for one relative with the average measures for a number of other relatives. For example, we might consider the covariance between offspring and the average parental scores, rather than scores for a single parent. Or we could turn it around and look at the covariance between a single parent and the average of all of that parent's offspring. In general, the expectations for such averaged relationships are the same as those discussed above for relatives considered one at a time. However, some preconditions must be met (Falconer, 1981).

What about epistasis? We noted earlier that, in addition to additive effects of alleles across loci (V_A), there is also nonadditive genetic variance. Although some of this nonadditive variance is due to interactions between alleles at a locus (V_D), the rest is due to nonadditive interactions between alleles at different loci (V_I). Because identical twins are genetically identical, their phenotypic covariance includes all additive and nonadditive genetic variance. However, phenotypic covariance for other familial relationships (particularly those, such as full siblings and fraternal twins, that share variance due to dominance) includes only some of the variance due to epistatic and dominance interactions (Falconer, 1981). Fortunately, this complexity turns out empirically to be less important than it might seem. We shall see that additive genetic variance accounts for the majority of genetic variance in most behavioral characters for which such information is available.

Table 9.9 summarizes the genetic and environmental components of variance responsible for the phenotypic covariance of relatives. For example, the phenotypic covariance between fraternal twins includes half of the additive genetic variance ($\frac{1}{2}V_A$), one-fourth of the nonadditive genetic variance due to dominance ($\frac{1}{4}V_D$), and environmental influences common to members of fraternal twin pairs ($V_{E_{c(DZ)}}$). In contrast, identical twins' covariance includes all additive and nonadditive genetic variance, as well as environmental influences common to members of identical twin pairs ($V_{E_{c(MZ)}}$). In Chapter 10, such differences in the components of covariance will be used to estimate the various components of genetic and environmental variance.

HERITABILITY

The concept of heritability is frequently misunderstood. However, if properly defined and employed, heritability is a useful concept. It is simply a statistic that describes the ratio of genetic to phenotypic variance — the proportion of observed variance in a population that can be explained by genetic variance. In other words, heritability describes the extent to which genetic differences among individuals in a population make a difference phenotypically. The environmental contribution to phenotypic variance is directly analogous to heritability. Unfortunately, there is no generally accepted word to express the proportion of individual differences unexplained by genetic factors. Of the various terms that have been proposed, we shall use a word suggested by Fuller and Thompson (1978): *environmentality*.

For any behavior, we are likely to observe a wide range of individual differences. These phenotypic differences may be caused by environmental experiences, as well as genetic differences. One important aspect of behavioral genetics involves partitioning phenotypic variability into parts due to genetic and environmental differences. Heritability and environmentality are descriptive statistics that quantify the magnitude of these parts relative to the total phenotypic variance.

What Heritability Is Not

Heritability Is Neither Constant Nor Immutable

Heritability describes a situation involving a particular phenotype in a population with a certain array of genetic and environmental factors at a given time. Heritability does not indicate an eternal truth concerning the phenotype, for it can vary from population to population and from time to time. It is a population parameter analogous to the population mean and variance. If the population changes, its parameters will change accordingly.

If genetic variance or environmental variance changes, heritability (and environmentality) can change. A relatively unexplored benefit of the concept of heritability is that it can describe changes in the mix of genetic and environmental factors in various populations, times, or developmental stages.

Heritability Does Not Refer to One Individual

Heritability is a descriptive statistic that applies to a population. If we say that height has a heritability of 0.80, that means that 80 percent of the variation in height observed in this population at this time is due to genetic differences. It obviously does not mean that an individual who is 5 feet tall grew to the height of 4 feet as the result of genes and that the other 12 inches were added by the environment. However, if an individual from this population were 10 inches taller than average, one could estimate (rather imprecisely) that 80 percent of this deviation was due to genetic effects and that 20 percent was due to environmental influence. The same reasoning, of course, applies to behavioral traits.

Heritability Is Not Absolutely Precise

Some people object that heritability implies a high degree of precision. Heritability, as we have said, is a descriptive statistic; like all descriptive statistics, it involves error. Correlations, for example, involve a range of error that is partially a function of the size of the sample from which the estimate is made. As in the case of other descriptive statistics, however, we can estimate the extent of error involved in heritability estimates (Klein et al., 1973).

What Heritability Is

In 1940 J. L. Lush defined heritability as "the fraction of the observed variance which was caused by differences in heredity," a useful alternative to the old nature–nurture dichotomy:

> Furthermore, it gradually came to be recognized that the question whether the nature or the nurture, the genotype or the environment, is more important in shaping man's physique and his personality is simply fallacious and misleading. The genotype and the environment are equally important, because both are indispensable. . . . The nature–nurture problem is nevertheless far from meaningless. Asking right questions is, in science, often a large step toward obtaining right answers. The question about the roles of the genotype and the environment in human development must be posed thus: To what extent are the *differences* observed among people conditioned by the differences of their genotypes and by the differences between the environments in which people were born, grew and were brought up? (Dobzhansky, 1964, p. 55)

Or as R. C. Roberts (1967) has stated: "We need to know how much of the total variation (in a population) is due to various genetic causes, for it is axiomatic that the importance of a source of variation is proportional to the contribution it makes to the total variation."

Heritability is the proportion of phenotypic variance that is attributable to genotypic variance:

$$\text{heritability} = \frac{V_G}{V_P}$$

In the numerical example presented in Table 9.6, both the genetic and environmental variances are equal to 2.0. In the simplest case, when there is no correlation or interaction between genetic and environmental factors, the phenotypic variance is 4.0. In this case, heritability is 0.5, meaning that 50 percent of the phenotypic variance is explained by genetic variance. The other 50 percent of the phenotypic variance is caused by environmental variance. Thus,

$$\text{environmentality} = \frac{V_E}{V_P} = \frac{2}{4} = 0.5$$

Tables 9.7 and 9.8 indicate that correlations or interactions between genetic and environmental factors will increase phenotypic variance. These effects can also contribute to the phenotypic resemblance between relatives, and thus can have an effect on behavioral genetic analyses (Plomin, DeFries, and Loehlin, 1977).

Lush (1949) later distinguished between two types of heritability. *Broad-sense heritability* (h_B^2) is the type of heritability we have been discussing. It is the proportion of phenotypic differences due to all sources of genetic variance, regardless of whether the genes operate in an additive or nonadditive manner. *Narrow-sense heritability* (h^2), on the other hand, is the proportion of phenotypic variance due solely to additive genetic variance. If V_G refers to all genetic variance and V_A refers to additive genetic variance,

$$h_B^2 = \frac{V_G}{V_P} \text{ and } h^2 = \frac{V_A}{V_P}$$

Narrow-sense heritability is particularly interesting in the context of selective breeding studies, where the important question is the extent to which offspring will resemble their parents. As we noted earlier, additive genetic variance involves the extent to which characters "breed true." On the other hand, broad-sense heritability is important in many other contexts. The most important situation involves the relative extent to which individual differences are due to genetic differences of any kind. We can obtain the appropriate answer by assessing broad-sense heritability. In addition to these descriptive functions, broad-sense and narrow-sense heritability can be used in a predictive way. Heritability predicts genotypic values of individuals, given the mix of genetic and environmental variance in a population at a particular time.

Path Analysis

The concept of heritability can also be presented by the analysis of *paths* — the statistical effect of one variable on another independent of other variables. For some, it is easier to understand the concept of heritability visually in a path model rather than strictly in algebraic terms.

We can construct a path model of the effects of genetic and environmental factors on a behavioral phenotype, as in Figure 9.10. This is the same as the statement $P = G + E$. The "paths" in this case express the extent to

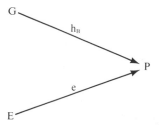

Figure 9.10

Path model of the genetic and environmental components of the phenotypic value (P). See text for explanation.

which genetic and environmental deviations cause phenotypic deviations. Thus, h_B is the path by which genetic deviations (G) from the population mean cause phenotypic deviations. In fact, the h_B path is the proportion of the phenotypic standard deviation (S_P) caused by the genetic standard deviation like S_G:

$$h_B = \frac{S_G}{S_P}$$

Remembering that the standard deviation is the square root of variance, you can see that the h_B path is, in fact, the square root of broad-sense heritability. Similarly, e is the square root of environmentality. Path analysis, which was introduced by the geneticist Sewall Wright (1921) over 60 years ago, has recently become popular for describing complexities of multifactorial models (Li, 1975; Loehlin, 1987) in both the social and biological sciences, as discussed later.

Multivariate Analyses

We have been focusing on the genetic–environmental analysis of only one behavior—a univariate (one variable) approach. However, several behaviors can be measured for each individual and subjected to multivariate quantitative genetic analysis. If two characters (X and Y) are measured for each individual in a population and a correlation is observed, this phenotypic correlation may be due to either genetic or environmental factors. Among the genetic causes, pleiotropy is the most interesting, since it results in permanent correlations between characters. Genetic correlations can also result from temporary linkages due to recent admixtures of populations or nonrandom mating. However, these linkages are soon broken up by recombination. Thus, pleiotropy is the most useful way of conceptualizing the genetic correlation between behaviors.

It is easy to visualize how environmental effects may give rise to correlations between characters, such as height and weight. A favorable diet, for example, may result in higher height and weight, whereas an unfavorable diet may be accompanied by depressed values for both characters. At the psychological level, the phenotypic correlations among measures of specific cognitive abilities may be due to environmental influences, such as the intellectual environment of the home or the quality of schooling, which affect various specific cognitive abilities in a similar way.

Why would we want to know the extent to which genetic and environmental factors contribute to the phenotypic correlation between two behaviors? When we study behaviors one at a time, many show genetic influence, but it is highly unlikely that each of these is influenced by a completely different set of genes. If the same genes affect different behaviors, we can observe a correlation among the behaviors. The same reasoning applies to environmental influences: They may affect several behaviors, producing correlations among them. Thus, the importance of multivariate genetic–environmental analysis lies in its potential for revealing the genetic and environmental bases of phenotypic covariance. Multivariate quantitative genetic analysis is one of the most important advances in behavioral genetics during the past decade (DeFries and Fulker, 1986).

Using the example of specific cognitive abilities again, the studies discussed in Chapters 11, 12, and 13 suggest substantial genetic influence for each of the specific cognitive abilities. However, it is possible that one set of genes influences all of these mental abilities. The phenotypic correlation among specific cognitive abilities may be largely due either to their genetic correlation or to a single set of environmental influences. Of course, there may also be several independent sets of genetic or environmental influences.

In summary, multivariate genetic–environmental analysis asks some important questions: Are independent gene systems involved, or do genetic influences overlap for some or all of the behaviors? Do various environmental factors make independent contributions, or are broad environmental influences responsible for the behaviors? Although there have been few multivariate behavioral genetic analyses, they suggest that genetic and environmental factors are neither very broad nor very narrow. More surprisingly, genetic and environmental correlations are correlated—that is, the structure of genetic influences seems to be similar to the structure of environmental influences (DeFries, Kuse, and Vandenberg, 1979; Fulker, 1979; Loehlin and Nichols, 1976; Martin and Eaves, 1977). Although most of us would probably predict different patterns of genetic and environmental influence, the possibility of similar genetic and environmental structures is reasonable.

Just as quantitative genetics can be applied to the variance of a single behavior, it can also be applied to the correlation between two behaviors (DeFries, Kuse, and Vandenberg, 1979). In fact, any behavioral genetic method that can partition the variance of a single behavior can also be applied to the partitioning of the covariance between two behaviors. Path analysis provides an easy way to visualize this analysis. Figure 9.11 extends the path analysis of a single behavior (see Figure 9.10) to the analysis of the correlation between two phenotypic characters, P_x and P_y. Just as the variance of a single character (P_x) is due to an environmental path (e_x) and a genetic path (h_x), the

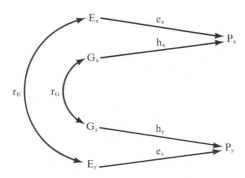

Figure 9.11
Path diagram of the phenotypic correlation between two characters (P_x and P_y) measured on an individual as a function of the genetic correlation (r_G) and the environmental correlation (r_E).

phenotypic correlation between X and Y ($r_{P_xP_y}$) may be caused by an environmental chain of paths ($e_xe_yr_E$) and a genetic chain of paths ($h_xh_yr_G$), where r_E and r_G are environmental and genetic correlations, respectively. These chains of paths are phenotypically standardized, and thus add up to the phenotypic correlation ($r_{P_xP_y}$):

$$r_{P_xP_y} = h_xh_yr_G + e_xe_yr_E$$

Genetic and environmental chains are especially useful for investigating the causes of phenotypic correlations between characters. These chains provide standardized measures of the genetic and environmental contributions to phenotypic resemblance. However, both genetic or environmental correlations (i.e., r_G and r_E by themselves) are also informative. The genetic correlation provides a measure of the extent to which two characters are influenced by the same genes. Likewise, the environmental correlation measures the extent that two characters are affected by the same environmental influences. Whether one estimates genetic chains, genetic correlations, or both depends on the purpose of the investigator (Plomin and DeFries, 1979).

This merely sums up in more precise terms that the phenotypic correlation between two behaviors may be due to genetic or environmental influences. The phenotypic correlation by itself, however, does not provide a useful index of the importance of the genetic and environmental chains. Even when the phenotypic correlation between two behaviors is negligible, there may be substantial genetic and environmental chains of influence between the two behaviors if the genetic and environmental chains work in opposite directions — that is, if one is positive and the other negative. For example, the same genes may affect specific cognitive abilities, leading to a positive genetic correlation. However, environmentally, one might develop a few abilities to the exclusion of the others, leading to a negative environmental correlation. Moreover, even if two behaviors are both substantially heritable, the phenotypic correlation between them may be environmental in origin. For example, verbal ability and spatial ability are phenotypically correlated, and both show substantial heritability. However, it is possible that completely different sets of genes influence the two abilities. In other words, their genetic correlation could be zero. If this were the case, the environmental chain would be solely responsible for the phenotypic correlation.

Genetic and environmental chains can be estimated by methods analogous to those used to estimate heritability. In Table 9.9 we presented the genetic and environmental components of variance that contribute to the phenotypic covariance between relatives for a single character. When we consider the phenotypic covariance between two characters rather than for one, we need to introduce a new concept, *cross-covariance*. Rather than studying the covariance of character X in parents and character X in offspring, we consider the cross-covariance of character X in parents and character Y in offspring. Phenotypic cross-covariance between parents and offspring may be due to their genetic and environmental similarity. In fact, the components of cross-covariance between relatives are the same as those listed in Table 9.9. This should not be surprising in view of the relationship between the univariate and multivariate analyses as just described. Thus, the phenotypic cross-covariance for characters X and Y, for parents and offspring, involves half of the additive genetic covariance, as well as common environmental influences. Phenotypic cross-covariances for identical twins include all genetic sources of covariance in addition to shared environmental influences. The use of familial resemblance in univariate analyses of variance and in multivariate analyses of covariance is outlined in the following sections.

RESEMBLANCE OF RELATIVES REVISITED

In Table 9.9 we described the genetic and environmental components of covariance for different family relationships. These relationships include both genetic and environmental components of variance. If we could find a relationship or combination of relationships that included only genetic variance, we could easily obtain the heritability statistic by dividing the genetic variance by the phenotypic variance. Determining such relationships is the essence of quantitative genetic methods, which will be discussed in the following chapters.

Univariate Analysis

Regressions and correlations are useful because they are merely covariances divided by variances. If the covariance consists solely of the genetic component of variance, then the correlation between relatives estimates heritability. In this case, the correlation is found by dividing the genetic variance by the phenotypic variance. This is the definition of heritability.

Consider identical twins who have been separated from birth. As shown in Table 9.9, identical twins share all genetic variance, plus common environmental influences. However, if they have been separated from birth, they do not have a common postnatal environment. Thus, their phenotypic covariance estimates V_G, and the correlation between them directly estimates heritability, V_G/V_P. The important thing to remember is that identical twins are genetically identical, whether or not they share environments. If they do not share environments, their correlation estimates heritability. Path analysis presents a picture of this idea.

Each identical co-twin's phenotype is caused by genetic and environmental influences, as shown in Figure 9.12. However, identical co-twins have the same genotype. A useful feature of path analysis is the ability to trace the components of a correlation by following the paths. For identical twins reared together, one chain of paths from the phenotype of one identical twin to the phenotype of the co-twin is $(h_B)(h_B)$, or h_B^2. Another chain is (e) $(r_{E_{MZ}})$ (e) or

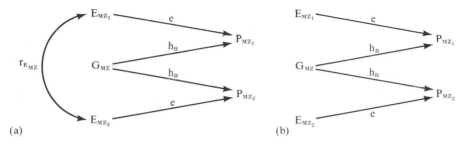

Figure 9.12
Path diagram for identical twins reared (a) together and (b) apart in uncorrelated environments.

$e^2(r_{E_{MZ}})$. The correlation between two phenotypes, given an appropriate path model, is the sum of the chains of paths. Thus, the correlation between identical twins reared together is $h_B^2 + e^2 (r_{E_{MZ}})$, which means that the correlation includes broad-sense heritability and environmental influences shared by MZ twins. This statement merely reiterates that identical twins reared together have genetic and environmental factors in common. However, as shown on the right half of Figure 9.12, identical twins who do not share environmental influences (that is, where $r_{E_{MZ}} = 0$) share only the genetic paths. Thus, their correlation directly estimates h_B^2 (broad-sense heritability).

It is in this sense that correlations can be used to imply causation, despite the revered rule to the contrary. Data from experiments are in fact analyzed in terms of correlations — more specifically, regressions — using the computational short-cut of analysis of variance. Just as we do not hesitate to impute causation to correlations from experiments, correlations in behavioral genetic research can also imply causation, because its methods involve quasi-experiments in which the effect of different degrees of genetic and environmental resemblance is assessed. For example, if identical twins reared apart correlate for some trait, how can such a correlation be explained other than by hereditary influence?

Another issue is that familial correlations represent components of variance; they are not squared to determine variance. Earlier in this chapter, we indicated that a correlation between X and Y is squared to determine the amount of variance in Y that can be predicted by X. In the case of a familial correlation, such as the correlation between identical twins reared apart (Figure 9.12), the correlation is not squared because at issue is the percent of variance common to the twins, rather than the percent of variance of twins' scores that can be predicted by the co-twins' scores (Jensen, 1971). The correlation for identical twins reared apart estimates heritability directly. For the same reason, the correlation for genetically unrelated children adopted together into the same adoptive family estimates the proportion of the variance due to shared environmental influences.

Multivariate Analysis

We have just seen in Figure 9.12 that the univariate correlation for pairs of identical twins reared in uncorrelated environments estimates h_B^2. Figure 9.13 extends this relationship of two characters, X and Y. We indicated earlier that the cross-covariance for trait X in one relative and trait Y in another has the same components of covariance as the univariate situation summarized in Table 9.9. Similarly, cross-correlations for two characters for relatives have the same relationship to univariate familial correlations. Thus, as shown in Figure 9.13, the cross-correlation for X and Y for separated identical twins is equivalent to the genetic chain discussed earlier:

$$r_{P_{x_1}P_{y_2}} = h_x h_y r_G$$

In other words, if the phenotypic correlation between two traits is due entirely to their genetic correlation, then the cross-correlation for pairs of separated identical twins should be similar to the phenotypic correlation between X and Y observed within individuals. Of course, we do not need to find separated identical twins in order to conduct multivariate quantitative genetic analyses. As we have said, any behavioral genetic analysis of the variance of a single character can be applied to the correlation among characters.

MODEL FITTING

Single correlations, such as the correlation for identical twins reared apart or the correlation for genetically unrelated children adopted together, may be quite sufficient for many purposes for estimating quantitative genetic parameters. Other behavioral genetic designs involve the comparison of two correlations, such as the twin method, which compares correlations for identical and fraternal twins, and the results of these studies can also be interpreted just by examining the correlations and calculating quantitative genetic parameters directly. However, even for such simple designs, fitting an explicit quantitative genetic model to observed data has many advantages over just calculating

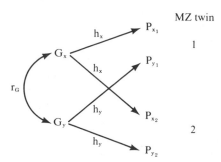

Figure 9.13
Path diagram of the phenotypic correlation between two characters (P_x) and (P_y) measured on a pair of identical twins reared apart in uncorrelated environments.

quantitative genetic parameters. An elementary approach to model fitting for simple designs such as the twin design is described in Chapter 12 (DeFries and Fulker, 1985). Model fitting becomes particularly critical for the interpretation of data from various sources and designs—for example, when family, twin, and adoption data are included in the same analysis. Raymond B. Cattell (1953, 1960) has long argued for the need for this multimethod approach, which he called Multiple Abstract Variance Analysis (MAVA). When designs become more complicated, simply examining correlations is not adequate for estimating parameters.

In quantitative genetics, model fitting basically involves solving a series of simultaneous equations in order to estimate genetic and environmental parameters that best fit observed familial correlations (Jinks and Fulker, 1970; Eaves, Last, Young, and Martin, 1978). The major advantages of such model-fitting analyses include the following: Models make assumptions explicit; model fitting tests the fit of a particular model with its set of assumptions; it is able to analyze data for several different familial relationships simultaneously; it provides appropriate estimates of quantitative genetic parameters and errors of estimate given the assumptions of the model; and it makes it possible to compare the fit of alternative models. Model fitting is as appropriate for the analysis of quantitative genetic analyses of animal designs (e.g., Hahn et al., 1987; Jinks and Broadhurst, 1974; Mather and Jinks, 1982) as for human designs, although the following examples refer to human designs.

A complete description of model fitting is beyond the scope of this primer; however, an excellent introduction to the topic that includes behavioral genetic examples is available (Loehlin, 1987; see also Connell and Tanaka, 1987). In order to provide an example of a simple model-fitting analysis, consider the model implied by the path diagrams in Figure 9.12. The model contains two observations—correlations for identical twins reared apart and for identical twins reared together—and two *latent* variables—G and E. The goal is to solve the model for the unknown paths e and h. As explained in relation to Figure 9.12, expectations can be written for the correlations for identical twins reared together (MZT) and for identical twins reared apart (MZA):

$$r_{MZT} = h_B^2 + e_c^2$$
$$r_{MZT} = h_B^2$$

This model provides two equations with two unknowns; the assumptions of the model were mentioned earlier, such as the assumption that MZA do not share correlated environments, that is, that selective placement is unimportant for MZA. Given observed correlations for MZT and MZA, the equations can be solved using simple algebra in order to estimate the genetic and environmental parameters. The MZA correlation estimates heritability and the shared environment component is estimated as the extent to which the MZT correlation exceeds the MZA correlation.

Model-fitting analyses usually involve many more equations and parameters, in which case it becomes necessary to solve the equations using procedures other than algebra. For example, equations could be added for fraternal twins reared together, nontwin siblings, and parents and offspring. Parameters could also be added to estimate assortative mating, to distinguish additive and nonadditive genetic variance, and to estimate parameters separately by gender or by age. It is important to have more equations than unknowns in the model because such overdetermined designs make it possible to test the fit of the model to the data, in addition to estimating the parameters; if there are more unknowns than equations, the parameters cannot be estimated. The goodness-of-fit index is usually χ^2 (chi square), which is a widely used statistic that indicates the statistical significance concerning the fit between expectations and observed data. The main analytic procedure used to solve complex series of equations is called *maximum-likelihood model fitting*. Such an approach employs computer programs that maximize the fit between the model and the data by finding the set of parameter estimates that yield the smallest possible discrepancies with the data. One of the most widely used maximum-likelihood computer programs is LISREL, which stands for LInear Structural RELations (Jöreskog and Sörbom, 1984). A special issue of the journal *Behavior Genetics* has recently focused on LISREL applications to twin analyses (Boomsma, Martin, and Neale, 1989).

As an example of model fitting involving several groups, consider an analysis involving twins, adoptive siblings, and parents and offspring. Familial correlations for the trait Sociable as measured by a self-report questionnaire called the Thurstone Temperament Schedule are presented in Table 9.10. In glancing at the correlations, it is easy to see that some genetic

Table 9.10

Correlations for the trait Sociable in two twin studies and an adoption study

Pairing	Correlation	Number of pairs
1. MZ twins: Michigan	0.47	45
2. DZ twins: Michigan	0.00	34
3. MZ twins: Veterans	0.45	102
4. DZ twins: Veterans	0.08	119
5. Father-adopted child	0.07	257
6. Mother-adopted child	−0.03	271
7. Father-natural child	0.22	56
8. Mother-natural child	0.13	54
9. Adopted-natural child	−0.05	48
10. Two adopted children	−0.21	80

SOURCE: From Loehlin, 1987.

influence is suggested by the pattern of correlations: For example, identical twins (MZ, monozygotic) are more similar than fraternal twins (DZ, dizygotic), and adoptive parents and their adopted children do not resemble each other as much as nonadoptive parents and offspring. Shared environmental influence does not appear to be important because adoptive relatives do not resemble each other. We could estimate each of these parameters from each informative comparison and average the estimates; for example, averaging the four correlations for adoptive relatives (lines 5, 6, 9, and 10) yields an average correlation (weighted by sample size) of $-.01$, indicating that shared environment accounts for essentially none of the variance for the Sociable scale. However, model fitting is much more informative, because as mentioned earlier, it tests an explicit model using all of the data simultaneously and it can compare alternative models.

A path diagram of a model of causal paths that might underlie such patterns of correlations is shown in Figure 9.14. The trait P is a score on the Sociable scale measured on both individuals of each pair. The model proposes that correlations between relatives on the trait are due to three independent sources: additive effects of genes, G; nonadditive effects of genes due to

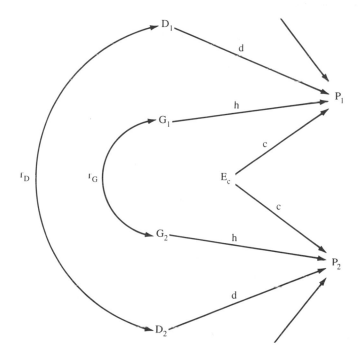

Figure 9.14

Path model of genetic and environmental sources of correlation between two individuals. G = additive genetic effect, D = nonadditive genetic effect, E_c = shared environment, P = phenotypic score; 1, 2 = two individuals. (Adapted from Loehlin, 1987.)

dominance, D; and the environment common to pair members, E_c. The residual arrow to P allows for effects of nonshared environment unique to each individual, and, in all but the MZ pairs, for genetic differences as well. The assumed genetic correlations (r_G and r_D) for the different types of relatives follow from Table 9.9: The correlation for first-degree relatives for additive genetic variance is .50, and for MZ twins it is 1.0. For nonadditive genetic variance due to dominance, genetic correlations are 1.0 for MZ twins, .25 for genetically related siblings, including DZ twins, and .00 for parents and offspring. The genetic correlations are all zero in the case of genetically unrelated relatives. These genetic correlations assume that assortative mating does not occur for the trait and that adopted children are randomly placed in adoptive homes with respect to the trait. Because the data set includes no genetically related relatives adopted apart (such as identical twins reared apart), all relatives can also resemble each other for reasons of shared environment.

From this model, Table 9.11 shows equations for the correlations between test scores of members of the relatives whose data are listed in Table 9.10. The model keeps open the possibility that shared environmental influence differs in magnitude for identical twins, siblings, and parents and their offspring by allowing different c paths for these relationships. The model, however, does not discriminate between environmental relationships of parents and adopted or natural children, between mothers and fathers' relationships with their children, or between DZ twins and adoptive siblings; models could be constructed to permit these comparisons. There are 10 observed correlations described in Table 9.11 and five unknowns (h, d, and the three c parameters).

Five models with one to four unknowns were analyzed using model-fitting procedures, with results shown in Table 9.12. A "null model" that tests whether all 10 correlations are equal can confidently be rejected because the model does not fit the data. A χ^2 of 36.02 with 9 degrees of freedom (10 groups minus 1 parameter estimated) indicates a probability (p) value of less than .001. This p value means that the data significantly depart from this null

Table 9.11

Equations for correlations between pairs of individuals in different relationships

Relationship	Table 9.10 pairings	Equation
MZ twins	1,3	$h^2 + d^2 + c_1^2$
DZ twins	2,4	$0.5h^2 + 0.25d^2 + c_2^2$
Parent, adopted child	5,6	c_3^2
Parent, natural child	7,8	$0.5h^2 + c_3^2$
Adoptive siblings	9,10	c_2^2

SOURCE: From Loehlin, 1987.

Table 9.12
Solutions of equations in Table 9.11 for various combinations
of parameters

Model	χ^2	df	p
1. all rs equal (null)	36.02	9	<0.001
2. h^2 only	9.09	9	0.44
3. $h^2 + c^2$	8.47	8	0.40
4. $h^2 + d^2$	7.60	8	0.48
5. $h^2 + c_1^2 + c_2^2 + c_3^2$	2.60	6	0.85

SOURCE: From Loehlin, 1987.

model; fewer than one in a thousand times would such a result be observed by chance. None of the other models departs significantly from the data. None-theless, the fit of the various models can be compared directly when models are nested — that is, when one model includes a subset of the parameters in another model. In this case, the difference in χ^2 for the two models is itself distributed as χ^2 and may be tested for significance. Model 2, which includes only additive genetic variance, is not improved significantly by adding any of the other parameters. That is, the fit is not improved significantly by adding a single shared environment parameter (model 3, which equates the three c parameters); the χ^2 difference between models 2 and 3 is 0.62, which with one degree of freedom yields a probability value greater than .40. Similarly, adding the dominance parameter (model 4) does not significantly improve the fit of model 2 (χ^2 difference = 1.49, df = 1, $p > .20$). Adding the three separate environmental parameters for parent and child, siblings and MZ twins yields a marginally better fit, although still not significantly better than model 2 (χ^2 difference = 6.49, df = 3, $p < .10$).

The rule of parsimony would suggest that the simplest model is best — model 2, which includes only additive genetic variance. However, because model 5 marginally improves the fit, Table 9.13 lists parameter estimates for both models 2 and 5. The single-parameter model 2 yields a narrow heritabil-ity estimate of .40, which means that about 40 percent of the variance in the

Table 9.13
Parameter estimates for Table 9.12 solutions (from Loehlin, 1987).

Model				
Model 2	$h^2 = .40$			
Model 5	$h^2 = .34$	$c_{p\text{-}c}^2 = 0.02$	$c_{sib}^2 = -0.13$	$c_{MZ}^2 = 0.11$

NOTE: These parameter estimates are lower than those presented by Loehlin (1987), which were corrected for unreliability of measurement.
SOURCE: From Loehlin, 1987.

Sociable scale is due to additive genetic factors. Because the shared environment parameter is not included in this model, the rest of the variance is presumed to be due to nonshared environmental factors. The parameter estimates for model 5 are especially interesting in that they suggest that environmental effects on sociability are almost independent for parents and children, positively shared by identical twins, and negatively related for other siblings. This result suggests the following story: Parents have little effect on their children's sociability; identical twins share environmental effects, perhaps because of their considerable physical similarity; and other siblings, including fraternal twins, contrast themselves in terms of sociability. Additional research is needed to replicate this finding and to pin down the processes by which it occurs.

The results of model-fitting analyses can only be used to test the fit of alternative models; as in tests of any scientific hypothesis, model fitting cannot prove that a particular model is correct. Another caveat should be mentioned. Although model fitting is sophisticated and elegant (the state of the art in behavioral genetic analysis), it has the disadvantage of being complex and sometimes seems to be a black box from which parameter estimates magically appear. We should not stand too much in awe of model fitting or allow it to obfuscate the basic simplicity of most behavioral genetic designs. For example, the twin design discussed in Chapter 12 estimates genetic influence on the basis of the difference between identical and fraternal twin correlations. If the identical twin correlation does not exceed the fraternal twin correlation for a particular trait, there is no genetic influence, and model-fitting approaches must come to that conclusion or there is something wrong with the model. Model fitting can be best viewed as providing refined analyses of the basic data of behavioral genetics—resemblance for relatives who vary in genetic and environmental relatedness. For this reason, the introduction to behavioral genetic methods and results in the remaining chapters focuses on the basic methods and data and merely refers to model-fitting analyses.

Model fitting was developed by quantitative geneticists (Wright, 1921), for one reason, because quantitative genetics provides a strong theory of the origins of individual differences that permits the development of explicit models that can be tested in a rigorous manner. Quantitative genetics as a theory is the focus of the following section.

QUANTITATIVE GENETICS AS A THEORY OF INDIVIDUAL DIFFERENCES

Individual differences are substantial for any behavioral characteristic studied sensitively enough to detect them. Moreover, the behavioral issues of greatest relevance to society are issues of individual differences. For example, although it would be interesting to know why human infants are natural language users, society is more concerned about individual differences: Why are some children delayed in their use of language? Do differences in language development make a difference later in life, for example, in reading ability? Can language disabilities be prevented or remediated? Societal interest in the behavioral sciences comes from the study of things that make a difference: the description, prediction, explanation, and alteration of individual differences. Nonetheless, most research in the behavioral sciences ignores individual differences, the "very standard deviation." Three reasons for this neglect of individual differences include historical happenstance, the greater statistical demands of individual differences research, and concerns about studying inequality, as discussed elsewhere (Plomin, DeFries, and Fulker, 1988).

Another reason for the neglect of individual differences is that such research often appears atheoretical. Without a theory, data gathering can lead to a collection of inconsequential facts. Researchers interested in individual differences tend to describe phenomena empirically and stay close to their data; seldom does this research go beyond description to explanation. However, quantitative genetics provides the basis for a general theory of the etiology of individual differences of a scope and power rarely seen in the behavioral sciences. The theory is unlike most contemporary theories in the behavioral sciences in that it is not limited to a particular substantive domain. It is analogous in this respect to learning theory, which does not predict what is learned but rather the processes by which learning occurs. Quantitative genetics is also formally similar to learning theory in that neither specifies molecular mechanisms. That is, learning theory does not predict which neurotransmitters are involved in the learning process, nor does it even predict that neurotransmitters are involved at all. Although quantitative genetics is based on the proposition that variation in DNA leads to phenotypic variation, it does not specify which genes are responsible for phenotypic variance. Quantitative genetic theory can be summarized in terms of 10 propositions in order to emphasize the descriptive, predictive, and especially the explanatory power of the theory.

Genetic differences among individuals can lead to phenotypic individual differences. The fundamental tenet of quantitative genetic theory is that genetic differences among individuals can lead to phenotypic differences, even for complex traits (such as behavior) that are highly polygenic and influenced as well by nongenetic factors. Genetic influence does not imply genetic determinism in the sense of a direct or close relationship between genes and their effects on behavior. The pathways from gene expression through cells, tissues, and organs to behavior are likely to be very complex. Genetic influence means only that genetic differences among individuals relate to behavioral differences observed among them; no specific genetic mechanisms or gene-behavior pathways are implied. A related issue is the "new" molecular genetics, the explosion of knowledge that has come from recombinant DNA techniques developed during the past decade, described in Chapter 7. New molecular concepts such as split genes and transposable genes do not vitiate the theory of quantitative genetics, which is based upon the postulates of classical Mendelian genetics. Quantitative genetics represents the "bottom line" of genetic variability as it affects phenotypic variability. That is, quantitative genetic methods assess the total impact of inherited genetic variability of any kind, regardless of its molecular source.

To the extent that heredity is important, the expected phenotypic similarity among relatives is a function of their genetic similarity. The essence of the theory is its quantification of the degree of resemblance expected for different types of family relationships. For example, as described in relation to Table 9.9, parents and offspring share half of the additive genetic variance and none of the nonadditive genetic variance that affects a trait. If additive genetic variance completely explains the variance of a trait, the expected phenotypic correlation for parents and their offspring is .50; if additive genetic variance is unimportant for a particular trait, additive genetic variance will not contribute to their similarity. Quantitative genetic methods spring directly from differences in the degree of hereditary resemblance for different types of relatives. If heredity is important, full siblings should be more similar than half-siblings and even more similar than adoptive siblings, who are unrelated genetically. Similarly, in the classic twin design, identical twins should be more similar than fraternal twins. Quantitative genetic theory offers refinements of these expectations of familial resemblance such as nonadditive genetic variance (both dominance and epistasis), assortative mating, and genotype–environment correlation and interaction.

The genetic component of variance for a particular trait can be decomposed into additive and nonadditive components of variance. As indicated in Table 9.9, additive and nonadditive genetic variance contributes differentially to the similarity of relatives. For example, even though they are first-degree relatives sharing half of the additive genetic variance, parents and offspring share no nonadditive genetic variance, whereas full siblings also share a quarter of the nonadditive genetic variance due to dominance. Most importantly, identical twins share all genetic variance, including all nonadditive genetic variance due to dominance and epistasis. Thus, to the extent that nonadditive genetic variance is important for a particular trait, quantitative genetic theory predicts that identical twins will resemble each other but first-degree relatives will not. This distinction is important because, as discussed in later chapters, some behaviors, especially some personality traits, appear to be influenced by epistatic genetic factors.

Environmental differences among individuals can lead to phenotypic individual differences. This proposition seems more nearly a definition because no one needs to be convinced that environmental differences can make a difference in behavior. However, it is an important feature of quantitative genetic theory that it recognizes both genetic and environmental components of variance; as we shall see in later chapters, behavioral genetic research provides evidence for the importance of environmental as well as genetic influence. Because of its balanced approach, recognizing both genetic and environmental influences, quantitative genetics also suggests novel ways to study environmental influences, as described in later chapters. It should be mentioned that quantitative genetics employs a much broader definition of environment than is usual in psychology. "Environment" in this context literally means "nongenetic," in the sense that it is that portion of variance that cannot be accounted for by heredity. This definition of environment thus includes biological factors, such as anoxia at birth, prenatal effects of drugs, and even environmental influences on DNA itself, as well as traditional environmental factors, such as childrearing, school environments, and peers.

Quantitative genetic parameters will change when genetic and environmental sources of variance change. As mentioned earlier, quantitative genetics is a theory of populations and its parameters will change as populations change. Populations with different blends of genetic and environmental influences will yield different estimates of quantitative genetic parameters. In addition, historical change can produce different results across time in the same population. For example, as described in the previous chapter, changes in gene frequency are expected as a result of natural selection and migration, and changes in genotypic frequences occur when assortative mating and inbreeding change. Precise predictions can be made about the effects of such population changes.

Environmental change across generations, cohort effects, can also be important (e.g., Elder, 1985). The dramatic changes in our society even during the past three decades—such as the widespread availability of radio and television and educational opportunities—are all likely to change the environmental contribution to behavioral development. Furthermore, the contribution of genetic and environmental factors can change during development. For example, one might expect that the relative importance of environmental variation increases during childhood as children experience increasingly diverse environments. Genetic effects also change during development. Such age-to-age changes in genetic and environmental influences are the focus of a new subdiscipline: developmental behavioral genetics (Plomin, 1986a).

The environmental component of variance for a particular trait can be decomposed into two subcomponents, one shared by family members and the other not shared. As indicated earlier, environmental variance can be decomposed into two components, E_c and E_w. Common or shared environmental influences (E_c) are those that make members of a family similar to each other; the remainder of the environmental variance, the portion not shared by family members, is called *within, independent,* or *nonshared* (E_w). This distinction in quantitative genetic theory has led to one of the most important discoveries made concerning environmental influence: Shared environmental influences are generally of little importance in the development of personality, psychopathology, and perhaps intelligence after childhood (Plomin and Daniels, 1987). Given that environmental variance is important but shared environment is not, critical environmental influences must be of the nonshared variety, making children in the same family as different from one another as are pairs of children selected randomly from the population. The importance of this finding lies in the fact that much previous environmental research has been misguided: Environmental factors relevant to differences in behavioral development lie, not *between* families but *within* families. This important finding is discussed at length in later chapters.

Heredity can affect measures of the environment and the relationship between environmental measures and behavioral measures. Measures of the environment often indirectly assess behavior. For example, the most widely used environmental measure in studies of cognitive development is the Home Observation for Measurement of Environment, or HOME (Caldwell and Bradley, 1978), an observation/interview measure that assesses dimensions such as maternal responsiveness, involvement, and restrictiveness. Such parental behaviors could be influenced by genetic factors that affect the parent or the child. Even aspects of the physical environment of the home can be influenced genetically. For example, the frequently used item, "books on the shelf," could reflect genetic influence on parents' reading interest and ability. Clearly, self-report-questionnaire measures of the environment might be filtered through genetically influenced characteristics of the individual. For example, several studies have shown that perceptions of the warmth of one's family environment is influenced by genetic factors, whereas perceptions of the control exercised in the family environment is not (Plomin et al., 1988).

In addition to genetic and environmental "main effects," phenotypic variance may be due to genotype–environment interaction. Genotype–environment interaction denotes an interaction in the statistical sense of a conditional relationship: The effect of environmental factors depends upon genotype. For example, emotional and unemotional children might not differ in adjustment when reared in a stable supportive environment; however, in an unstable environment, behavioral problems might erupt for emotional children but not for those who are less emotional. In other words, genotype–environment interaction refers to nonlinear combinations of genetic and environmental influences. A well-known discussion of the interface between heredity and environment asserted that the "organism is a product of its genes and its past environment" (Anastasi, 1958, p. 197). The point was that there could be no behavior without both environment and genes. Although no one would deny this truism, it does not mean that we cannot study the effects of environment or of heredity. As explained earlier, behavioral genetics does not apply to *an* organism or *the* organism; its focus is on variance, differences among individual organisms in a population. Even though there can be no behavior without both genes and environment, we can study the relative contributions of genes or environment to individual differences in a population for a particular behavior. Environmental differences can occur when genetic differences do not exist (for example, differences observed within pairs of identical twins). Also genetic differences can be expressed in the absence of environmental differences (for example, differences among members of a genetically heterogeneous population of mice reared in the same controlled laboratory environment). Although it is difficult to assess the overall importance of genotype–environment interaction for phenotypic variation, rearing inbred strains of mice in various environments, for example, or adoption studies can be used to isolate specific interactions between genotypes and measures of the environment. Research to date has not uncovered many genotype–environment interactions for mice or men. For example, in a series of studies involving thousands of mice, few significant and consistent interactions were found (Henderson, 1967, 1970, 1972; see review by Fuller and Thompson, 1978). Similarly, a recent review of analyses of human data found few examples of genotype–environment interaction (Plomin, DeFries, and Fulker, 1988). However, the search for genotype–environment interaction, especially for human behavior, has just begun.

Phenotypic variance may also be due to genotype–environment correlation. Genotype–environment correlation literally refers to a correlation between genetic deviations and environmental deviations as they affect a particular trait. (See Table 9.7 and the associated discussion.) In other words, it describes the extent to which individuals are exposed to environments as a function of their genetic propensities. For example, given that verbal ability is significantly heritable, parents who are gifted in verbal ability provide their children with both genes and an environment conducive to the development of verbal ability (Meredith, 1973). Three types of genotype–environment correlation have been described (Plomin, DeFries, and Loehlin, 1977). As in the previous example of verbal ability, *passive* genotype–environment correlation occurs because children share heredity as well as environmental influences with members of their family and can thus passively inherit environments correlated with their genetic predispositions. *Reactive*, or evocative, genotype–environment correlation refers to experiences of the child that derive from reactions of other people to the child's genetic propensities. *Active* genotype–environment correlation occurs, for example, when children actively select or even create environments that are correlated with their genetic propensities. This has been dubbed niche building or "niche picking." Genotype–environment correlations can be negative as well as positive—for personality, for example, it has been suggested that "society likes to 'cut down' individuals naturally too dominant and to help the humble inherit the earth" (Cattell, 1973, p. 145). Adoption studies can be used to isolate specific genotype–environment correlations for human behavior (Plomin, DeFries, and Loehlin, 1977). An important spur for research in this field is Scarr and McCartney's (1983) theory based on the three types of genotype–environment correlation as processes by which genotypes transact with environments during development. A recent review of research on genotype–environment correlation concluded that it represents a promising area for future behavioral genetic research (Plomin, DeFries, and Fulker, 1988); new model-fitting analyses of genotype–environment correlation for IQ data suggest that the passive type of genotype–environment correlation may be important (Loehlin and DeFries, 1987).

Correlations among traits can be mediated genetically as well as environmentally. As indicated earlier, during the past decade, behavioral genetics has been extended from univariate analysis of the variance of traits considered one at a time to multivariate analysis of the covariance among traits, one of the most important advances in quantitative genetic theory (DeFries and Fulker, 1986). Moreover, multivariate analysis applied to longitudinal data assesses genetic and environmental sources of the correlation between a trait at one age and at a later age (Plomin, 1986b). The genetic correlation indicates the extent to which genes that affect the trait at one age overlap with genetic effects at a later age; the extent to which the genetic correlation is less than 1.0 indicates that age-to-age genetic change has occurred. Not surprisingly, such analyses of height and weight show that stability from childhood to adulthood is largely mediated by genetic factors (Plomin, DeFries, and Fulker, 1988); more of a surprise are recent analyses that suggest that genetic effects on IQ in early childhood are substantially the same as genetic effects on IQ in adulthood (DeFries, Plomin, and LaBuda, 1987).

For these reasons, quantitative genetics is a powerful theory of individual differences. At its most general level, the theory provides an expectation that individuals will differ in complex behavioral traits. In this way, quantitative genetics organizes a welter of data on individual differences so that they are no longer viewed as imperfections in the species type or as nuisance error in analysis of variance, but rather as the quintessence of evolution.

An attractive feature of quantitative genetic theory is that, in philosophy of science jargon, the theory is progressive. That is, it leads to new predictions that can be verified empirically; in addition, potential problems with the theory are examined rather than ignored. Newer views of the philosophy of science attempt to return empirical evidence to its role as judge of scientific truth (e.g., Gholson and Barker, 1985). According to this view, successful theory maximizes empirical successes and minimizes conceptual liabilities. To the extent that this newer view becomes accepted in the behavioral sciences, quantitative genetics is likely to be seen as the basis for a powerful theory of the origins of individual differences.

SUMMARY

After a brief overview of statistics (mean, variance, standard deviation, covariance, correlation, and regression), the single-gene model of quantitative genetics is described. Genotypic value, additive genetic value, and dominance deviation are defined. The full quantitative genetic model is a polygenic extension of the single-gene model. We provide a hypothetical example, which also includes genotype–environment correlation and interaction.

Quantitative genetic methods estimate genetic and environmental components of variance from the phenotypic covariance of various types of relatives that differ in genetic relatedness or in environmental relatedness. The covariance of relatives can be used to estimate within-family and shared environmental influences, as well as heritability. Heritability, either in its narrow sense ($h^2 = V_A/V_P$) or its broad sense ($h_B^2 = V_G/V_P$), is the proportion of phenotypic variance attributable to genotypic variance. Environmentality

($e^2 = V_E/V_P$) is the proportion of phenotypic variance that is attributable to environmental (nongenetic) variance. Quantitative genetic methods are usually applied to the variance of a single character (the univariate approach), but they are equally applicable to the study of the genetic and environmental etiology of correlations among several characters (the multivariate approach). Testing explicit models represents an important advance in quantitative genetics.

Quantitative genetics provides a powerful theory for the study of individual differences in behavior. The following chapters describe the methods of behavioral genetics in greater detail and apply these methods to behavioral examples.

CHAPTER·10

Quantitative Genetic Methods: Animal Behavior

$\mathbf{I}$n Chapter 3 we discussed methods used to investigate single-gene influences on animal behavior. These methods include analyses of strain distributions, Mendelian crosses, and recombinant inbred strains. In this chapter, we shall consider more general methods of genetic analysis that can begin to untangle genetic and environmental factors, regardless of whether the behavior is influenced by a single gene or by many genes. These methods, like quantitative genetic theory, discussed in the previous chapter, can be understood at different levels. They can be viewed simply as experiments in which genetic factors are manipulated to determine whether genes can influence behavior. Or they can be coupled with quantitative genetic theory to make estimates of the relative contribution of genetic and environmental factors. The three basic methods are family studies, strain studies, and selection studies. In this chapter, examples primarily rely on mouse research; however, extensive behavioral genetic research has also been conducted using these methods with *Drosophila* (Ehrman and Parsons, 1981; Grossfield and Ringo, 1984) and the rat (Blizard and Fulker, 1981).

Although research on our own species is usually more interesting to students beginning to learn about behavioral genetics, nonhuman research is of vital importance to the field. The main reason is experimental control—both genetic and environmental. In terms of genetics, we can control breeding, and this has led to the powerful techniques for genetic analysis that are discussed in this chapter, such as inbred strains and selected lines. The feasibility of these breeding techniques is enhanced by the very short generation intervals for laboratory animals like *Drosophila* and mice. As we shall see in this chapter, an inbred strain whose members are genetically identical permits a degree of genetic replication across studies and decades that is unparalleled by any other research strategy. Selection studies in the laboratory have produced selected lines that differ for nearly all genes that affect a particular behavior. These selected lines have served as important animal models for further analysis of the physiological processes that intervene between genes and behavior.

Environmental control in the laboratory is of two kinds. First, background variability in experiences are roughly controlled so that genetic effects can be seen more clearly. Second, specific features of the environment can be manipulated; this kind of control is invaluable in understanding how heredity interacts with the environment. An important strength of behavioral genetics research is that it has developed both nonhuman and human research simultaneously and, at its best, in parallel.

FAMILY STUDIES

Relatives who share genes ought to be similar for a particular character, assuming that genes influence that character. As indicated in the last chapter (Table 9.9) and summarized in the first column of Table 10.1, genes contribute differentially to various family relationships. We showed that half-siblings share one-fourth of the additive genetic variance and no dominance variance. Parents share half of the additive genetic variance and no dominance variance

Table 10.1
Phenotypic resemblance of relatives, assuming no environmental covariance

Relatives	Genetic components of covariance	Regression (b) or correlation (r)	Relationship to heritability
One parent and offspring	$\frac{1}{2}V_A$	$b_{op} = \frac{1}{2}V_A/V_P$	$2b_{op} = h^2$
Midparent and offspring	$\frac{1}{2}V_A$	$b_{o\bar{p}} = \frac{1}{2}V_A/\frac{1}{2}V_P$	$b_{o\bar{p}} = h^2$
Half-siblings	$\frac{1}{4}V_A$	$r_{HS} = \frac{1}{4}V_A/V_P$	$4r_{HS} = h^2$
Full siblings	$\frac{1}{2}V_A + \frac{1}{4}V_D$	$r_{FS} = (\frac{1}{2}V_A + \frac{1}{4}V_D)/V_P$	$h^2 \leq 2r_{FS} \leq h_B^2$

with their offspring. Full siblings share half of the additive genetic variance and one-fourth of the dominance variance. We can compare these different relationships to determine the extent to which their genetic similarity predicts their phenotypic similarity.

With laboratory animals, environments can be controlled to some extent, so that environmental contributions to phenotypic similarity may often be ignored. If we assume that environments are controlled, the components of covariance include only genetic ones. In the previous chapter, we showed that correlations and regressions are merely covariances divided by variances. We also defined heritability as genetic variance divided by phenotypic variance. The third column of Table 10.1 indicates that the genetic covariances can be divided by an appropriate variance to obtain a regression or correlation. The last column shows the relationship between regression or correlation and heritability.

For example, the genetic component of covariance between scores of offspring and one parent is half of the additive genetic variance. The regression between parent and offspring divides their covariance (which is $\frac{1}{2}V_A$) by the phenotypic variance of the parent (the variable from which we are predicting offspring scores). Thus, the phenotypic regression of offspring on their parents estimates half of the narrow-sense heritability, assuming that shared environment is unimportant. Therefore, to estimate narrow heritability, we double this regression:

$$2\left(\frac{\frac{1}{2}V_A}{V_P}\right) = \frac{V_A}{V_P} = h^2$$

As noted in the previous chapter, the covariance between average parental (midparent) and offspring scores has the same components as the covariance between one parent and offspring. However, as indicated in Table 10.1, the regression of offspring on midparent estimates h^2, not $\frac{1}{2}h^2$. The reason is that the variance of midparent scores is half that of single-parent scores when mating is random. The regression takes the covariance component ($\frac{1}{2}V_A$) and divides it by the variance of the midparent scores, which is $\frac{1}{2}V_P$. Thus, this regression directly estimates h^2. The latter method of estimating heritability has several advantages, including the fact that it is unbiased by assortative mating (Falconer, 1981). Remember that these estimates of genetic influence are based on the assumption that environmental influences have been controlled so that they do not contribute to correlations between family members.

For parents and offspring, the preferred statistic is regression (Falconer, 1981). For siblings and twins, however, correlations are preferable. Actually, a special kind of correlation (intraclass) is used, rather than the usual interclass correlation. The intraclass correlation is used so frequently that we shall briefly describe it.

The *interclass* ("between-class") correlation assumes that there are two distinct characters, such as variables X and Y, and this is not the case for the relationship between siblings. The *intraclass* ("within-class") correlation takes into account all possible pairings of siblings within a family. However, if we randomly assign one sib to one arbitrary "class" and the other sibling to another arbitrary "class" and then compute the usual interclass correlation, the answer is much the same as the intraclass correlation.

The calculation of an intraclass sibling or twin correlation uses mean squares from a one-way analysis of variance in which the groups are the pairs. (Mean squares are the sum of squared deviations from the mean divided by the appropriate degrees of freedom.) The point of the analysis of variance is to determine the variance between pairs of siblings and the total variance, because the intraclass correlation is the extent to which total variance is due to variance between pairs. For this purpose, mean squares from the analysis of variance are used. Mean squares within pairs indicate differences within pairs; however, mean squares between pairs contain a component of variance due to differences within pairs as well as twice a component of variance due to differences between pairs. The component of variance between pairs indicates the extent to which members of sibling pairs resemble each other compared to other pairs. Total variance is the sum of components between and within pairs. For this reason, the intraclass correlation can be calculated from analysis of variance mean squares between (MSB) and within (MSW) as follows: $(\text{MSB} - \text{MSW}) \div (\text{MSB} + \text{MSW})$. Summarizing the previous discussion of components of variance in these mean squares indicates that the intraclass correlation represents the proportion of total variance that is due to variance between pairs:

$$\text{Intraclass correlation} = \frac{\text{MSB} - \text{MSW}}{\text{MSB} + \text{MSW}} = \frac{V_w + 2V_b - V_w}{V_w + 2V_b + V_w} = \frac{2V_b}{2V_t} = \frac{V_b}{V_t}$$

An additional advantage of the intraclass correlation is that it permits the computation of a correlation when there are more than two siblings in a family. For details, see Haggard (1958) or other intermediate-level statistics books.

The correlation among half-siblings must be multiplied by 4 to estimate h^2. The consequence of this is that any sampling variability will also be multiplied by 4. Thus, estimates of h^2 based on half-sibling correlations tend to be imprecise except when sample sizes are large. In animal breeding research, for example, where records of hundreds of progeny artificially sired by hundreds of males are available, this method has been very useful.

Doubling the correlation of full siblings will overestimate h^2, if dominance variance (V_D) occurs. However, it will underestimate h_B^2 because full siblings share only $\frac{1}{4} V_D$. Doubling the correlation does not yield $(V_A + V_D)/V_P$, which is broad-sense heritability; rather, it yields $(V_A + \frac{1}{2} V_D)/V_P$.

These statistical procedures are the technical basis for the simple point with which we begin. The family study method is based simply on the fact that genetically related individuals ought to be similar phenotypically for any behavior that is influenced by genes. Moreover, genes contribute to varying extents for different family relationships (such as parents and their offspring, full siblings, and half-siblings). By comparing such family relationships, we can determine the contribution of genetic similarity to observed phenotypic similarity. An example will help to clarify the family study method.

Open-Field Behavior in Mice

As an example of the use of these methods, we may again consider open-field behavior in mice. DeFries and Hegmann (1970) tested 72 males and 144 females of the F_2 generation derived from C57BL/6 × BALB/c crosses, and then mated each male with two females. In this way, the resulting 128 litters (841 mice) included full siblings and also half-siblings related through their father.

The heritability of open-field behavior may be estimated in several different ways from these data, as indicated in Table 10.2. The regression of offspring on midparent estimates a narrow-sense heritability of 0.22, while the half-sib correlation estimates a narrow-sense heritability of 0.16. Of these two estimates, which are quite similar, the parent–offspring regression is more accurate because half-sibling correlations are less reliable. However, the full-sibling correlation suggests a much higher heritability. Recall that full-sibling correlations are due both to additive and to nonadditive components of genetic variance. But even this may not tell the whole story. The rest of the answer may lie in the fact that the full siblings were reared in the same litter by the same mother, thus sharing prenatal and postnatal environmental influences. Earlier, we assumed that common environmental sources of covariance among relatives could be safely ignored in laboratory animals in controlled environments. However, in the case of full siblings, the environment is not controlled and evidently contributes substantially to their phenotypic similarity.

Although other animal research—such as studies of alcohol preference in mice (Whitney, McClearn, and DeFries, 1970), locomotor activity in *Drosophila* (Connolly, 1968), and avoidance learning in swine (Willham, Cox, and Karas, 1963)—has used the family study method, there are surprisingly few such studies, considering the obvious applicability of this technique

Table 10.2
Phenotypic resemblance of relatives for open-field activity

Relatives	Regression or correlation	Estimate of heritability
Midparent and offspring	0.22	0.22
Half-siblings	0.04	0.16
Full siblings	0.37	0.74

SOURCE: After DeFries and Hegmann, 1970.

for estimating heritability. However, there are many examples of family studies of human behavior, and these will be described in the next chapter.

Multivariate Analyses

In the previous chapter, we indicated that quantitative genetic methods can be applied to both multivariate and univariate problems. In other words, we can analyze the phenotypic correlation among characters, as well as the variance of characters taken one at a time. In terms of open-field behavior, a negative correlation (about -0.40) is usually observed between open-field activity and defecation — that is, mice who run around a lot in the open field do not leave many mementos of their travels. We know that both open-field activity and defecation are influenced by genes to some extent, but are any of the same genes involved? In other words, is the phenotypic correlation between activity and defecation due to genetic or environmental factors?

Hegmann and DeFries (1970) addressed this issue using data from the study described above. Cross-correlations for parents and offspring (for example, correlations between activity scores in parents and defecation scores in offspring) in the genetically segregating F_2 and F_3 generations were used to estimate genetic correlations, as discussed in the previous chapter. We indicated that familial cross-correlations include the same components of covariance as univariate familial correlations. Thus, cross-correlations for parents and offspring include additive genetic covariance, as well as environmental influences shared by parents and offspring. However, we can assume that the controlled laboratory setting has weakened such environmental deviations. Environmental correlations were obtained from the genetically invariant parental inbred strains and their F_1 cross. Phenotypic correlations observed between activity and defecation within the inbred or F_1 individuals must be caused by environmental factors, because these individuals do not vary genetically.

Table 10.3 shows that three of the four genetic correlations between open-field activity and defecation scores are large and negative. This indicates that many of the same genes that influence open-field activity pleiotropically affect defecation as well. Analysis of recombinant inbred strains provides additional evidence of a large negative genetic correlation between activity and defecation (Blizard and Bailey, 1979). Later in this chapter we shall consider a selective breeding study that also substantiates this finding. Selection for open-field activity resulted in a correlated response for open-field defecation.

Table 10.3 also shows that the pattern of genetic correlations among the single-day, open-field measures is mirrored by that of the environmental correlations. For example, the highest genetic correlation (0.94) is between day 1 and day 2 activity, and the highest environmental correlation (0.59) is also between these two measures. Such similarity between the genetic and environmental correlations would be rather surprising to those who might

Table 10.3
Genetic correlations (above diagonal) and environmental correlations (below diagonal) of single-day, open-field behavioral scores of mice

	Day 1 activity	Day 2 activity	Day 1 defecation	Day 2 defecation
Day 1 activity	—	0.94	−0.51	−0.89
Day 2 activity	0.59	—	−0.10	−0.76
Day 1 defecation	−0.30	−0.25	—	0.20
Day 2 defecation	−0.21	−0.44	0.34	—

SOURCE: After Hegmann and DeFries, 1970.

expect that some phenotypic correlations are caused solely by genetic factors and others solely by environmental influences. One might argue that the similarity between the genetic and environmental correlation was caused partially by a confounding of genetic and environmental influences. The genetic correlations contained possible environmental influence shared by parents and offspring. Nonetheless, the results of this study tend to be the rule rather than the exception in multivariate studies of both nonhuman and human behavior and suggest that genetic correlations and environmental correlations are correlated.

In the previous chapter we suggested that this result is really not surprising, at least in retrospect. Hegmann and DeFries (1970) wrote: "From the standpoint of biological efficiency, it would seem most reasonable that correlated characters should respond similarly to both genetic effects and environmental deviations" (p. 285). What could cause the similarity of genetic and environmental correlations? A hypothetical example proposed by DeFries, Kuse, and Vandenberg (1979) involves the metabolic pathway from tyrosine to norepinephrine, as illustrated in Figure 10.1. Dopamine and norepineph-

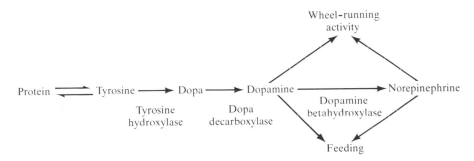

Figure 10.1
Metabolic pathway of the relationship between brain amines and rodent behavior. (From DeFries, Kuse, and Vandenberg, 1979.)

rine are neurotransmitters that relate positively to wheel-running activity and negatively to eating. A gene substitution that increases the activity of the tyrosine hydroxylase (TH) or dopa decarboxylase (DD) enzymes will result in more wheel running and less eating. In other words, mutations that affect TH or DD activity will have pleiotropic effects on wheel running and eating, resulting in a negative correlation between the two behaviors. But how could environmental influences do the same thing? Diet may serve as an example. A diet rich in tyrosine might result in more wheel running and reduced feeding. Changes in room temperature could also do this, since lower temperatures might decrease wheel running and increase feeding. Thus, at least at the biochemical level, genetic and environmental correlations may be similar.

STRAIN STUDIES

Animals derived from intense inbreeding, such as brother–sister matings over many generations, eventually become homozygous at all autosomal loci, such that animals of the same sex are genetically identical. The origin of such inbred strains of mice is described in Box 10.1. Because different inbred strains become homozygous for different alleles, the strains are genetically different from one another. This fact can be used to determine whether genetic differences affect behavior. If different strains are reared in similar environments (for example, standard laboratory cages, food, temperature, and lighting), behavioral differences will reflect genetic differences (although prenatal and postnatal parental influences may also affect the behavior). Various numbers of genes may be involved. Many genes are usually responsible for complex behaviors. Inbred strains can often be distinguished on the basis of their coat color (see Figure 10.2), and they also differ behaviorally.

For example, two widely studied inbred strains pictured in Figure 10.2 are BALB/c and C57BL/6. Figure 10.3 shows their average scores for open-field activity and defecation. The C57BL/6J mice are much more active and defecate much less than the BALB/cJ mice. The mean activity and defecation scores of derived F_1, backcross, and F_2 and F_3 generations are also shown. Note that there is a strong relationship between the average behavioral scores and the percent of genes obtained from the C57BL/6 parental strain. From left to right, these percentages are 0 percent (BALB), 25 percent (B_1), 50 percent (F_1, F_2, and F_3), 75 percent (B_2), and 100 percent (C57BL). Such a large strain difference suggests a role for genetic influence. Later in this section, we shall discuss more sophisticated uses of data from such strain studies.

Over a thousand behavioral investigations involving genetically defined mouse strains were published between 1922 and 1973 (Sprott and Staats, 1975). From 1974 through 1978, there were over 650 studies of this type (Sprott and Staats, 1978, 1979), and this pace continued into the 1980s (Sprott and Staats, 1980, 1981). Studies such as these have demonstrated that

Box 10.1

Inbred Strains of Mice

The common house mouse (*Mus musculus*) has become the most widely used animal in behavioral genetics research. Part of the reason for its popularity is that although it has the behavioral complexity characteristic of mammals, it has a short breeding time. Another important reason is the diversity of inbred strains that have been developed. As discussed in Chapter 8, brother–sister inbreeding over 20 generations will produce 98 percent homozygosity. Thus, individuals of an inbred strain are nearly identical genetically. Because of the increase in homozygosity, inbred strains are often susceptible to severe inbreeding depression, with a consequent drop in fertility. Many attempts to create inbred strains fail because the strains happen to "lock on" to harmful recessive alleles, which are expressed in the homozygous condition.

The first known inbred strain was created in 1907 by 20 generations of brother–sister matings from a pair of mice, who were homozygous recessive for three genes relating to coat color. The genes were dilute, brown, and nonagouti. (The dominant gene for agouti produces grizzled-looking, banded colors in the hair.) The strain was given the acronym dba (later changed to DBA) to refer to these three recessive genes that make the strain easily identifiable. Animals of the DBA strain and some of the other strains commonly used in behavioral genetic research are shown in Figure 10.2.

In 1913 H. J. Bagg started an albino strain whose name (Bagg albino) was later shortened to BALB/c. In 1921 the BALB/c line was crossed to another albino stock to begin the A strain, which has been particularly useful in cancer research because of its susceptibility to tumors. Crosses between the BALB/c and DBA strains led to several C strains, such as C3H and CBA.

Two other commonly used strains were begun in 1921, when a female identified as C57 gave birth to both black and brown offspring. These were separately inbred to produce the C57BL (black) and C57BR (brown) strains. The ancestry of the inbred strains corresponds to known genetic differences among the strains. For example, in terms of genetic markers, C57BL strains differ to a greater extent from the A, BALB/c, C3H, and DBA than the latter strains differ among themselves (Roderick, 1980).

There are now over 100 inbred strains available for research. They are designated by the general strain name, followed by a slash and information about the particular laboratory responsible for the strain and other specific designations. DBA/1J, for example, refers to a subline of DBA maintained by the Jackson Laboratory in Bar Harbor, Maine. And DBA/2IBG refers to a different subline of DBA maintained by the Institute for Behavioral Genetics in Boulder, Colorado.

genetic variance is nearly ubiquitous—almost all behaviors chosen for investigation showed strain differences. Although the strain comparison method now tends to be overshadowed by more sophisticated genetic analyses, it still provides a simple and highly efficient test for the presence of heritable variation. For example, strain comparisons have demonstrated considerable genetic variance for such characters as olfaction in mice (Wysocki, Whitney,

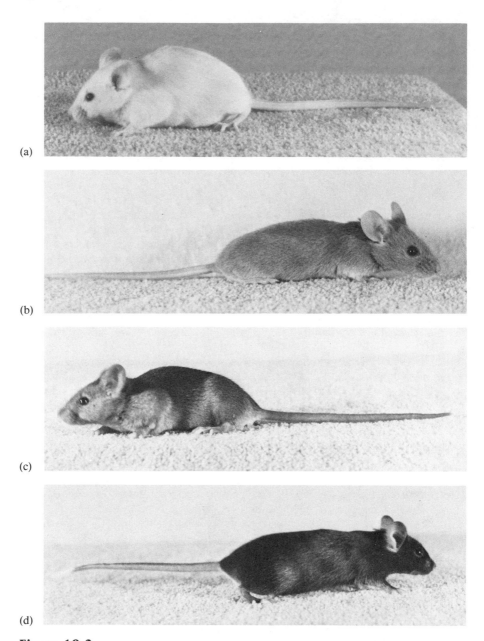

(a)

(b)

(c)

(d)

Figure 10.2
Four common inbred strains of mice. (a) BALB/cJ. (b) DBA/2IBG. (c) C3H/2IBG.
(d) C57BL/6IBG. (Courtesy of E. A. Thomas.)

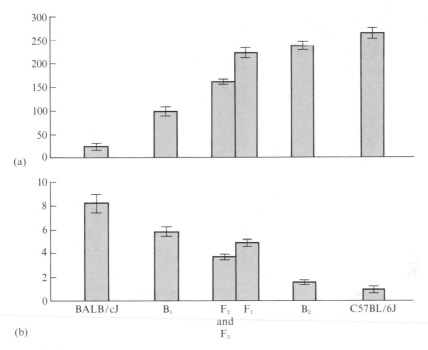

Figure 10.3
Mean open-field (a) activity and (b) defecation scores (± twice the standard error) of BALB/cJ and C57BL/6J mice and their derived F_1, backcross (B_1 and B_2), F_2, and F_3 generations. (From "Response to 30 generations of selection for open-field activity in laboratory mice" by J. C. DeFries, M. C. Gervais, and E. A. Thomas. *Behavior Genetics*, 8, 3–13. Copyright © 1978 by Plenum Publishing Corporation. All rights reserved.)

and Tucker, 1977) and fruit flies (Hay, 1976), taste perception in rats (To-bach, Bellin, and Das, 1974), reinforcing value of alcohol (Elmer, Meisch, and George, 1987), EEG correlates (Maxson and Cowen, 1976), developmental patterns of seizure susceptibility (Deckard et al, 1976) in mice, and performance in various learning situations by mice (Anisman, 1975; Padeh, Wahlsten, and DeFries, 1974; Sprott and Stavnes, 1975) and by *Drosophila* (Hay, 1975).

Genetic Effects on Learning

Because behavioral scientists studying animal behavior frequently focus on learning, we shall give a few examples of the widespread differences among mouse strains in learning situations. These data suggest caution in generalizing findings beyond the particular strain studied.

Genetic differences in learning have been found nearly every time they were studied. (See reviews of mouse research by Bovet [1977], of fly research by McGuire [1984], and of rat research by Hewitt, Fulker, and Broadhurst [1981]). Genetic differences have been shown in mice for active avoidance learning, passive avoidance learning, escape learning, bar pressing, reversal learning, discrimination learning, maze learning, and even heart rate conditioning. Active avoidance learning will serve as an example. This type of learning is usually studied in an apparatus known as a "shuttle box," which has two compartments and an electrified floor. (See Figure 10.4.) An animal is placed in one compartment; a light is flashed on, followed by a shock (delivered by an electrified grid on the floor) that continues until the animal moves to the other compartment. Animals learn to avoid the shock by moving to the other compartment as soon as the light comes on.

Actually, only some animals learn to avoid the shock. Before experimenters became aware of genetic causes of learning differences, they were puzzled by the wide range of differences in their genetically haphazard subjects. They believed that the differences would disappear if they could only measure learning with enough precision. There used to be a joke called "the Harvard law of animal behavior": When stimulation is precisely and repeatedly applied in a highly controlled setting, the animal will react exactly as it pleases (Scott, 1958). The far right side of Figure 10.5 shows that avoidance learning scores for random-bred (heterogeneous stock) Swiss mice range from near zero to greater than 50 percent. Most learning experiments have used rodents that are genetically heterogeneous, but of unknown heritage. The data for the six inbred strains in Figure 10.5 indicate that the inbreds are much

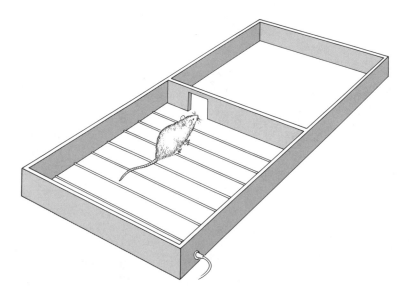

Figure 10.4
Shuttle box used to study avoidance learning in mice. (From *The Experimental Analysis of Behavior* by Edmund Fantino and Cheryl A. Logan. W. H. Freeman and Company. Copyright © 1979.)

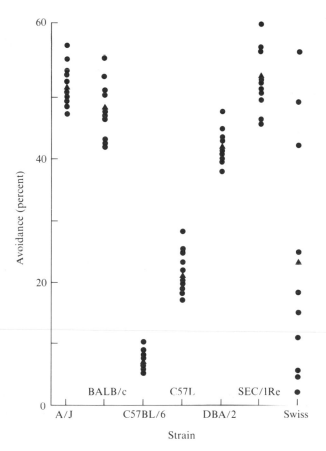

Figure 10.5
Avoidance learning in six strains of inbred mice and a random-bred strain (Swiss).
Each point represents the performance of a mouse during five sessions. The
triangles represent the mean. (From "Strain differences and learning in the mouse"
by D. Bovet. In A. Oliverio, ed., *Genetics, Environment, and Intelligence.*
Copyright © 1977 by Elsevier/North Holland Biomedical Press. All rights reserved.)

more homogenous than the random-bred mice. Moreover, the substantial
differences that exist among the inbred strains point to the influence of
genetic differences on active avoidance learning. Four of the inbred strains
learn to avoid shock over 40 percent of the time. The C57L strain, however,
avoids shock only about 20 percent of the time, and the C57BL/6 strain
avoids shock on less than 10 percent of the trials.

Some mice perform even more poorly, as indicated by the day-to-day
performances shown in Figure 10.6. The CBA strain avoids fewer than 5
percent of the shocks. This figure also illustrates differences in the rate of
learning. By the third day of training, the DBA mice greatly accelerate in
performance, whereas the BALB/c mice keep plodding along, improving their

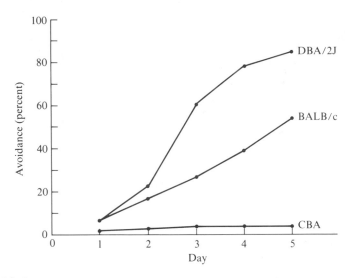

Figure 10.6
Avoidance learning (five days, 100 trials per day) for three inbred strains of mice. (From "Genetic aspects of learning and memory in mice" by D. Bovet, F. Bovet-Nitti, and A. Oliverio. *Science*, 163, 139–149. Copyright © 1969 by the American Association for the Advancement of Science.)

performance but at a slower rate. Figure 10.7 shows that such strain differences are not peculiar to aversive learning (that is, learning to avoid an unpleasant event such as shock). Performances of the same three strains on an appetitive task—learning to run through a maze to obtain food—show the same pattern. The DBA/2J strain learned quickly, the CBA animals were slow (although they learned a bit this time), and the BALB/c strain was intermediate. However, strains fastest in one learning situation are not always faster in another (Padeh, Wahlsten, and DeFries, 1974).

Another classical question in learning concerns the effect of massed trials versus trials distributed over longer periods of time. There is no universal answer to this question. In some situations, C3H inbred mice appear to learn well only when their practice is massed. DBA mice, on the other hand, perform much better when the learning trials are spread out (Bovet, Bovet-Nitti, and Oliverio, 1969).

In summary, inbred strains differ in their performance on various learning tasks. These results suggest that genetic differences affect learning. The next section indicates that inbred strains provide important information about the environment as well as heredity.

Environmental Effects on Behavior

Inbred strains are useful in demonstrating environmental effects. If mice of a single inbred strain, reared in different environments, differ in behavior, environmental factors are implicated. Studies of this kind have investigated

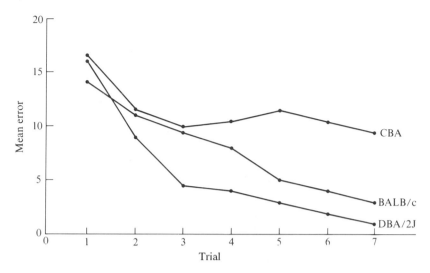

Figure 10.7
Maze-learning errors (Lashley III maze) for three inbred strains of mice. Each point represents the mean errors of 16 mice given one daily trial for ten days. (From "Genetic aspects of learning and memory in mice" by D. Bovet, F. Bovet-Nitti, and A. Oliverio. *Science*, 163, 139–149. Copyright © 1969 by the American Association for the Advancement of Science.)

the behavior of inbred strains under various environmental conditions, such as "enriched" environments with playthings and lots of room for running, crowded environments, and environments made stressful by shock. The results of such studies indicate that such environmental circumstances can influence behavior.

A few experiments have studied the joint effects of genotypic and environmental differences. One of the best-known studies of this type used selectively bred lines rather than inbred strains (Cooper and Zubek, 1958). These researchers worked with rats that had been selectively bred to run through a maze with few errors ("maze bright") or with many errors ("maze dull"), as described in the next section. Rats from these two lines were reared under one of two conditions from weaning at 25 days of age to 65 days. One condition was "enriched," in that the cages contained many movable toys. For the other condition, "impoverished," gray cages without the movable objects were used. Animals reared under these conditions were compared to maze-bright and maze-dull animals reared in a normal laboratory environment (in a different experiment).

The results of testing these animals for maze-running errors (Figure 10.8) showed that there is a large difference between the two lines after rearing in the normal environment. This is not surprising because the lines were selectively bred for maze-running differences in this environment. The enriched condition had no effect on the maze-bright animals, but it substantially improved the performance of the maze-dull rats. On the other hand, an impoverished environment was extremely detrimental to the maze-bright

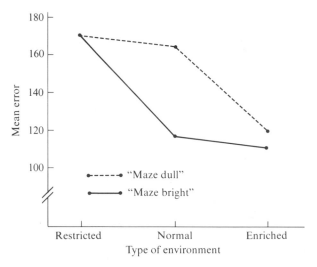

Figure 10.8
Genotype–environment interaction. Maze-running errors for maze-bright and maze-dull rats reared in restricted, normal, and enriched environments. (From Cooper and Zubek, 1958.)

rats, but had little effect on the maze-dull ones. In other words, there is no simple answer concerning the effect of deprived and enriched environments in this study. It depends on the genotype of the animal. This example illustrates genotype–environment interaction, the differential response of genotypes to environments.

Other studies of learning have also found a significant interaction between genotype and environment. However, an important study by Norman Henderson (1972) suggests that the effect of early experience may be mediated by temperamental characteristics, such as curiosity and level of motivation. For example, in the next section we shall see that the maze-bright and maze-dull animals used in the R. M. Cooper and J. P. Zubek study differ more in terms of such temperamental characteristics than they differ in learning ability per se. Henderson studied six inbred strains of mice reared in either standard or "enriched" environments. Previous studies of early experience had relied on learning tasks that may be influenced by motivational and exploration differences, such as learning to run through a maze in order to obtain food and learning to avoid a shock with a certain cue (such as a light). Henderson tested the mice in two escape-learning tasks (escape from water and escape from shock), which provided uniformly high motivation and minimized curiosity. As in other studies, learning proved to be substantially influenced by genetic factors, as evidenced by large strain differences. Unlike other studies, rearing environment and genotype–environment interaction appeared to have little effect. Henderson argued that learning, independent of such temperamental characteristics as curiosity and motivation, may not show rearing effects or genotype–environment interaction.

Prenatal and Postnatal Factors

Strain differences may be due to prenatal and postnatal environmental influences, as well as to genetic differences between the strains. For example, differences in activity between BALB/c and C57BL/6 mice may be caused by some prenatal or postnatal rearing difference between the BALB/c and C57BL/6 mothers. Although such "environmental" differences may ultimately be based on maternal genetic differences between the two strains, it is useful to separate such influences from more direct genetic differences.

The most efficient test is simply a *reciprocal cross* between the two strains. This involves crossing BALB/c males with C57B/6 females, and comparing the offspring to the offspring of BALB/c females and C57BL/6 males. Although their mothers are from different strains, the hybrid offspring in these two groups have the same genotypes. If either prenatal or postnatal maternal effects are important, then the genotype of the mother should make a difference in the pups' behavior.

In the large study of open-field behavior mentioned earlier (DeFries and Hegmann, 1970), hybrid offspring were obtained from reciprocal crosses between BALB/c and C57BL/6 mice. Figure 10.9 depicts mean daily open-field activity of the two inbred strains and their reciprocal-cross offspring. As we saw in Figure 10.3, the C57BL/6 mice are much more active than the BALB/c mice. The hybrids tend to be more like the C57BL mice, indicating

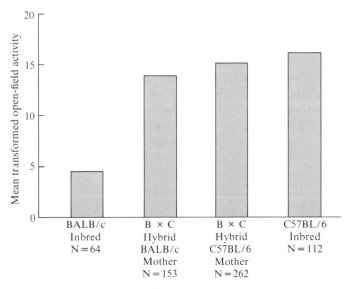

Figure 10.9
Mean transformed open-field activity of two inbred strains of mice and their reciprocal crosses. (Data from DeFries and Hegmann, 1970.)

some dominance for open-field activity. Moreover, the hybrids reared by a BALB/c mother are nearly as active as the hybrids reared by C57BL/6 mothers. The slight maternal effect, although statistically significant, is insubstantial compared to the large differences between the two inbred strains. A similarly slight maternal effect was also found for open-field defecation. Thus, although a small maternal effect is indicated, it is clear that it cannot begin to account for the large strain difference in open-field behavior.

Cross-Fostering

The slight maternal effect indicated by the above results of the reciprocal-cross method may be either prenatal or postnatal in origin. Postnatal influences can be isolated by a technique called *cross-fostering*. At birth, pups are transferred (cross-fostered) to mothers of a different strain. If the pups reared by a foster mother behave like the pups of the foster mother's strain, then postnatal influences are implicated. However, if the cross-fostered pups are like non-fostered peers, then the particular strain difference is not influenced by postnatal factors.

Despite the reasonableness of assuming the possibility of a maternal influence, neither reciprocal-cross nor cross-fostering studies have shown many important maternal effects, at least for rodents. However, one study (Reading, 1966) found a slight, but significant, effect of cross-fostering for open-field activity in BALB/c and C57BL/6 mice. Figure 10.10 illustrates the results of this study. C57BL/6 mice were not affected by cross-fostering, but the BALB/c mice were slightly more active when cross-fostered to a C57B/6 mother. As before, the maternal effect is quite small when compared to the overall difference between the two strains. The C57BL/6 mice cross-fostered to BALB/c mothers were still nearly 40 percent more active than the BALB/c mice cross-fostered to C57BL/6 mothers. Thus, the strain difference between BALB/c and C57BL/6 mice does not seem to be caused to any major extent by postnatal environmental influences. A similar conclusion emerges from a study of maternal effects on locomotion in which some maternal effect was found for 21-day-old C57BL/6 pups cross-fostered to a BALB/c mother, but not at 75 days of age; the effect was not found for BALB/c pups cross-fostered to C57BL/6 mothers, nor was it found for reciprocal crosses (Le Pape and Lassalle, 1984).

Ovary Transplants

An elegant technique to determine maternal effects involves transplanting the ovaries from one female to another. Although this sounds like science fiction, the operation is relatively easy, and the technique has been used in experimental embryology since 1909 (Palm, 1961). It certainly has elements of science fiction, as evidenced by the title of an early article, "Offspring from Unborn Mothers" (Russell and Douglas, 1945).

Ovary transplants have unique advantages over other techniques for assessing maternal effects. Reciprocal crosses involve hybridization, and the

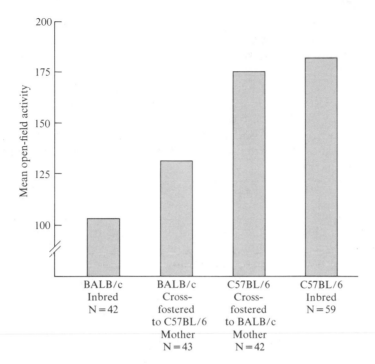

Figure 10.10
Mean open-field activity of two inbred strains of mice and reciprocal cross-fosterings. (Data from Reading, 1966.)

prevalence of hybrid vigor makes it likely that hybrids are less susceptible to environmental influences (Falconer, 1960). Cross-fostering studies require transferring a litter shortly after birth to a different mother. Ovary transplantation avoids these problems. However, just like skin grafts, ovary transplants require histocompatibility between donor and recipient. Thus, transfers between females of the same strain can be made with no difficulty. But this does not help us understand the role of prenatal influences. Fortunately, hybrids are compatible with both of their inbred parents. By transferring ovaries from two inbred strains to genetically identical hybrid mothers, we can determine whether inbred strain differences are due to the genotype of the offspring or that of their mother.

The first behavioral study using ovary transplants was conducted by DeFries et al. (1967). Ovaries of BALB/c and C57BL/6 females were transplanted to F_1 (BALB/c × C57BL/6) females whose own ovaries were removed. The operation is quite simple because the ovary is a self-contained unit. A small incision is made in the abdominal wall and each ovary with its surrounding fat pad is removed. The ovaries, within their transparent capsules, are then simply transferred to the ovarian cavity of the recipient female, whereupon the fat pads graft to the abdominal wall. In just a few days, the donor's ovaries will release ova.

The hybrid females with BALB/c ovaries were mated with BALB/c males, thus producing BALB/c inbred fetuses (from BALB/c sperm and BALB/c eggs) carried by a hybrid mother. The offspring of these females were designated B/H (a BALB/c offspring carried by a hybrid mother). Similarly, hybrid females with C57BL/6 ovaries were mated with C57BL/6 males. These pups were called C/H. Also, inbred pups reared by their own unoperated mother (B/B and C/C) were used for comparison purposes. This procedure is outlined in Figure 10.11.

Open-field activity of these four groups of offspring is illustrated in Figure 10.12. As we have often seen previously, BALB/c (B/B) inbreds are considerably less active and defecate more than the C57BL/6 (C/C) inbreds. More importantly, the activity scores of offspring reared prenatally and post-natally by foster hybrid mothers (B/H and C/H) are essentially the same as those of the inbreds. These results strikingly refute the hypothesis of a mater-nal effect for open-field activity. Similar results were obtained for open-field defecation. However, the study did demonstrate the effect of maternal envi-ronment on body weight, a character previously recognized as being suscepti-ble to maternal effects.

Despite the effectiveness of ovary transplantation, we are aware of only one other study that has used it to separate pre- and postnatal environmental influences from genetic influences on behavior (Roubertoux and Carlier,

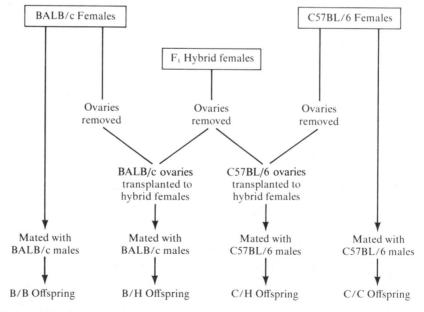

Figure 10.11
The ovary-transplant experiment of DeFries and associates. (DeFries et al., 1967.)

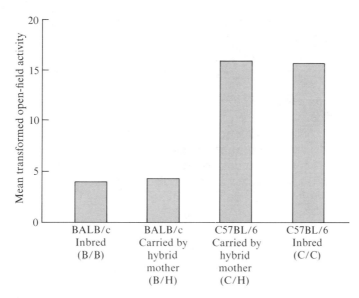

Figure 10.12
Mean transformed open-field activity of two inbred strains of mice and inbreds carried by hybrid mothers. See text for explanation of ovary transplantation. (Data from DeFries et al., 1967.)

1988). In the case of open-field behavior, it provides the best case that the maternal effect is not an important source of the difference between the BALB/c and the C57BL/6 strains.

Classical Analysis

In addition to demonstrating that genetic differences between strains produce striking behavioral differences, strain studies can also be used to estimate heritability. Because this method involves the analysis of parental, F_1, and F_2 generations used by Mendel, it has been referred to as the *classical analysis*. When highly inbred strains are crossed to produce F_1 hybrids, individuals within each of the three populations are genetically identical to each other (isogenic). Each hybrid mouse is heterozygous, but all are heterozygous in the same way. Thus, variability within these three populations must be caused by nongenetic factors. The phenotypic variance observed in each of these populations therefore provides an estimate of the environmentally caused variance:

$$V_E = V_{P_1} = V_{P_2} = V_{F_1}$$

where V_{P_1} is the phenotypic variance observed in parental strain 1, etc.

In the F_2 generation, genes assort and the individuals differ from each other genetically as well as environmentally. Because the F_2 individuals differ genetically and environmentally, their phenotypic variance represents all sources of variability:

$$V_{F_2} = V_G + V_E$$

Thus, the variance of the F_2 individuals is equivalent to the total phenotypic variance. Heritability is the proportion of phenotypic variance due to genetic variance. Given that phenotypic variance includes both genetic and environmental variance, we can estimate genetic variance as $V_{F_2} - V_E$. The estimate of V_E can come from the phenotypic variance of the inbred strains or of the F_1 generation. The best estimate is the pooled parental and F_1 variances. To estimate heritability, we divide this estimate of genetic variance by the total phenotypic variance of the F_2 population:

$$h_B^2 = \frac{V_{F_2} - V_E}{V_{F_2}} = \frac{V_G}{V_G + V_E} = \frac{V_G}{V_P}$$

Classical analysis estimates broad-sense heritability, because any genetic differences—those caused by nonadditive as well as additive genetic effects —contribute to the genetic variance in the F_2 population. More sophisticated methods, such as model fitting, are now routinely employed to analyze such data (e.g., Mather and Jinks, 1971; Jinks and Broadhurst, 1974).

Open-field behavioral data from BALB/cJ and C57BL/6J mice and their F_1 and F_2 crosses (DeFries and Hegmann, 1970) will again serve as an example. For activity, the variance of the F_2 population was 16.1. The pooled variance for the genetically invariant populations (two parental inbred strains and their F_1) was 9.6. The variance of the F_2, which includes both genetic and environmental variance, is significantly greater than the estimate of V_E, thus suggesting the influence of genes. We can estimate broad-sense heritability as follows:

$$h_B^2 = \frac{V_G}{V_P} = \frac{V_{F_2} - V_E}{V_{F_2}} = \frac{16.1 - 9.6}{16.1} = 0.40$$

This estimate of broad-sense heritability suggests that about 40 percent of the phenotypic variance in the segregating F_2 population is due to genetic differences. This means, of course, that the majority of variance is due to environmental differences. These same data suggest a heritability for open-field defecation of 0.29 ($V_{F_2} = 0.55$, $V_E = 0.39$).

In the previous section, parent–offspring regression suggested that the narrow-sense heritability of open-field activity was 0.22. We have just arrived at a broad-sense heritability estimate of 0.40. The difference between broad- and narrow-sense heritability is due to nonadditive genetic variance, as discussed in Chapter 9. However, it may be that the classical analysis will overestimate genetic variance because, in practice, it usually begins with parental strains that differ markedly. For example, the BALB and C57BL strains differ substantially in open-field activity. Segregation of blocks of genes with positive genotypic values, and blocks of genes with negative genotypic values within the F_2 may result in overestimates of the genetic variance. Not only do BALB and C57BL differ substantially in terms of open-field behavior, they also differ more than most inbred strains genetically, as suggested by their different heritage (Box 10.1). BALB and C57BL differ on about half of the polymorphic genetic markers studied in mice, which is a greater genetic difference than is found between other strains (Ward, 1985). Incidentally, work with genetic markers provides some confirmation of the quantitative genetic results just discussed. Across several inbred strains, an index of genetic differences is able to account for about 13 percent of the variance in open-field activity and the c locus for albinism explains an additional 8 percent of the variance (Ward, 1985).

Diallel Design

The classical analysis just described considers generations derived by crossing two inbred strains. This analysis can be extended to all possible crosses between several strains. A *diallel design* compares several inbred strains and all possible F_1 hybrid crosses. It was first called "the method of complete intercrossing" (Schmidt, 1919). Like the classical analysis of two inbred strains, diallel analysis provides information concerning genetic variance, heterosis, and maternal effects. Environmental dimensions can also be added (Henderson, 1967; Hyde, 1974). However, the diallel method is more efficient because it can assess most genetic parameters using only F_1 generations — that is, without waiting for F_2 animals. Moreover, the diallel method includes the genetic variance of several inbred strains and is thus less limited in its conclusions.

As an example, let us once again consider open-field activity in mice. Henderson (1967) conducted a diallel cross involving 1,440 mice of four strains. Table 10.4 summarizes the open-field activity data. Mean scores for the four inbred strains are listed on the diagonal. As we have seen, the C57BL/10 mice are more active than the BALB/c mice, and the C3H/He strain is the least active. The other scores are those obtained by the reciprocal crosses between the strains. Comparing the scores above the diagonal to those below, it is apparent that maternal effects are not important. For example, there is no maternal effect for the BALB/c $\times$ C57BL/10 cross. With a BALB/c mother, the hybrids had an average activity score of 49. With a C57BL/10 mother, their average score was 50. These data generalize the conclusion concerning maternal effects to the other strains and their crosses. Analyses of variance for these four strains suggested broad-sense heritabilities of 0.40 for open-field activity and 0.08 for defecation. As in the two-strain comparison between BALB and C57BL discussed in the previous section, estimates of genetic variance may be inflated compared to other behavioral genetic designs by the inclusion of strains that differ markedly. As indicated in Table 10.4,

Table 10.4
Diallel analysis of four inbred mouse strains for open-field activity

Maternal strain	Paternal strain			
	C57BL/10	DBA/1	C3H/He	BALB/c
C57BL/10	**56**	47	49	50
DBA/1	45	**55**	29	40
C3H/He	46	22	**21**	24
BALB/c	49	42	21	**35**

SOURCE: After Henderson, 1967.

two high-scoring strains (C57BL and DBA) and two low-scoring strains (C3H and BALB) were studied; a more representative sampling of strains would yield lower estimates of heritability.

In summary, studies comparing inbred strains are useful, not only for determining the influence of genes on behavior, but also for studying the role of the environment. In combination with manipulated environmental experiences, inbred strain studies can provide information concerning genotype–environment interaction as well as prenatal and postnatal maternal influences.

SELECTIVE BREEDING STUDIES

Both natural and artificial selection are effective only to the extent that the traits selected are under the influence of heredity. Animal breeders have successfully bred for behavioral characters throughout recorded history, long before there was any understanding of why it worked, and selection remains an important part of animal husbandry today. In addition to commercial applications, such as production of poultry, dairy cattle, and animals bred for meat, the results of artificial selection are apparent in our pets. Because dogs are such a familiar example of genetic variability, we shall digress for a moment to discuss breeds of dogs.

Dogs

Despite the tremendous variety of dogs, they are all members of a single species. Dogs have successfully been selected for behavior and morphology for the past ten thousand years. Everyone has seen their vast morphological differences, such as the forty-fold difference in weight between a Chihuahua and a Saint Bernard. However, behavior has clearly been as important in their selection. J. P. Scott and J. L. Fuller (1965) describe some very old accounts of breeds. In 1576, the earliest English book on dogs classified breeds primarily on the basis of behavior. Terriers, for example, were bred to creep into burrows to drive out small animals. Another book, published in 1686, described the behavior for which spaniels were originally selected. They were bred to creep up on birds and then spring to frighten the birds into the hunter's net. With the advent of the shotgun, different spaniels were bred to point rather than to crouch, although cocker spaniels still crouch when frightened. However, the author of the 1686 work was more concerned about the personality of spaniels: "*Spaniels* by Nature are very loveing, surpassing all other Creatures, for in *Heat* and *Cold*, *Wet* and *Dry*, *Day* and *Night*, they will not forsake their *Master*" (cited by Scott and Fuller, 1965, p. 47).

Behavioral classification of dogs continues today. The American and British kennel clubs still classify dogs on the basis of their behavioral function, rather than their genetic similarity. The selection process can be quite fine-tuned. For example, in France, where dogs arc uscd chiefly for farmwork, there are 17 breeds of shepherd and stock dogs specializing in aspects of this work. In England, dogs have been bred primarily for hunting, and there are 26 recognized breeds of hunting dogs. Breeding, however, has had some mixed blessings. By selecting for certain characteristics, in many cases breeders have accidently selected for defects. For example, many breeds, such as German shepherds, are bred for a "downhill carriage" in which the shoulders are higher than the hips. Though this gives the dogs a powerful appearance, it also makes them more likely to have problems with their hip joints and may result in lameness.

An extensive behavioral genetics research program on breeds of dogs was conducted over two decades by Scott and Fuller (1965). They studied the development of pure breeds and hybrids of the five representative breeds pictured in Figure 10.13: basenjis, beagles, cocker spaniels, Shetland sheep dogs, and wire-haired fox terriers. The selection history of each of these strains is quite different. Basenjis were recently brought from Africa, where they were bred as general-purpose hunting dogs. They were used to drive small game into a net, to track, and to flush birds. Cocker spaniels are descended from bird dogs and are now primarily house pets. Terriers are aggressive scrappers, as their selection history described above would suggest. Unlike the other breeds, Shetland sheep dogs have not been bred for hunting, but rather for performing complex tasks under close direction from their masters. They were originally small dogs, which were crossed with the large Scotch collie, and then crossed again with smaller breeds. They are currently being bred as small dogs with a collielike appearance.

These breeds are all about the same size, but they differ markedly in behavior. Before we generalize about the behavior of the breeds, we should note that breeds of dogs are not genetically invariant, like inbred strains of mice. Considerable genetic variability exists within breeds, despite the substantial genetic differences between them. Average behavioral differences among the breeds reflect their breeding history. Spaniels are very people-oriented and nonaggressive. Terriers are considerably more aggressive. Basenjis

Figure 10.13
J. P. Scott with the five breeds of dogs used in his experiments with J. Fuller. Left to right: wire-haired fox terrier, American cocker spaniel, African basenji, Shetland sheep dog, and beagle. (From *Genetics and the Social Behavior of the Dog* by J. P. Scott and J. L. Fuller. Copyright © 1965 The University of Chicago Press. All rights reserved.)

seldom bark and are very fearful of people until they are a couple of months old, at which time they can be rapidly tamed. Shetland sheep dogs are very responsive to training. In short, Scott and Fuller found behavioral breed differences just about wherever they looked—in the development of social relationships, emotionality, and trainability, as well as many other behaviors and physical characteristics. In addition, more sophisticated genetic analyses such as backcrosses, were conducted in order to shed light on the genetic architecture of dog breeds.

Heritability

Quantitative genetic considerations can predict the success of selection. At the most basic level, if heritability is zero, selective breeding will completely fail to produce the desired results. If heritability is 1.0, selection will quickly succeed. Intermediate levels of heritability will yield partial success.

More specifically, narrow-sense heritability is used to estimate the response to selection because it involves only additive genetic variance. (See Chapter 9.) Selection will succeed only to the extent that additive genetic variance is present. Consider a behavior such as open-field activity in a genetically heterogeneous group of mice. If we breed mice high in activity in the open field, and if activity is heritable, their offspring will have an average activity score more like their parents' average than that of the average of the base population. However, if open-field activity is not heritable, the offspring of these highly active animals will have an average activity score the same as that of the base population.

We need to introduce two terms to describe this logic. The first is the *selection differential*, the difference between the mean of the selected parents and that of the base population. (See Figure 10.14.) The other is the *response to selection* (or gain), the difference between the mean of the offspring of the selected animals and that of the base population. If heritability is zero, then the response to selection will be zero, no matter how severe the selection differential. In fact, the response to selection (R) is a simple function of heritability (h^2) and the selection differential (S):

$$R = h^2S$$

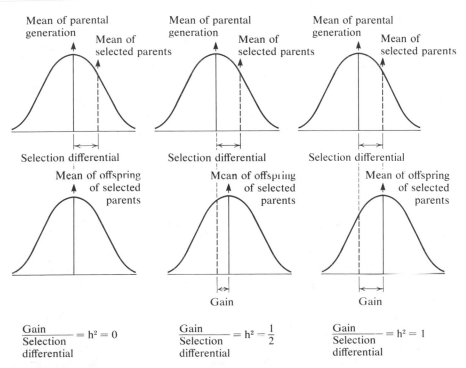

Figure 10.14

Relationship among selection differential, response to selection or gain, and realized heritability. (From *Heredity, Evolution, and Society*, 2nd ed., by I. Michael Lerner and William J. Libby. W. H. Freeman and Company. Copyright © 1976.)

Thus, as heritability approaches 1.0, the response to selection will approach the selection differential. In other words, the offspring of the selected parents will yield an average score similar to their parents' average score. Thus, if we know the heritability of a character, we can predict the response to selection for a given selection differential, although precise predictions are complicated by dominance, pleiotropy, linkages, and differing initial allelic frequencies (Wright, 1977). This relationship is especially important for evolutionary biology, where the character being selected is relative fitness. (See Box 10.2.)

By rearranging terms, we see that

$$h^2 = \frac{R}{S}$$

Thus, a selection study can be used to estimate heritability retrospectively, the so-called realized heritability. The relationship between these variables is illustrated in Figure 10.14. In the first example, heritability is zero because the mean of the offspring is just like the mean of the unselected base population. The middle example indicates a response to selection ("gain" in Figure 10.14) that is half the selection differential; so that heritability is 0.50. In the third example, illustrating a heritability of 1.0, the response to selection is the same as the selection differential.

Another important feature of selection studies is that they indicate the extent to which behavior is affected by many genes. For example, if a single gene were responsible for behavioral variation, selection should yield a substantial separation of high and low lines in a single generation. On the other hand, if many genes affect a trait, selection should continue to be successful for many generations, pushing the high and low lines further apart as more genes that affect the trait are sorted out into the two lines. We shall see that selection studies provide the best available evidence that behavioral traits are highly polygenic.

Maze Running in Rats

Selective breeding experiments were employed early in the history of behavioral genetics. In 1924, Edward Tolman reported the results of two generations of selection for maze learning by rats. Tolman saw the genetic approach, and selective breeding in particular, as a tool for "dissecting" behavioral characteristics:

> The problem of this investigation might appear to be a matter of concern primarily for the geneticist. Nonetheless, it is also one of very great interest to the psychologist. For could we, as geneticists, discover the complete genetic mechanism of a character such as maze-learning ability—i.e., how many genes it involves, how these segregate, what their linkages are, etc.—we would necessarily, at the same time, be discovering what psychologically, or behavioristic-

ally, maze-learning ability may be said to be made up of, what component abilities it contains, whether these vary independently of one another, what their relations are to other measurable abilities, as, say, sensory discrimination, nervousness, etc. The answers to the genetic problem require the answers to the psychological, while at the same time, the answers to the former point the way to those of the latter. (Tolman, 1924, p. 1)

As his own contribution toward this end, Tolman began with a diverse group of 82 rats, which were assessed for learning ability in an enclosed maze. Using as a criterion for selection "a rough pooling of the results as to errors, time, and number of perfect runs," nine male and nine female "bright" rats were selected and mated with each other. Similarly, nine male and nine female "dull" rats were selected to begin the "dull" line. The offspring of these groups constituted the first selected generation. These animals were then tested in the maze, and selection was made of the brightest of the bright and the dullest of the dull. These two groups of selected animals were mated among themselves (brother × sister) to provide the second selected generation of "brights" and "dulls."

The results were quite clear in the first generation. The bright parents produced bright offspring, and the dull parents produced dull. Due to the completeness of the data presented by Tolman (1924), it is possible to estimate that the realized heritabilities for the three characters selected were as follows: errors, 0.93; time, 0.57; and number of perfect runs, 0.61. However, due to the small sample size, these estimates are subject to large standard errors. The difference between "brights" and "dulls" decreased in the next generation, primarily because of a drop in the efficiency of performance of the bright line. These second-generation results were, of course, disappointing, and Tolman examined various possible explanations. In the first place, the maze he had used was not a particularly reliable measuring instrument. Second, the mating of brother with sister may have led to inbreeding depression.

To facilitate further investigation, an automatic, self-recording maze was developed by Tolman in collaboration with Robert Tryon. With the new maze, which provided superior control of environmental variables and proved to be highly reliable, Tryon began the selection procedure again, starting with a large, heterogeneous "foundation stock" of rats collected from several different laboratories. Tolman's energies were spent developing his theory of learning, and he did no further actual experimentation in behavioral genetics. Nevertheless, he made a continuing contribution to the field by insisting on the importance of heredity in his well-known H.A.T.E. (heredity, age, training, endocrine plus drug and vitamin conditions) list of individual-difference variables.

In Tryon's experiment, selection was based on the total number of entrances into blind alleys from days 2 to 19, following a preliminary run of eight days to acquaint subjects with the maze. As in the Tolman study, deliberate inbreeding was practiced. The results are shown in Figure 10.15.

Box 10.2

Fundamental Theorem of Natural Selection

Calculations of the selection differential must consider not only the means of the selected parents and the base population, but also the possibility that some of the selected parents may produce more offspring than others. For example, suppose the highest-scoring selected parents did not produce as many offspring as other selected parents. Unless we weighted the parental mean by the number of offspring, the calculated selection differential would overestimate the expected mean of their offspring. Although this may appear to be a small point, it serves to introduce a very important topic, the fundamental theorem of natural selection. This theorem states that changes in evolutionary fitness require additive genetic variance.

The derivation of the fundamental theorem of natural selection can be understood if we refer again to the problem of differential reproduction by selected parents in a selective breeding study. In order to adjust for the problem, a weighted selection differential is used that weights the parental mean by the number of their offspring. A mating pair's relative fitness is the ratio between the number of progeny (f_i) contributed to the next generation by that pair, and the average number of progeny of all selected pairs ($\bar{f}$). Thus, we can obtain the weighted selection differential (S) by multiplying the deviation of each pair's midparent value from the population mean ($X_i - \bar{X}$) by that pair's relative fitness ($f_i/\bar{f}$). We then sum these values for all selected pairs and divide by the number of pairs to get an average. In more concise algebraic terms:

$$S = \left(\frac{1}{n}\right) \Sigma \left[\frac{f_i}{\bar{f}} (X_i - \bar{X})\right]$$

For example, if the population mean for some character were 10.0 and the mean of the selected population was 15.0, the unweighted selection differential would be 5.0. However, if the highest-scoring selected animals reproduced less than the lower-scoring selected animals, then the weighted selection differential (that is, weighted by each pair's relative fitness) would be less than 5.0.

The above equation facilitates an important derivation obtained by Fisher (1930) many years ago. Instead of considering mean values for some trait (X) as above, think about reproductive or relative fitness, the most important characteristic in an evolutionary sense. We described relative fitness as $f_i/\bar{f}$, which we can symbolize as F. If we use the above equation to find a weighted selection differential for relative fitness, (S_F), we substitute as follows:

$$S_F = \left(\frac{1}{n}\right) \Sigma [F_i(F_i - \bar{F})]$$

This is equivalent to

$$\frac{\Sigma F_i^2 - [(\Sigma F_i)^2]/n}{n} = V_{P_F}$$

This, in turn, is equivalent to the phenotypic variance of F, relative fitness. (Readers who have studied statistics will recognize the form of this equation as the computational shortcut for computing a variance.)

As described in the text, the response to selection is a function of heritability and the selection differential. For relative fitness, the response to selection (R_F) is the change in relative fitness per generation:

$$R_F = h_F^2 S_F$$

Because narrow-sense heritability is equivalent to the ratio of additive genetic variance to phenotypic variance, and, as seen above, S_F equals the phenotypic variance of F, this equation reduces to

$$R_F = (V_{A_F}/V_{P_F})(V_{P_F})$$
$$= V_{A_F}$$

In other words, change in relative fitness per generation is equal to the additive genetic variance. If additive genetic variance for characters related to fitness is present in a species, then there is room for further selection. However, when characters related to fitness have been selected as severely as possible (meaning that there can be no more response to selection), no additive genetic variance will remain. We would, therefore, expect heritability to be low for major components of fitness, such as fertility (Falconer, 1981). However, the converse is not necessarily true. Characters with low heritability are not necessarily important components of fitness. They may be low in heritability because environmental influences are very important.

The fundamental theorem of natural selection also implies that most of the genetic variance of fitness characters should be nonadditive. Because dominance–recessiveness is responsible for inbreeding depression and hybrid vigor, we would also expect to find considerable inbreeding depression and hybrid vigor, as well as low heritabilities, for fitness characters in stable populations (Bruell, 1964a,b, 1967).

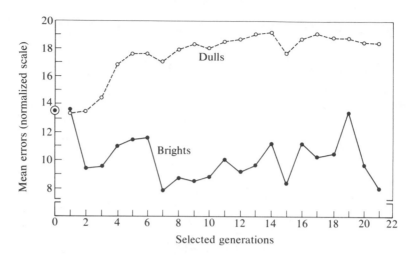

Figure 10.15
The results of Tryon's selective breeding for maze brightness and dullness. (From "The inheritance of behavior" by G. E. McClearn. In L. J. Postman, ed., *Psychology in the Making.* Copyright © 1963. Used with permission of Alfred A. Knopf, Inc.)

There is a fairly consistent divergence between the bright and dull lines through generation 7, at which time there is practically no overlap between the distributions of the two groups. The dullest bright rats were about equal in performance to the brightest dull rats. Little or no additional response to selection was observed in the later generations. Tryon provided sufficient information for Patrick Tyler (1969) to calculate the realized heritability at 0.21.

A general problem in the interpretation of selective breeding studies is evident in the results of Tryon's experiment. The problem is that we may think that we are selecting for one characteristic, when, in fact, we have selected for another. In Tryon's study, we might like to think that he selected for the rat equivalent of intelligence, as the terms "bright" and "dull" connote. But there are many other possibilities. All we know is that in a certain maze, given a certain testing procedure, the bright animals made fewer errors than the dull animals. This difference could be caused by a perceptual or motor defect in the dulls. Although such peripheral hypotheses have not been substantiated by research, a central hypothesis that competes with the "rat intelligence" notion is that the dull animals are more "emotional," or easily frightened, than the bright animals. This hypothesis was suggested by L. V. Searle (1949), who obtained scores for the dull and bright animals on 30 measures of learning and emotionality. He found that the brights performed better than the dulls on only two of five measures of maze learning. His results also indicated that the dulls might be more emotional than the brights. A later selection study (Thompson and Bindra, 1952), using a different maze, also resulted in successful selection for errors in maze running, but the differences between the selected lines in this study did not seem to be caused by emotional or motivational differences.

Although Tryon's selection experiment is a classic in experimental behavioral genetics, the design suffers from several inadequacies that have been perpetuated in some more recent selection research. As indicated, deliberate inbreeding was practiced by both Tolman and Tryon. Although a secondary objective of these studies was to produce highly inbred lines with uniform behavioral differences, inbreeding may impede the response to selection, which was the primary objective. Inbreeding results in a decrease in genetic variance within lines and thus a decrease in the potential selection response. In addition, inbreeding is almost always accompanied by a reduction in fertility.

Another inadequacy of this experimental design is the lack of an unselected control group. When such a group is included, it is possible to evaluate the effects of environmental changes across generations. In addition, the response to selection in the high and low lines can each be measured by their deviation from the mean of the control group. In this manner, it is possible to determine the degree of asymmetry of response to selection. Finally, selected and control lines should each be replicated. Since selection experiments involve considerable intergeneration variability, the reliability of the result can be indicated by the inclusion of replicate selected lines. Even more important is the fact that replicate lines are critical in analyzing genetic correlations among characters (see Chapter 9), the so-called correlated response to selection. Fortuitous correlations between the character under selection (for example, maze-running performance) and other characters of interest (such as temperament) may often occur when only one high line and one low line are maintained. However, if similar associations are noted in each of two or more replicates, the correlation is much more likely to indicate a causal (pleiotropic) relationship.

The selection study of open-field behavior described in the next section incorporated the refinements suggested above.

Open-Field Behavior

An exemplary study of mammalian behavior was conducted at the University of Colorado (DeFries, Gervais, and Thomas, 1978). Thirty generations of mice were selected for open-field activity, in a study involving the testing of more than 14,000 mice over a ten-year period. Selection began with an F_3 generation of a cross between BALB/c and C57BL/6 inbred strains. Selection was bidirectional—that is, both high- and low-active lines were selected. Also, a control line was maintained, and each of these three lines was replicated. Selection for the most or least active animals from each generation could lead to inbreeding because the selected animals could come from the same litter. In order to avoid inbreeding, a male and a female were selected from each litter and then mated at random within lines. Ten mating pairs were maintained for each of the six lines in each generation.

Figure 10.16 traces the selection progress over 30 generations. After 30 generations of selective breeding, there was a thirty-fold difference between the high and low lines. Measures on replicated lines indicate the reliability of this result. By the thirtieth generation (S_{30}), the low-active lines had nearly reached the bottom limit of zero activity scores. However, the high lines showed no sign of reaching a "selection limit" (Falconer, 1981). Moreover, the difference in open-field activity between the high- and low-active lines after 30 generations of selection exceeded the difference between the progenitor strains (BALB/c and C57BL/6) by a factor of three. These results suggest

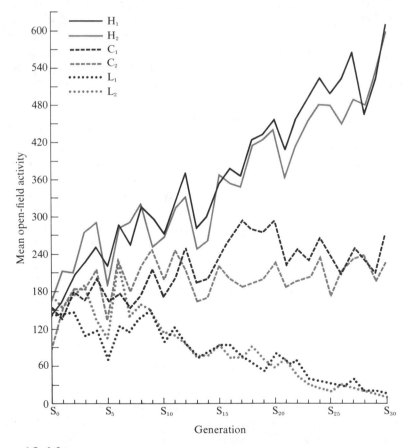

Figure 10.16
Mean open-field activity scores of six lines of mice: two selected for high open-field activity (H_1 and H_2), two selected for low open-field activity (L_1 and L_2), and two randomly mated within line to serve as controls (C_1 and C_2). (From "Response to 30 generations of selection for open-field activity in laboratory mice" by J. C. DeFries, M. C. Gervais, and E. A. Thomas. *Behavior Genetics*, 8, 3–13. Copyright © 1978 by Plenum Publishing Corporation. All rights reserved.)

that many genes affect open-field activity. The high-active mice now run the equivalent total distance of the length of a football field during two three-minute test periods. As indicated in Figure 10.17, there is no overlap in the distributions of open-field activity for the high and low lines.

Although the mice were selected for open-field activity, there was a correlated response to selection for defecation in the open-field apparatus.

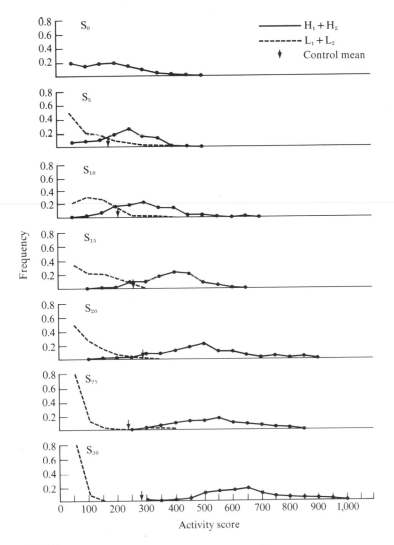

Figure 10.17
Distributions of activity scores of lines selected for high and low open-field activity. Average activity of controls in each generation is indicated by an arrow. (From "Response to 30 generations of selection for open-field activity in laboratory mice" by J. C. DeFries, M. C. Gervais, and E. A. Thomas. *Behavior Genetics*, 8, 3–13. Copyright © 1978 by Plenum Publishing Corporation. All rights reserved.)

(See Figure 10.18.) This suggests that open-field activity and defecation are genetically correlated. (See Chapter 9.) The average defecation scores of the low-active lines were about seven times higher than those of the high-active lines. The study of correlated responses to selection is an example of the multivariate approach discussed in Chapter 9. In this case, open-field activity and defecation seem to be influenced by many of the same genes. Also, as one would expect from our discussion of albinism and open-field behavior in Chapter 4, there are almost no albinos in the high-active lines, whereas the

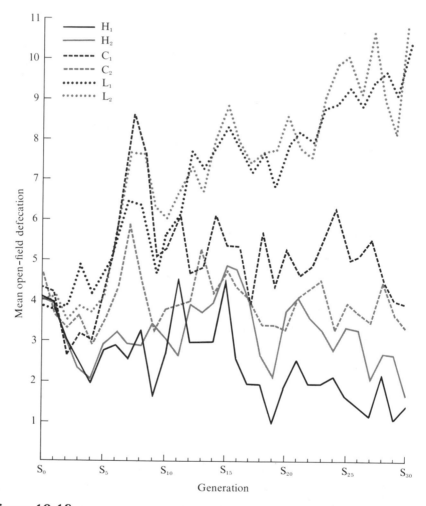

Figure 10.18

Mean open-field defecation scores of six lines of mice: two selected for high open-field activity (H_1 and H_2), two selected for low open-field activity (L_1 and L_2), and two randomly mated within line to serve as controls (C_1 and C_2). (From "Response to 30 generations of selection for open-field activity in laboratory mice" by J. C. DeFries, M. C. Gervais, and E. A. Thomas. *Behavior Genetics*, 8, 3–13. Copyright © 1978 by Plenum Publishing Corporation. All rights reserved.)

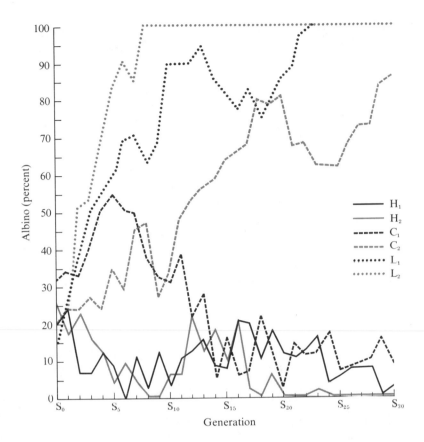

Figure 10.19
Frequency of albinism in six lines of mice: two selected for high open-field activity
(H_1 and H_2), two selected for low open-field activity (L_1 and L_2), and two
randomly mated within line to serve as controls (C_1 and C_2). (From "Response to
30 generations of selection for open-field activity in laboratory mice" by J. C.
DeFries, M. C. Gervais, and E. A. Thomas. *Behavior Genetics*, 8, 3–13. Copyright
© 1978 by Plenum Publishing Corporation. All rights reserved.)

low-active lines are completely albino. (See Figure 10.19.) It is as if the
continued selection pressure has scooped up all those alleles that make for less
activity, including the albino one, and concentrated them in the low-active
lines, while concentrating in the high lines those alleles that make for more
activity.

It is possible to use these data to estimate the realized heritability of
open-field activity. As previously discussed, selection—either artificial or
natural—is based on additive genetic variance. Thus, narrow-sense heritability estimates from parent–offspring regressions should agree with heritability
estimates from selection studies. After five generations of selection for open-
field activity, the pooled estimate of narrow-sense heritability was 0.26, which
is remarkably close to the parent–offspring estimate of 0.22 discussed earlier.

This study demonstrates that marked behavioral changes can occur during selective breeding when heritability is considerably less than 1.0.

Other Behavioral Selection Studies

Selection has been successful for many other behaviors in the laboratory. One study focused on a socially relevant behavior, sensitivity to the effects of alcohol (McClearn, 1976). Mice, like humans, differ in their reaction to alcohol. When mice are injected with the mouse equivalent of several martinis, they will "sleep it off" for varying lengths of time. Sleep time in response to ethanol injections was measured by the length of time it took an injected mouse to right itself in a cradle like that pictured in Figure 10.20. The selection progress is pictured in Figure 10.21. After 15 generations of selective breeding, the long-sleep animals sleep for an average of two hours. Many of the short-sleep mice are not even knocked out, and their average sleep time is only about 10 minutes. Figure 10.22 indicates that there is no overlap between the long-sleep and short-sleep lines. In fact, although one would hesi-

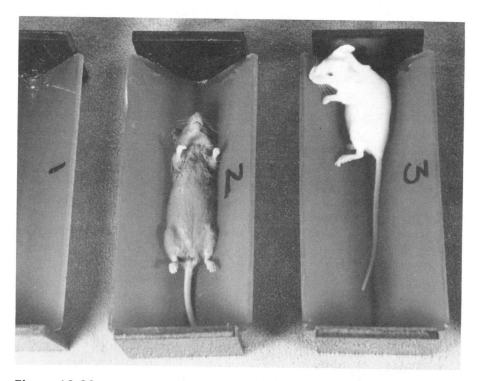

Figure 10.20
The "sleep cradle" for measuring loss of righting response after ethanol injections in mice. (a) A long-sleep mouse still on its back sleeping off the ethanol injection. (b) A short-sleep mouse that is just about to right itself. (Courtesy of E. A. Thomas.)

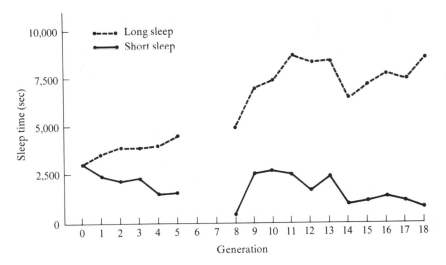

Figure 10.21
Results of ethanol sleep-time selection study. Selection was suspended during
generations 6 through 8. (From McClearn, unpublished.)

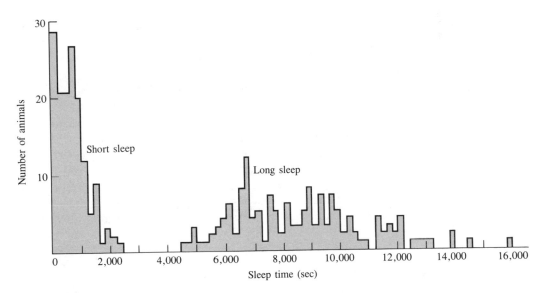

Figure 10.22
Distributions of ethanol sleep time after 15 generations of selection. (From
McClearn, unpublished.)

tate to say it while inebriated, the shortest of the long sleepers sleep longer than the longest of the short sleepers. As in the case of open-field activity, successful selection occurred despite the fact that heritability from these data is only moderate, about 0.25, a finding that has been confirmed in a classical Mendelian cross between the high and low lines (Dudek and Abbott, 1984). Again, the steady divergence of the lines over 18 generations indicates that many genes affect this behavior.

Selection has been successful for many other mouse behaviors, such as alcohol preference, learning, susceptibility to audiogenic seizures, and wildness and tameness. Recent selection research on mice includes studies of susceptibility to alcohol withdrawal seizures (Crabbe et al., 1985); aggressive behavior (Ebert and Hyde, 1976; Ebert, 1983); nerve conduction velocity (Hegmann, 1975), nest building (Lynch, 1981); rearing, which is a component of exploratory behavior (van Abeelen and van Nies, 1983); and seizure susceptibility (Chen and Fuller, 1976; Deckard et al., 1976). In rats, widely used selected lines are the Maudsley Reactive (MR) and Nonreactive (MNR) strains selected, respectively, for low and high defecation in the open field (Blizard, 1981). The Roman high and low avoidance lines (RHA and RLA) were selected for active avoidance learning in the shuttle box (Figure 10.4; Bignami, 1965; Driscoll and Battig, 1982); two other selection studies of avoidance conditioning in rats have been conducted (Bammer, 1983; Brush, Froehlich, and Baron, 1979). Successful selective breeding has also been accomplished for spontaneous activity, activity in the open field, and saccharine preference. Aggressiveness and aspects of mating behavior have been selected in chickens. Quail have been selectively bred for mating ability (Sefton and Siegel, 1975). Geotaxis, phototaxis, mating speed, and activity have been selected in *Drosophila* (including studies by Polivanov [1975], Watanabe and Anderson [1976], and Kekic and Marinkovic [1974]). Several selection studies involving conditioning in the blowfly, *Phormia regina*, have been successful (Hirsch and McCauley, 1977).

This success in selective breeding is another indicator of the considerable genetic variation that exists for behavior. More importantly, such studies generate groups of animals ideal for research on the physiological mechanisms underlying behavior. Peter Broadhurst (1975, 1976), for example, has reviewed the large number of experiments conducted from 1964 to 1974 with his Maudsley Reactive and Nonreactive rats selected for open-field defecation. Broadhurst (1978) has also reviewed selective breeding studies that involved selection for pharmacological differences. These studies are a part of the fast-developing area, *behavioral pharmacogenetics* (see also Horowitz and Dudek, 1983; Petersen, 1982).

SUMMARY

Three basic methods are used to assess polygenic influences on animal behavior. *Family studies* determine whether genetic similarity is evidenced phenotypically for different family relationships, such as parents and their offspring, full siblings, and half-siblings. There are surprisingly few family studies of animal behavior. *Strain studies* compare the behavior of inbred strains. Prenatal and postnatal environmental effects (especially maternal effects) can be studied using reciprocal crosses, cross-fostering, and ovary transplantation. Classical analysis of inbred strains and their F_1 and F_2 crosses can be used to estimate heritability. The diallel method compares several strains mated in all possible combinations. The third method is *selection*. Open-field behavior in mice has been used throughout this chapter as an example of the utility of the various methods. All of them converge on a narrow-sense heritability estimate of about 0.25.

CHAPTER·11

Family Studies
of Human Behavior

Genes influence human behavior in the same way that they affect any phenotype. They control the production of proteins, which interact in physiological systems, thus affecting behavior indirectly. Obviously, we cannot (nor do we wish to) selectively breed people, assign them to controlled environments, or produce genetically identical groups analogous to inbred strains of mice. We can, however, find naturally occurring situations in which (1) genetic influences are controlled or randomized so that the effects of the environment can be studied and (2) environmental influences are controlled so that the effects of genes can be studied. Although studies of complex human behavior lack the experimental control of studies of nonhuman animals, there is a silver lining in this particular cloud. From the beginning, human behavior has been studied in the context of genetic and environmental variation as it exists naturally in populations. Thus, human behavioral geneticists have provided information concerning the relative contributions of genetic and environmental factors to observed variation, given the mix of genetic and environmental influences in the real world. In contrast, little is known about the full range of environmental and genetic variation in mice in natural populations because inbred strains are typically studied in the controlled environment of the laboratory.

In this and the next two chapters, we shall consider three basic methods of studying the relative influences of genetic and environmental factors on human behavior: family studies, twin studies, and adoption studies. Throughout the chapters, cognitive behavior and psychopathology will be used as examples of complex behaviors influenced by both genetic and environmental factors. Far more behavioral genetic data are available for cognition and psychopathology than for other human behaviors. In Chapter 1 we discussed three perspectives on behavior. The first recognizes no important differences between or within species, the second accepts differences between but not within species, and the third recognizes differences between and within species. The third view not only represents that of behavioral genetics, but it also characterizes the orientation of much research in the areas of cognitive behavior (particularly IQ and specific cognitive abilities), personality, and psychopathology. Other aspects of behavioral science have been slower to take advantage of behavioral genetics methods because their predominant orientation has been closer to the first or second perspectives. There are, however, clear signs of change.

COMPONENTS OF COVARIANCE IN FAMILY STUDIES

Genetic similarity applies to human family relationships as it does to the nonhuman ones discussed in the previous chapter. (See Table 10.1.) Genetically related individuals should be similar phenotypically to the extent that genes influence a particular behavior. Thus, in addition to testing specific genetic models, as described in Chapter 4, family studies have been useful in demonstrating familial resemblance or familiality, the *sine qua non* for establishing genetic influence. With laboratory animals, we could assume that environments are controlled, and we could test the adequacy of this assumption by reciprocal crosses, cross-fostering, and ovary transplantation. For human families, this assumption cannot be made. Thus, for human behavior, phenotypic similarity for different family relationships can only be viewed as compatible with a genetic hypothesis, not as conclusive proof of genetic influence.

In this sense, family studies provide an upper-limit estimate of heritability. (See Chapter 9.) That is, genetic influence is usually no greater than the degree of familiality (except when nonadditive genetic variance is important or when environmental factors are negatively correlated for family members). If family studies show no familial resemblance, the genetic hypothesis is disconfirmed. The finding of no familiality would also suggest that there are no familial environmental influences making family members similar to one another. Such environmental influences shared by family members are known as shared or between-family (common-family) environment, as discussed in Chapter 9. When behavioral scientists talk about environment, they usually mean between-family influences that make family members similar to one another and different from other families. For example, if families differ in child-rearing practice or income, these environmental factors may underlie observed phenotypic similarity among family members. All other environmental influences are called nonshared or within-family environment. These factors create differences among members of the same family. Although within-family environmental influences have been given much less attention than between-family influences, human behavioral genetics studies point to a major role for this type of environmental effect, particularly for personality and psychopathology (Plomin and Daniels, 1987). Heredity, of course, creates differences between as well as within families. This is one explanation for the fact that like sometimes, but not always, begets like. Although hereditary factors predict differences within families, we know next to nothing about possible within-family environmental factors.

As described in our discussion of nonhuman research in the previous chapter, we can study various family relationships such as parents and their offspring, full siblings, and half-siblings. Half-siblings are individuals with only one parent in common. This situation occurs, for example, when divorced individuals have children from their previous marriage, as well as from their current marriage. In human research, we can also consider twin relationships. The genetic relationship between fraternal twins is just like that between any full siblings—as shown in Table 9.9, they share $\frac{1}{2}V_A + \frac{1}{4}V_D$. They also share between-family environmental influences. Identical twins have the same genes because they come from the same fertilized egg. Thus, their genetic covariance includes all additive and nonadditive genetic variance ($V_A + V_D$) plus between-family environmental influences.

In the following sections, we shall review family studies of IQ and schizophrenia. Until recently, family studies on these two dimensions have dominated the research scene. As described in Chapter 14, attention has turned to other aspects of cognition and psychopathology—especially specific cognitive abilities and affective disorders. In addition, many other domains have been studied, and brief summaries of this behavioral genetic research are also included in Chapter 14.

INTELLIGENCE QUOTIENT (IQ)

There is much more to cognition than IQ, but more familial data are available for IQ than for any other behavior. This is clearly not the place for an extended discussion of the nature of IQ or its application. (See, for example, Jensen, 1980; Vernon, 1979; Willerman, 1979a.) Behavioral genetics studies have begun to consider specific cognitive abilities, rather than a single measure of general cognitive functioning, and data from this research will also be presented. However, behavioral genetics studies have just begun to open the door on the complexity of cognitive functioning. Future studies are likely to make use of theoretical approaches, such as Jean Piaget's, as well as the theories and instrumentation of experimental psychology, in order to increase our understanding of cognitive processes related to the perception of stimuli, attention, and memory storage and retrieval.

Many studies of familial resemblance for IQ were conducted in the late 1920s and the 1930s. Although these studies individually suffered from various difficulties, such as small sample size or bias in sampling (see McAskie and Clarke, 1976), the combined weight of their evidence demonstrates substantial familial influence. Figure 11.1 summarizes the results of family studies published prior to 1980 (Bouchard and McGue, 1981). The average correlations weighted by the size of each sample are 0.42 for parents and their offspring, and 0.47 for full siblings. These are first-degree relatives who share about half of the additive genetic variance, as explained in Chapter 9. The IQ correlations for half siblings and cousins are 0.31 and 0.15, respectively. Half-siblings are second-degree relatives and cousins are third-degree relatives whose coefficients of genetic relationship are 0.25 and 0.125, respectively.

The correlation of 0.47 for siblings may be due to genetic or environmental similarity, or both. If the phenotypic similarity for siblings were not influenced by genetic factors, then all of the phenotypic variance would obviously be caused by environmental (more technically, nongenetic) factors. The correlation of 0.47 would indicate that half of this variance is shared by siblings (between-family environment, as discussed in Chapter 9), and the other half is due to environmental influences not shared by siblings (within-family environment). If genetic similarity were responsible for all of the phenotypic similarity, then broad-sense heritability would be 1.0, meaning that all of the phenotypic variance can be accounted for by genetic differences. Twin studies and adoption studies, reviewed in the next two chapters, suggest that the answer is in the middle. The phenotypic variance for sibs is due both to genetic variance and to environmental variance between and within families.

The summary data in Figure 11.1 suggest a correlation of 0.42 between parents and their offspring. Taken at face value, this correlation suggests a similar conclusion to those based on the sibling data. On the other hand, single parent–offspring correlations estimate only the upper limit of half of the heritability, and they may also be inflated by assortative mating.

Data from the largest and most recent family study of cognition (De-Fries et al., 1976, 1979) provide a better estimate of parent–offspring similarity, and suggest substantially less familial resemblance. The focus of the study, conducted in Hawaii, was specific cognitive abilities. But the test battery included a shortened version of Raven's Progressive Matrices, regarded as one

---

Figure 11.1

Family studies of IQ. A summary of correlation coefficients compiled by Bouchard and McGue from various sources. The horizontal lines show the median correlation coefficients, and the arrows show the correlation expected if IQ were entirely due to additive genetic variance. (From "Familial studies of intelligence: A review" by T. J. Bouchard, Jr., and M. McGue. *Science*, 212, 1055–1059. Copyright © 1981 by the American Association for the Advancement of Science.)

	Range of correlations	Number of correlations	Number of pairings	Weighted average
Single parent–offspring reared together		32	8,433	0.42
Siblings reared together		69	26,473	0.47
Half-siblings		2	200	0.31
Cousins		4	1,176	0.15

Range of correlations: 0.0 0.10 0.20 0.30 0.40 0.50 0.60 0.70 0.80 0.90 1.00

of the best single measures of culture-fair abstract reasoning. For 830 Caucasian families, the mean single parent–offspring correlation between parents (aged 35 to 55) and their children (aged 13 to 33) for this test was 0.26. This is lower than the median correlation of 0.42 reported in Figure 11.1. We can estimate the upper limit of heritability by doubling this correlation, thus estimating that narrow-sense heritability should be no greater than 0.52. Bear in mind that this estimate may include genetic variance due to assortative mating. As mentioned in Chapter 10, the best upper-limit heritability estimate is the regression of offspring on midparent (which is not affected by assortative mating). In the Hawaii family study of cognition, this regression was 0.52 (DeFries et al., 1979). The correspondence between this value and the estimate based on single parent–offspring correlations suggests that assortative mating did not affect the single-parent estimate.

Another measure of general cognitive ability from the Hawaii study was a composite score based on the common variance of 15 tests of specific cognitive abilities. This composite, called the first principal component, correlates highly with scores on a standard IQ test. The mean single parent–offspring correlation for this general cognitive factor in the Caucasian families was 0.35. The regression of offspring on midparent, which would probably be the best single estimate of the upper limit of narrow-sense heritability for general cognitive ability, was 0.60.

This same study also provides a good example of the possibility that such estimates can differ in various populations. In addition to the 830 families of European ancestry, the study also included 305 families of Japanese ancestry. For Raven's Progressive Matrices, the offspring–midparent regression was 0.24, compared to 0.52 in the Caucasian families. For the general composite, it was 0.42 rather than 0.60. Thus, there appears to be less familial resemblance (due to genetic or environmental influences) between parents and offspring in these families of Japanese ancestry than in the Caucasian families. Large sex-by-generation and ethnic group-by-generation interactions were found, suggesting the influence of cultural factors on these group differences in Hawaii (DeFries et al., 1982).

The estimated upper limit for narrow-sense heritability in the Hawaii study of cognition is about 0.60, when the estimate is based on midparent–offspring similarity in Caucasian families. For families of Japanese ancestry, it is lower. Both of these estimates are lower than the estimate based on the median correlation of 0.42 between parents and offspring in Figure 11.1.

The large-scale Hawaii study yielded a mean correlation of 0.31 for 455 pairs of Caucasian siblings. This correlation suggests that the upper limit of broad-sense heritability is 0.62, which is similar to the estimate based on parent–offspring regression. Even more interestingly, the correlation for 147 pairs of siblings of Japanese ancestry was 0.33. This suggests that genetic and environmental influences shared by siblings is about the same for Caucasian and Japanese groups, even though the familial influence shared by parents and offspring may be greater in Caucasian than in Japanese families.

Thus, both the parent–offspring and sibling data from the Hawaii family study of cognition point to less familiality than do the older data.

Subsequent chapters will show a similar pattern of results. For example, recent twin and adoption IQ data suggest a lower heritability than that suggested by older data (Plomin and DeFries, 1980). Although it is possible that environmental or genetic changes in the population underlie the differences between newer and older data, restriction of range in newer studies appears to be a factor (Caruso, 1983), and methodological variations may also be important. For example, some of the older studies tested families as a unit, so that any differences in test administration would increase familial correlations. In contrast, tests in the Hawaii family study were administered to many families at the same time in the same testing facility, using highly standardized procedures.

SCHIZOPHRENIA

At least 15 percent of the U.S. population is affected by mental disorders during any one year (Regier et al., 1978). One of the major kinds of severe mental disorders, called psychoses, is schizophrenia. Schizophrenia is characterized by long-term thought disorders, hallucinations, and disorganized speech. It has an incidence of about 1 percent in the general population.

Although early man viewed insanity in demonic terms, the familial nature of insanity was generally recognized by the sixteenth century. At the end of the nineteenth century, E. Kraepelin suggested two major types of insanity—schizophrenia and manic-depressive psychosis (depression alternating with manic elevations of mood). Family studies in the twentieth century have been important in specifying the extent of familial transmission, as well as in studying the heterogeneity of psychopathology. Although schizophrenic individuals are more likely than average to have schizophrenic relatives, they are no more likely to have manic-depressive relatives, suggesting that the familial causes of schizophrenia are distinct from those involved in manic-depressive disorder. The use of family studies to determine the heterogeneity of mental disorders has become even more important in recent years, and we shall consider these advances later.

Psychopathology introduces new problems for quantitative genetic analysis. Mental illness has not been studied in a quantitative way—that is, as the extreme of a normal distribution of mental functioning. "Either–or" sorts of data have been compiled in the past, although that is changing. Because of the qualitative classification of mental illnesses, the usual quantitative analyses are not appropriate. The incidence of psychopathology in relatives of an affected index case is compared to the incidence in the general population or to a control group. Familial resemblance is indicated when the incidence in the relatives is greater than the incidence in the general population. However, good incidence figures are not easy to obtain. Incidence is different from prevalence. Prevalence is the number of individuals in a population who are affected at a particular time, regardless of age. Incidence figures, on the other hand, may be viewed as a lifelong prevalence estimate. The age of risk for schizophrenia, for example, is about 15 to 45 years of age. Persons younger than 45 still have a chance of becoming schizophrenic. Also, older individuals may have been hospitalized for a mental disorder, but are no longer. Incidence figures take these cases into account. A special incidence figure used in family studies is called a *morbidity risk estimate*, which is an estimate of the risk of being affected. Because comparable estimates of population incidence relevant to a study's particular diagnosis are rarely available, control groups of individuals relatively free of psychiatric disorders are typically studied. However, a better approach is to select controls with a treated psychiatric condition that differs from that of the probands in order to control for referral biases and to explore the specificity of transmission of the disorder (Tsuang, Winokur, and Crowe, 1980).

Even though mental disorders are often viewed as discontinuous characters, they may well have a quantitative genetic basis. However, the discontinuous method of measurement (qualitative classification) makes it difficult to estimate heritability. We can, however, assume that an all-or-none disorder has an underlying continuous liability. Individuals having a liability above a certain threshold value are assumed to be affected. Those whose liability is below this threshold value are not affected. Liability is thus an unobserved construct that is presumed to be a continuous function of both genetic and environmental deviations. If there is such a threshold for a relatively rare character, concordance even for identical twins can be high only if heritability is very high (Smith, 1974). The heritability of the liability can be estimated, although several assumptions must be made (Falconer, 1965; Reich, Cloninger, and Guze, 1975).

A brief summary of family studies of schizophrenia follows in order to illustrate the usefulness of the family study method. Other aspects of psychopathology are reviewed in Chapter 14.

Although there is variability in strictness of diagnosis (Taylor and Abrams, 1978), the risk of schizophrenia in the general population is about 1 percent (Gottesman and Shields, 1982). The risks for various relatives of schizophrenics are summarized in Table 11.1. It is clear that schizophrenia runs in families. In 14 studies of over 8,000 schizophrenics, the risk for parents of schizophrenics is over five times that for the general population. It is likely that a lower rate for parents than for other first-degree relatives is found because schizophrenics are less likely to marry and those who do marry have relatively few children. Siblings of schizophenics provide the most stable data. They are approximately ten times more likely to become schizophrenic than an individual chosen randomly from the population. (See Table 11.1.) This rate increases to 16.7 percent for those siblings who have an affected parent as well as an affected sibling.

In seven studies of children of schizophrenics, their median risk was nearly 13 percent. It is interesting that the risk is no greater when the mother is schizophrenic than when the father is schizophrenic, which suggests that maternal influences are not critical. When both parents are schizophrenic, the risk is three times as great. The average weighted risk for first-degree relatives of schizophrenics is 8.4 percent, more than eight times the risk for individuals

Table 11.1
Median morbidity risk estimates for
the general population and for
relatives of schizophrenic index cases

	Percent
General population	1.0
First-degree relatives	
Parents of schizophrenics	5.6
Siblings of schizophrenics	10.1
Children of schizophrenics	12.8
Second-degree relatives	3.3
Third-degree relatives	2.4

SOURCE: After Gottesman and Shields, 1982.

chosen randomly from the population. One use of familial resemblance for schizophrenia is to select children at risk for schizophrenia in order to study the development of the disorder; this is called the *high-risk* design, described in Box 11.1. However, it is important to note that over 90 percent of schizophrenics do not have a schizophrenic first-degree relative.

As shown in Table 11.1, the risk for second-degree relatives of schizophrenics, such as grandchildren (who share about one-fourth of the grandparents' segregating genes), is about three times that for the general population. Third-degree relatives (who share about one-eighth of the segregating genes) are twice as likely to be schizophrenic, compared to the general population. Recent family studies yield similar results (Baron et al., 1985; Kendler, Gruenberg, and Tsuang, 1985).

The search for etiologically distinct syndromes has become the major focus of family studies in psychopathology. This work is multivariate in the sense that it asks whether the familial factors that affect one type of psychopathology also affect another. For example, as mentioned earlier, schizophrenic individuals are no more likely to have manic-depressive relatives, indicating that the familial origins of schizophrenia differ from the origins of manic-depressive disorder. However, attempts to identify etiologically distinct subtypes of schizophrenia on the basis of symptoms have not been successful. That is, the classical subtypes of schizophrenia—paranoia and catatonia—do not generally breed true in family studies (Farmer, McGuffin, and Gottesman, 1984). This is seen most dramatically in a follow-up of the famous Genain quadruplets, who were concordant for schizophrenia but showed quite variable symptoms (DeLisi et al., 1984). One approach to the problem of heterogeneity divides schizophrenics on the basis of a family history for schizophrenia (Murray, Lewis, and Reveley, 1985), an approach that has been useful for other disorders, such as alcoholism. However, there are problems with this approach (Eaves, Kendler, and Schulz, 1986). The new molecular genetic techniques described in Chapter 7 might be useful in this

Box 11.1

High-Risk Studies of Psychopathology

Psychiatric research during the past 20 years has extensively utilized the "high-risk" family study in which children whose biological parents had been diagnosed as mentally ill are studied longitudinally. The goal of this research is to follow the developmental course of psychopathology before the occurrence of overt disturbances to gain a better idea of the causes of psychopathology, as well as to find ways to intervene in the developmental process. It is quite possible, however, that the genes that affect adult disorders show no effects in childhood. For example, onset of schizophrenia typically occurs in early adulthood, and it is possible that genes involved in schizophrenia do not show their effects until late in adolescence.

Fifteen long-term studies of children at risk for schizophrenia have cooperated in the Risk Research Consortium, which includes 1,200 children with a least one schizophrenic parent and 1,400 normal control subjects (Watt et al., 1984). In early childhood, few differences are found between risk and control children, although by the middle childhood years more differences emerge. Although few solid conclusions can be drawn from this research, the two best documented findings are that high-risk subjects show impaired performance on complex measures of sustained attention and a lack of smooth eye movements during a visual pursuit task (Erlenmeyer-Kimling and Cornblatt, 1987).

Only about 13 percent of the children at risk are expected to become schizophrenic, which weakens the comparison between the children at risk and the control children. The oldest children in these studies are just now entering the age of onset for schizophrenia, so in the next few years it will be possible to determine which of the at-risk children are in fact schizophrenic; the earlier data for these children can then be studied in a more refined search for early indices of schizophrenia.

Over a dozen high-risk studies of children of patients with depression are also underway (Orvaschel, 1983). Preliminary reports indicate results similar to those for schizophrenia: Some differences can be found between children at risk and control children, but the differences are neither great nor consistent from study to study.

context if linkage were found for a particular type of schizophrenia. This approach could begin to break down the heterogeneity of schizophrenia on the basis of causes rather than symptoms. However, it should be noted that although most researchers anticipate progress based on splitting schizophrenia, some argue that we need to go in the opposite direction, lumping various schizophrenic disorders into a spectrum of schizoid disorders (Gottesman and Shields, 1982; Baron and Risch, 1987).

Another use of family studies is to test single-gene hypotheses, as described in Chapter 3. Extensive analyses of this type for schizophrenia have found little support for the hypothesis of a single major locus, and multifactorial polygenic models are generally accepted (Faraone and Tsuang, 1985).

Family studies have established that familial influences play a role in schizophrenia. Twin studies and adoption studies are needed to determine the relative roles of genetic and environmental factors. However, the family studies carry an important message. First-degree relatives are about 50 percent similar genetically, but fewer that 10 percent of the first-degree relatives of schizophrenics become schizophrenic. Thus, the results of family studies suggest that the development of schizophrenia may be substantially affected by nonshared environmental influences.

SUMMARY

Family studies determine whether or not there is familial resemblance for the behaviors being investigated. If familiality is observed, it may be due either to genetic or environmental influences. Family studies do not indicate whether observed familial resemblance is caused by shared environment or shared heredity, although they can provide an upper-limit estimate of heritability (or of between-family environmental influences). Substantial familiality has been found in family studies of IQ and schizophrenia. Twin and adoption studies — described in the next two chapters — are needed to determine the relative importance of genetic and environmental influences underlying familial resemblance.

CHAPTER·12

Twin Studies

The study of twins has a long history. In 1875 Francis Galton, the father of behavioral genetics, reported a study of life histories of two groups of twins. One group consisted of pairs who were similar at birth and the other group was dissimilar at birth. His idea was to assess the ability of the environment to make initially similar twins different and to make initially different twins similar. These comparisons led Galton to make the oft-quoted statement that "there is no escape from the conclusion that nature prevails enormously over nurture" (1876, p. 406). Galton was aware that there were two types of twins, one type that developed from a single fertilized egg (which we now call identical, or monozygotic) and another type that was derived from two separately fertilized eggs (called fraternal, or dizygotic). However, Galton did not suggest that evidence for genetic influence could be adduced by comparing identical and fraternal twins, the essence of the twin method. The twin method was not put to use until 50 years later (Merriman, 1924).

By the beginning of the twentieth century, most biologists were convinced of the existence of two types of human twins (Wilder, 1904). Nonetheless, E. L. Thorndike, who objectively measured specific cognitive abilities in twins in 1905, did not believe there were two types. He used twins to test environmental hypotheses, and found that younger and older twin pairs showed about the same degree of similarity for the cognitive measures. The first behavioral twin study to approach the modern method was conducted in 1924 by C. Merriman, who called the two types of twins "duplicates" and fraternals. He first showed that the similarity in IQ between the opposite-sex twins in his sample was the same as that between nontwin siblings (a correlation of 0.50). He then found that the same-sex twins had a much higher correlation for IQ (about 0.87). He concluded that the greater similarity in IQ between twin pairs in the group of same-sex twins was caused by the inclusion of all the "duplicate" twins in this group. Merriman then tried to separate the two types of same-sex twins using physical similarity as a criterion. This, as we shall see, is a reasonably accurate method for diagnosing zygosity (i.e., number of zygotes). Twenty-two same-sex pairs "resembled each other closely enough to frequently cause confusion of identity." The correlation on the Stanford-Binet IQ test for these twins was 0.99 in this small sample.

While most mammals have large litters, primates, including *Homo sapiens*, tend to have single offspring. However, all types of primates occasionally have multiple births. The embryology of twinning has been thoroughly analyzed for humans (Bulmer, 1970). In the case of single births, the embryo is suspended in a sac made of two membranes formed within the placenta. The innermost layer of the sac, the amnion, is surrounded by an outer layer, the chorion. Fraternal twins have completely separate chambers inside the womb; they develop from two different zygotes (dizygotic). When the embryos implant close to one another in the uterus, their placentas sometimes fuse, but they each still have a separate amnion and chorion. (See Figure 12.1.)

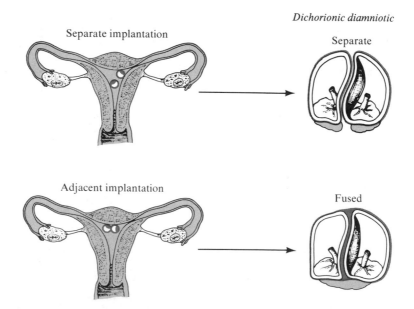

Figure 12.1
Fraternal twinning may result in either separate or fused placenta. (From "Twin placentation and some effects on twins of known zygosity" by G. Corney. In W. E. Nance, ed., *Twin Research, Part B: Biology and Epidemiology*, Copyright © 1978 by Alan R. Liss, Inc. All rights reserved.)

Identical twins result from a single zygote (monozygotic) that divides for unknown reasons. They share accommodations in the womb to some extent, depending on when the zygote splits. It used to be thought that identical twins always have the same chorion, but we now know that about a third of them have separate chorions and amnions, as do fraternal twins (Corney, 1978). These identical twins develop from zygotes that split before implantation, which occurs at about five days after fertilization. (See Figure 12.2.) The other two-thirds of identical twins develop from zygotes that split after implantation. These twins develop within the same chorion. When separation occurs five to ten days after fertilization, as it usually does, there are two amnions. About 4 percent separate after ten days, and these share their amnion as well

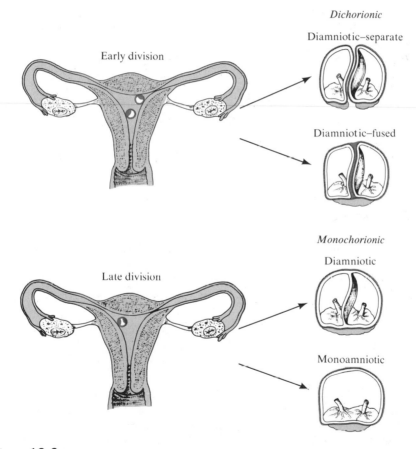

Figure 12.2
Identical twinning can result in separate or fused chorions and amnions, depending on the timing of the separation of the zygotes. (From "Twin placentation and some effects on twins of known zygosity" by G. Corney. In W. E. Nance, ed., *Twin Research, Part B: Biology and Epidemiology*, Copyright © 1978 by Alan R. Liss, Inc. All rights reserved.)

as chorion. Extremely late separation results in "Siamese" twins, individuals who are partially fused.

Twin births are not nearly as rare as most people would suppose. One in 83 deliveries in the United States is a twin birth. About 33,000 pairs are born each year in the United States, and about 2.4 million twin pairs are currently living in the United States. As many as 20 percent of fetuses may be twins, although both members survive in only about 5 percent of these cases. Many singleton births involve sole survivors of a twin pair; the unlucky member of such a pair is referred to as the "vanishing" twin.

The rate of twinning, especially fraternal twinning, is affected by maternal age and the number of previous offspring (Rao, 1978), as well as by fertility drugs. In a group of women who take fertility drugs, the rate of twinning can be as high as 50 percent (Wyshak, 1978). Also, in some countries, such as Nigeria, the rate of fraternal twinning is much higher than in the United States and Great Britain (Nylander, 1978). Finally, the tendency to produce fraternal twins may be inherited in some families (Nance et al., 1978), although heredity does not seem to be an important factor in identical twinning (Segreti, Winter, and Nance, 1978).

About one-third of the conceptions that produce twins result in opposite-sex fraternal twins, one-third in same-sex fraternal twins, and the remaining third in identical twins. After childhood, however, the proportion of identical twins is reduced to about one-quarter due to the decreased viability of this type of twin. Rates of identical twinning are remarkably similar for different maternal ages, but the fraternal-twinning rate fluctuates (Bulmer, 1970).

As discussed in the previous chapter, fraternal twins are 50 percent similar genetically on the average, like other first-degree relatives, while identical twins are identical genetically. By comparing the phenotypic similarity of identical twins with that of fraternal twins, we can conduct a natural experiment to investigate the effects of heredity and environment. Despite the twofold differences in genetic similarity, both types of twins share roughly the same environmental influences (an assumption that is discussed in detail later). If many pairs of twins are studied and identical twins are found to be no more similar than fraternal twins for a particular behavior, then genetic factors cannot be important in determining variability for that trait. In that case, the twofold greater genetic similarity of identical twins did not make them more similar phenotypically. On the other hand, if identical twins are significantly more similar than fraternal twins, the particular trait is influenced by genetic factors. Fraternal twins may be of the same sex or opposite sex, while identical twins are always of the same sex. Because of the sex chromosome differences between males and females, and because of the possibility that males and females are exposed to differing environments, twin studies are better natural experiments if we compare only same-sex fraternal twins and identical twins.

ZYGOSITY DETERMINATION

The twin method has often been subject to criticism. We shall discuss in detail two major issues: zygosity determination and the equal environments assumption. With respect to zygosity determination, a problem arose from the fact that some researchers made a subjective judgment of zygosity (whether the twins were identical or fraternal) and a subjective judgment of behavior. This method could have biased the results. It is now possible to perform highly accurate blood analyses by which the members of each pair of twins are compared for genetic markers in the blood. If any of these genetic markers are different, the twins must be fraternal; if all the markers are the same, the twins are identical. In addition, any errors in diagnosing the zygosity of twins will lower estimates of genetic influence based on the comparison of identical and fraternal twin correlations, because they will raise the fraternal twin correlation and lower the correlation for identical twins. Any polymorphic genetic markers could be used for the diagnosis of zygosity. Studies have traditionally used blood type systems, serum group systems, and red cell enzyme systems, but molecular genetic markers described in Chapter 7 could also be used. A new molecular genetic method involves a probe for a tandem repeat region of DNA, which occurs in several places in the genome; the number of repeats is highly variable for each region. (See Chapter 7 for a description of variable number tandem repeats.) When cut with a restriction enzyme and submitted to Southern blotting, DNA for each person yields a distinct DNA "fingerprint" of bands with different numbers of tandem repeats (Jeffreys, Wilson, and Thein, 1985). The pattern of bands is similar for genetically related individuals, but only identical twins have exactly the same DNA fingerprints, and this method thus provides an excellent test of zygosity without the need for analysis of many individual genetic markers.

We shall consider some new developments in diagnosis in detail because they have important implications for the ease of conducting twin research. As early as 1927, physical traits (such as hair and eye color) were used along with minimal blood analyses to diagnose zygosity (Siemens, 1927). By 1955 the analysis of blood groups began to be accepted as the best method (Smith and Penrose, 1955). However, as early as 1941, diagnoses based on blood markers and physical similarity were shown to be highly related (Essen-Möller, 1941).

In 1959 a detailed analysis of this relationship was reported by T. Husén in Sweden. At that time, blood analysis was much less extensive than it is now, and Husén was able to study only four markers. One of these was the ABO system. Identical twins, of course, must have the same blood type. However, about two-thirds of fraternal twin pairs can also be expected to have the same variant by chance. For this reason, Husén could achieve only 60 percent accuracy in classifying twins as identical if they had the same variant for the ABO blood system. However, the probability that a pair of fraternal twins will have the same variant for two blood systems is much less. For example, the accuracy of diagnosis goes up to 0.72 when we consider both the ABO and MNS blood group systems. With all four blood groups that Husén studied, the accuracy of diagnosis was 90 percent.

Husén found that evaluation of physical similarity is a good way of diagnosing zygosity. Even a subjective rating of similarity predicted well. For 25 pairs of identical twins, only five were rated as showing some "striking dissimilarities." Only one fraternal twin pair was rated "very similar." As a general rule, identical twins tend to be misclassified as fraternal twins more often than the opposite. Husén also found that fraternal twins are seldom mistaken for one another by family members and friends, while identical twins often are. However, just asking twins whether they are identical or fraternal is not a very accurate method (Husén, 1959; Carter-Saltzman and Scarr, 1977).

Even greater precision in diagnosing zygosity can be obtained by using specific physical characteristics, such as eye and hair color, in the same way that blood markers are used. For example, Husén found that 92 percent of identical twins had exactly the same eye color, as compared to 44 percent of fraternal twins. The comparisons for some other characteristics were 92 percent versus 38 percent for hair color; 96 percent versus 55 percent for

Table 12.1

Physical similarity and twin zygosity for children

	Percent of twins "exactly similar" (or "yes" responses by mothers)	
	Identical	Fraternal
"Is it hard for strangers to tell them apart?" (Asked of mothers.)	100	8
Eye color	100	30
Hair color	100	10
Facial appearance	49	0
Complexion	99	14
Weight	46	6
"Do they look alike as two peas in a pod?"	48	0
"Does either mother or father ever confuse them?"	79	1
"Are they sometimes confused by other people in family?"	93	1
Height	56	13
Number of pairs	181	84

SOURCE: Adapted from D. J. Cohen et al., "Reliably separating identical from fraternal twins," *Archives of General Psychiatry*, 1975, 32, 1373–1374.

complexion; 96 percent versus 16 percent for hairline pattern; and 93 percent versus 11 percent for position of teeth. For ten such characteristics, over 90 percent accuracy of diagnosis could be achieved.

These results are in accord with more recent analyses of data on adult (Cederlöf et al., 1961; Sarna et al., 1978; Magnus, Berg, and Nance, 1983) and adolescent (Nichols and Bilbro, 1966) twins, as well as data obtained when children were rated for similarity by their parents (D. J. Cohen, et al., 1973, 1975). The results of the study of children are described in Table 12.1, which lists ten physical similarity variables that were evaluated by mothers of identical and fraternal twins. All of these studies found an accuracy of over 90 percent in diagnosing zygosity from physical similarity. Because of the difficulty and expense of obtaining blood samples, particularly from young children, it is indeed fortunate that physical similarity can be used to diagnose zygosity.

EQUAL ENVIRONMENTS ASSUMPTION

Another issue that arose with respect to the twin method involved the equal environments assumption. The essence of the twin method is to compare the phenotypic similarities of identical twins and fraternal twins, and to be able to ascribe a finding of greater identical-twin similarity to their twofold greater genetic similarity. The method assumes that the degree of environmental similarity is about the same for the two types of twins. If the equal environments assumption is not correct (for example, if identical twins are treated more similarly than fraternal twins), then a finding of greater phenotypic similarity between identical twins might be due partially to greater environmental similarity. It could not be attributed to greater genetic similarity alone.

On the face of it, the equal environments assumption seems quite reasonable for many important variables. Both types of twins share the same womb at the same time. Both are reared in the same family at the same time. They are the same age and the same sex (if opposite-sex fraternal twins are excluded from the sample). Nonetheless, the major criticism of the twin method is that identical twins may be treated more similarly than fraternal twins. It should be noted that one could also argue that identical twins may be treated *less* similarly than fraternal twins, which would result in underestimates of genetic influence. For example, identical twins are so similar in appearance that their parents may accentuate slight behavioral differences. Also, the twins themselves may behave differently from one another in an attempt to forge separate identities.

Speculations such as these have been much more common than research designed to explore the issue. Fortunately, some empirical studies have been conducted concerning the question of the equal environments assumption. Such studies have been of three major types: the effects of labeling, direct assessments of environmental differences for identical and fraternal twins, and tests to determine whether such differential environments affect behavior.

The effect of labeling a twin pair as identical or fraternal has been studied using twins who were misclassified by their parents or by themselves. For example, when parents think that their twins are fraternal but they really are identical, will the mislabeled identical twins be as similar behaviorally as correctly labeled identical twins? Two studies (Scarr, 1968; Scarr and Carter-Saltzman, 1979) have suggested that the influence of such labeling is minimal. Twins whose zygosity is mistaken by their parents or by the twins themselves behave according to their true zygosity. For example, in a study of 400 pairs of adolescent twins in Philadelphia, the twins were simply asked if they were identical or fraternal; 40 percent were wrong about their zygosity. Table 12.2 presents the differences in performance on cognitive and personality tests between correctly and incorrectly classified twin pairs. Although the fraternal twins showed greater differences than the identical, the identical twins who thought that they were fraternal had only slightly greater differences than those who thought they were identical. The fraternal twins who thought that they were identical were no more similar than those who correctly thought that they were fraternal. We can conclude from this and an earlier study (Scarr, 1968) that labeling twins as identical or fraternal has little effect on their behavioral similarity.

The second approach to the question of the equal environments assumption involves measuring aspects of the environments of identical and fraternal twins to determine whether identical twins, in fact, experience more

Table 12.2

The effects of zygosity labeling: average absolute differences on cognitive and personality tests for identical and fraternal twins who were right and for those who were wrong about their zygosity

	Cognitive tests		Personality tests	
	Average difference within twin pairs	Number of pairs	Average difference within twin pairs	Number of pairs
Identical twins who were correct about their zygosity	0.66	89	0.81	98
Identical twins who were incorrect about their zygosity*	0.73	61	0.85	68
Fraternal twins who were correct about their zygosity	0.81	84	0.93	101
Fraternal twins who were incorrect about their zygosity*	0.81	49	0.99	64

*Includes pairs in which one or both twins were in error concerning their zygosity.
SOURCE: After Scarr and Carter-Saltzman, 1979.

similar environments. Some differences can be found (Vandenberg, 1976). For example, identical twins tend to be treated more similarly in terms of clothes, they study together more, and they share more friends (Wilson, 1934; Lehtovaara, 1938; Zazzo, 1960; Smith, 1965; Loehlin and Nichols, 1976). Evidence such as this has often been used to reject the twin method as biased. However, it is a mistake to assume that environmental differences such as these make a difference behaviorally. This approach does not go far enough. For example, the question can be asked whether parents create or respond to differences in their twins. This is the title of an article that reports an interesting observational study of parents' interactions with their twin children (Lytton, Martin, and Eaves, 1977). The answer to the question is that when parents treat identical twins more similarly than they would fraternal twins, parents are responding to differences in their children's behavior.

The third approach asks the more appropriate question: "Do observed differences in the environment of the two types of twins make a difference behaviorally?" The method was first suggested by John Loehlin and Robert Nichols (1976), who reported data from a large twin study of personality. The study included about 850 pairs of twins from a group of high school students who participated in the National Merit Scholarship Qualifying Tests in 1962 and 1965. The parents of the twins were asked to rate five environmental variables that had been found to differ for identical and fraternal twins. The results are presented in Table 12.3. Consistent with previous studies, the identical twins experienced slightly more similar environments than did the fraternal.

The important question is whether these environmental differences affect behavior. Because identical twins are identical genetically, differences within pairs of identical twins can be caused only by environmental factors. Some identical twin pairs are more similar behaviorally than others; and

Table 12.3

Mean scores for identical and fraternal twins on five items and their composite score concerning differential experience

Item	Score range	Identical		Fraternal	
		M	F	M	F
Dressed alike	1–3	1.7	1.5	2.0	1.8
Played together (age 6–12 years)	1–4	1.3	1.3	1.6	1.6
Spent time together (age 12–18 years)	1–4	1.8	1.7	2.2	2.0
Slept in same room	1–4	1.7	1.7	1.8	1.8
Parents tried to treat alike	1–5	1.9	2.0	2.3	2.2
Composite of above	6–23	9.7	9.6	11.6	10.8
Number of pairs		217	137	297	199

NOTE: A score of 1.0 indicates maximum similarity; larger numbers indicate less similarity.
SOURCE: After Loehlin and Nichols, 1976.

some identical twin pairs are exposed to more similar environments than others. Loehlin and Nichols tested the adequacy of the equal environments assumption by correlating identical twin differences for cognition and personality with the measures of similarity of environment. If similarity of environment makes a difference behaviorally, then identical twins whose environments are more similar should be more similar behaviorally. Only identical twins were included in this analysis because observed differences between them are clearly environmental. Fraternal twins, on the other hand, differ genetically as well as environmentally, so the investigators could not have been sure whether the genetic differences led to the environmental differences, or vice versa. Table 12.4 presents the correlations between behavioral differences and a composite index of differential environments. A positive correlation means that identical twins who were exposed to more similar environments were more similar in behavior. The low correlations provide strong support for the equal environments assumption of the twin method. Environmental variables that were unequal for identical and fraternal twins simply did not make a difference in cognition, personality, vocational interests, or interpersonal relationships.

What about the greater similarity of appearance of identical twins? The method used by Loehlin and Nichols can also be applied to this question. Although identical twins are usually quite similar in appearance, they vary in the amount they are mistaken for one another, as indicated in Table 12.1 (Cohen et al., 1975). For example, only 48 percent of the mothers of identical twins said that the twins "look alike as two peas in a pod." Thus, we can ask

Table 12.4

Test of the equal environments assumption: correlations between absolute differences for identical twins on behavioral measures and a composite measure of differential experience

	Correlation*	Number of pairs
Cognitive (average of 5 NMSQT subtests)	−0.06	276
Personality (average of 18 CPI scales)	0.06	451
Vocational interests (average of 12 VPI scales)	0.01	276
Interpersonal relationships (average of 6 types of relationships)	0.05	276

NOTE: NMSQT is the National Merit Scholarship Qualifying Tests; CPI is the California Psychological Inventory; and VPI is the Vocational Preference Inventory.
  *A positive correlation means that identical twins who were treated more similarly were more similar behaviorally.
SOURCE: After Loehlin and Nichols, 1976.

Table 12.5
Test of the equal environments assumption: rank order
correlations between absolute differences for identical twins
on behavioral measures and physical similarity

	Correlation*	Number of pairs
Stanford-Binet IQ test	0.05	47
WISC IQ test	−0.19	74
Personality test (average of 16 PF)	0.00	51

NOTE: 16 PF is Raymond Cattell's 16 Personality Factors test.
   *A positive correlation means that identical twins who were more similar
physically were more similar behaviorally.
SOURCE: Matheny, Wilson, and Dolan, 1976.

whether twins who are more easily mistaken for one another are more similar
in behavior (Matheny, Wilson, and Dolan, 1976; Plomin, Willerman, and
Loehlin, 1976). Matheny and associates compared absolute differences in IQ
scores and personality measures with ratings of physical similarity for blood-
diagnosed identical twins between the ages of three and thirteen. Their finding
of no significant correlations between behavioral differences and physical
similarity scores (see Table 12.5) suggests that the greater physical similarity
of identical than fraternal twins does not seriously affect the equal environments assumption of the twin method.

   Data such as these strongly support the reasonableness of the equal
environments assumption. Although other differences between the environments of identical and fraternal twins may yet be found, these results suggest
that such differential experiences will not necessarily affect the behavioral
similarity of the two types of twins.

   Another issue related to the equal environments assumption is that
fraternal twins tend to volunteer less often for twin studies than do identical
twins, perhaps because fraternal twins feel less "twinlike" than identical twins
(Lykken, McGue, and Tellegen, 1987). For this reason and others, twin
analyses should check that means and variances of identical and fraternal
twins are not significantly different.

REPRESENTATIVENESS

As in any experiment, generalizability is an issue. Are twins representative of
the general population? One way in which twins are different from the
nontwin population is that twins are three to four weeks premature on
average, compared to singletons, and they are about 30 percent lighter and 17
percent shorter at birth, a difference that disappears by middle childhood

(MacGillivray, Nylander, and Corney, 1975). In childhood, twins perform somewhat more poorly than singletons on verbal tests (Hay and O'Brien, 1983), perhaps because they retard each other's language learning (Savic, 1980). It appears that most of this deficit is recovered early in the school years (Wilson, 1983). If twins are different in means or variances from the population, the results of twin analyses might not completely generalize to the population at large. This is an empirical issue that should be examined before twin results are presumed to generalize to the population.

HERITABILITY

Some researchers prefer to test the significance of the difference between correlations for identical and fraternal twins and to leave it at that. We know, for example, that identical twins are significantly more similar in height and weight and in the total ridge counts of their fingerprints. Thus, genetic differences are implicated in the etiology of these traits.

It is useful in many contexts to go beyond this statement of statistical significance to estimate broad-sense heritability, although the cautions expressed in Chapter 9 must be kept in mind. Even if one chooses not to estimate heritability, it is important to know whether the pattern of correlations for identical and fraternal twins is reasonable, regardless of the significance of the difference between the correlations. For example, a correlation of 0.90 for identical twins and a correlation of 0.10 for fraternal twins is not consistent with an additive genetic hypothesis, because identical twins are only twice as similar in terms of additive genetic variance as fraternal twins. Twin studies frequently yield reasonable patterns of correlations. For example, in Table 12.6 we list the median correlations for height, weight, and total ridge count of fingerprints for adult twins (from Mittler, 1971). These data also agree with other familial information. For example, for total ridge count,

Table 12.6
Median correlations for identical and fraternal twins for height, weight, and total ridge count of fingerprints

	Correlation	
	Identical	Fraternal
Height	0.93	0.48
Weight	0.91	0.58
Total ridge count	0.96	0.49

SOURCE: After Mittler, 1971.

the sibling correlation is 0.51 and the single parent–offspring correlation is 0.46.

The components of covariance for various familial relationships were described in Chapter 10. Fraternal (DZ) twins have the same genetic components of covariance as full siblings (see Tables 9.9 and 10.1):

$$\text{Cov}_{\text{DZ}} = \tfrac{1}{2}V_A + \tfrac{1}{4}V_D + V_{E_{c(\text{DZ})}}$$

Because fraternal twins also share familial environment, we have referred to a component of common environment salient to fraternal twins (which may be greater than V_{E_c} for nontwin siblings). This is only a slightly more precise way of stating that the phenotypic covariance for fraternal twins can include about half of the genetic variance and some environmental variance. Identical (MZ) twins, on the other hand, covary completely genetically and also share a familial environment:

$$\text{Cov}_{\text{MZ}} = V_A + V_D + V_{E_{c(\text{MZ})}}$$

Thus, the difference between the phenotypic covariances of identical and fraternal twins is

$$\text{Cov}_{\text{MZ}} - \text{Cov}_{\text{DZ}} = \tfrac{1}{2}V_A + \tfrac{3}{4}V_D$$

assuming that the common environments are about the same for the two types of twins.

As described in Chapter 9, most researchers prefer to work with standardized covariances, known as correlations. A phenotypic correlation is a covariance divided by a variance. Thus, the fraternal twin correlation reflects the following components:

$$r_{\text{DZ}} = \frac{\tfrac{1}{2}V_A + \tfrac{1}{4}V_D + V_{E_{c(\text{DZ})}}}{V_P}$$

The identical twin correlation is represented as follows:

$$r_{\text{MZ}} = \frac{V_A + V_D + V_{E_{c(\text{MZ})}}}{V_P}$$

The difference between these two expressions is

$$r_{\text{MZ}} - r_{\text{DZ}} = \frac{\tfrac{1}{2}V_A + \tfrac{3}{4}V_D}{V_P}$$

Doubling this difference yields

$$2(r_{\text{MZ}} - r_{\text{DZ}}) = \frac{V_A + \tfrac{3}{2}V_D}{V_P}$$

As described in Chapter 9, broad-sense heritability is the proportion of phenotypic variance accounted for by additive and nonadditive genetic variance:

$$h_B^2 = \frac{V_A + V_D}{V_P}$$

Thus, doubling the difference between the identical and fraternal twin phenotypic correlations can somewhat overestimate broad-sense heritability because this estimate contains 1.5 times the dominance variance. Although we have not expressed epistatic variance, it, too, would lead to an overestimate of heritability. Another possible complexity is that assortative mating by parents will increase additive genetic variance shared by fraternal, but not by identical, twins. A larger additive genetic variance for fraternal twins will increase their phenotypic correlation and reduce the difference between the correlations for fraternal and identical twins. Thus, assortative mating can result in an underestimate of broad-sense heritability, although there are ways to adjust twin heritability estimates for the effect of assortative mating (Jensen, 1967). Yet another complexity is the possibility of genotype–environment interaction and correlation, as described in Chapter 9. Although there has been considerable research and theorizing about such problems, we can hold them in abeyance for this introduction because they are likely to have only small and counterbalancing effects on estimates of heritability from twin studies.

Doubling the difference between identical and fraternal twin correlations has been called Falconer's (1960, 1981) estimate of broad-sense heritability. Analyses of the correlations in Table 12.6 suggest that the broad-sense heritability is 0.90—i.e., 2(0.93 − 0.48) = 0.90—for height, 0.66 for weight, and 0.94 for total ridge count. As explained in Chapter 9, this is a shorthand way of saying that, of the individual differences in height for these samples and at this particular time, about 90 percent of the variation in height is due to genetic differences among people and about 10 percent is due to environmental differences.

Model-fitting procedures are increasingly used to analyze data from twin studies (e.g., Fulker, 1981). Model fitting for twin data was embedded in the example of model fitting presented in Chapter 9. (See Figure 9.14 and Tables 9.11, 9.12, and 9.13.) A recent issue of the journal *Behavior Genetics* is devoted to model-fitting analyses of twin data that test complex models that include, for example, reciprocal effects of sibling interactions, gender, and environmental exposure (Boomsma, Martin, and Neale, 1989). A simpler model-fitting approach that is easy to use for twin data and other simple designs involves linear regression rather than maximum-likelihood analyses, as explained in Box 12.1.

Box 12.1

Multiple Regression Model Fitting for Twin Data

A model-fitting approach that uses linear regression can be applied to twin data and other simple behavioral genetic designs (DeFries and Fulker, 1985). It begins with the regression of one twin's score on the other twin's score. However, the technique involves an analysis called *multiple regression* because one twin's score is predicted from other things in addition to the co-twin's score: the coefficient of relationship (1.0 for identical twins and 0.5 for fraternal twins) and a "dummy" interaction term that is the product of the twin's score and the coefficient of relationship. In multiple regression, the regression of the predicted variable on each predictor is calculated independently from each of the other predictors. This independent regression is called a partial regression because the effects of the other predictors are partialled out.

The point of all of this is that the partial regression for the interaction term in this multiple regression model indicates whether twin resemblance differs as a function of the coefficient of relationship, i.e., zygosity. In other words, a significant partial regression for the interaction indicates that twin resemblance differs significantly for identical and fraternal twins, which thus tests for genetic influence. In fact, this partial regression coefficient provides a direct estimate of h^2. In addition, in this multiple regression model, the regression of one twin's score on the other twin's score is also a partial regression in which the effects of the other predictors are partialled out. In other words, this partial regression represents twin resemblance independent of genetic influence. Thus, this partial regression estimates shared environmental influence.

The advantages of this multiple regression model-fitting analysis include the following: A model is tested, standard errors of estimates are provided, and additional parameters can easily be tested, such as age and gender (Ho, Foch, and Plomin, 1980). The approach can also be extended to include data other than twins (Zielenewski et al., 1987).

This multiple regression approach is especially useful in addressing one of the core issues in psychopathology: the relationship between the normal and the abnormal. When quantitative data are obtained, this multiple regression procedure can be used to assess the extent to which genetic and environmental factors that affect the disorder are the same as those that affect normal variability. In other words, the procedure tests whether the disorder is merely the extreme of a normal distribution of variability (DeFries and Fulker, in press).

ENVIRONMENTALITY

The twin method can also be described graphically (Nichols, 1965), as in Figure 12.3, which emphasizes the assumptions of the method. The rectangle represents the phenotypic variance of individuals for a particular trait. The first assumption is that the total variance of the identical and fraternal twins is

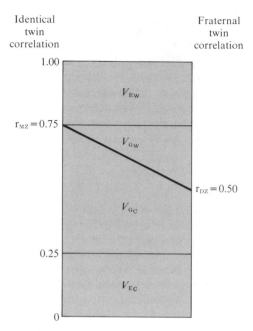

Identical twin correlation

Fraternal twin correlation

1.00

$V_{\mathrm{E_W}}$

$r_{\mathrm{MZ}} = 0.75$

$V_{\mathrm{G_W}}$

$r_{\mathrm{DZ}} = 0.50$

$V_{\mathrm{G_C}}$

0.25

$V_{\mathrm{E_C}}$

0

Figure 12.3
Pictorial representation of the twin method and its assumptions. (Adapted from Nichols, 1965.)

the same and that it does not differ from the variance in the general population to which the results will be generalized. Twin studies usually satisfy this assumption. Even in the few instances where there is a mean difference between twins and singletons in the population as a whole (one example is an average twin IQ of about 96), the variance of the two types of twins and the general population is the same. The variance in the rectangle has been broken down into environmental and genetic variance components using an example in which the correlation is 0.75 for identical twins and 0.50 for fraternal twins. On each side of the rectangle, the variance above the correlation represents differences within pairs, while the variance below the correlation stands for variance shared by members of a pair. Examination of the left side of the rectangle shows that identical twins share 75 percent of the variance, which also means that 25 percent of the variance is not shared. The differences within pairs of identical twins can only be environmental (except for somatic mutations and possible cytoplasmic differences at the time of separation). Such environmental influences are called nonshared, or within-family, environmental influences—in this case, they are within-pair environmental differences ($V_{\mathrm{E_w}}$). One assumption of the twin method is that these differences are the same for identical and fraternal twins (the "equal environments" assumption discussed above).

The right side of the rectangle shows that the difference between the correlation for identical and fraternal twins is genetic. This variance is sometimes referred to as within-family genetic variance ($V_{\mathrm{G_w}}$) because it is due to genetic influences that cause family members (in this case, fraternal twins) to differ genetically. Comparing the two sides of the rectangle reveals that the genetic covariance (V_{G_c}) for fraternal twins is half that for identical twins. Together, V_{G_w} and V_{G_c} types of genetic variance represent the extent to which phenotypic variance is due to genetic variance (i.e., $V_G = V_{G_c} + V_{G_w}$). In this example, the total genetic variance is 50 percent—double the difference between the correlations for identical and fraternal twins. Of course, twins and other family members can be similar environmentally as well as genetically. This is represented by the component labeled V_{E_c} which is calculated as $r_{MZ} - h_B^2$. Shared, or between-family, environmental influences were defined in Chapter 9 as environmental influences that make family members similar to one another and different from other families. We need to keep in mind that between-family environmental influence may be greater for twins than for nontwin siblings.

In this example, genetic variance accounts for half of the phenotypic variance. (Broad-sense heritability is 0.50.) Environmentality, the phenotypic variance due to environmental differences, is also 0.50. Half of this environmental variance, in turn, is due to differences within families and half to differences between families. Environmental variance is not always distributed evenly within and between families. As we shall see in the following section, some of the environmental variance relevant to cognition in childhood occurs between families. For psychopathology (and personality in general), however, environmental variance is due primarily to within-family influences.

OTHER TWIN DESIGNS

Three other twin study designs deserve attention: studies of families of identical twins, identical co-twin control studies, and studies of genetic and phenotypic similarity within pairs of fraternal twins. The study of adopted identical and fraternal twins is particularly powerful, and will be described in our discussion of adoption studies in Chapter 13.

Families of Identical Twins

An interesting set of family relationships revealed in families of identical twins has recently been subjected to behavioral genetics analysis. This type of analysis is called the families-of-identical-twins method (or the monozygotic half-siblings method) (Corey and Nance, 1978; Nance and Corey, 1976; Nance, 1976). Figure 12.4 illustrates the family configurations that occur when both members of an adult identical twin pair marry and produce offspring. These two families include relatives that share all of their genes (relationship A in Figure 12.4); half of their genes (relationships like B, C, and E); one-fourth of their genes (relationship F); and none of their genes (relationship D, if there is no assortative mating). This type of family study compares parents and offspring, as well as siblings. As in other family studies,

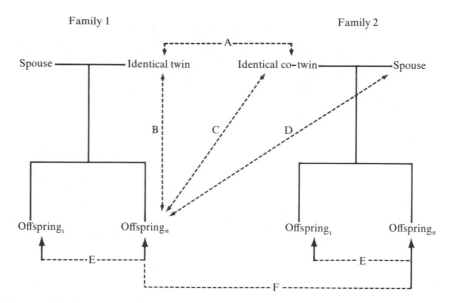

Figure 12.4
The families-of-identical-twins method. See text for explanation.

both heredity and family environment are shared in these relationships. The unique feature of this methodology involves the comparisons across the two families. In the families of male identical twins, for example, uncles and nephews/nieces (relationship C) are just as similar genetically as fathers and sons/daughters (relationship B). Also, cousins (relationship F) are half-siblings genetically because they have, in effect, the same parent from a genetic standpoint. These half-siblings are particularly interesting because they are reared in different families, thus avoiding the problems of other studies where half-siblings are separated by divorce, death, or desertion of one of their parents. The families-of-identical-twins method provides unique information concerning assortative mating, because genetically identical individuals have married different spouses.

These comparisons can be fitted to a model to ascertain genetic and environmental parameters (Nance, 1976). Although the model is complicated mathematically, the idea is simple. The most critical comparison is between the identical twin parents and their children versus their nephews/nieces. The similarity between identical twin parents and their children can be compared to the similarity between the same individuals and their nieces/nephews. Genetically, the degree of similarity is the same. But because the nieces/nephews are reared in a somewhat different home environment, fewer environmental influences are shared between the identical twin uncles and their nephews/nieces than between the identical twin fathers and their children. The comparison between the siblings (relationship E) and the half-siblings (relationship F) is also interesting, although the half-siblings are less similar than the full siblings environmentally, as well as genetically, because they are reared in different families. Another useful feature of this method is the opportunity to compare families of female identical twins and families of male identical twins to detect maternal effects and sex linkage. If there are maternal effects, half-siblings who are children of female identical twins should be more similar than half-siblings produced by male twins. For sex-linked characters, paternal half-brothers of male identical twins should be less alike than maternal half-brothers of female identical twins. Also, maternal half-sisters should be less similar than paternal half-sisters.

The basic problem with the families-of-identical-twins method is that the environments of the two families will be similar to the extent that genetically identical parents set up similar home environments. For example, suppose that the homes of the identical twin parents are essentially the same. Then identical twin parents and their nephews/nieces would be just as similar genetically and environmentally as parents and their offspring. It is possible to use path analyses to clarify not only this problem, but also the possible effects of assortative mating.

Studies have begun to apply the families-of-identical-twins method to behavioral traits. The earliest reports considered specific cognitive abilities, such as spatial ability (Rose et al., 1979a) and perceptual speed (Rose et al., 1979b). A recent study of the families of 74 pairs of identical twins found no evidence for genetic influence on an experimental measure of spatial ability called the rod-and-frame test (Tambs, 1987). Model-fitting approaches attempt to clarify this problem as well as the effects of assortative mating (Rose, Miller, and Fulker, 1981). However, data simulations suggest that the families-of-identical-twins design is not as powerful as some other behavioral genetic designs (Heath et al., 1985).

Identical Co-twin Control Studies

Co-twin control studies test the ability of environmental variables to alter phenotypes by exposing the members of identical twin pairs (co-twins) to differential environmental influences. The few (less than a dozen) co-twin control studies in the behavioral genetics literature fall into three categories.

The earliest and largest category includes studies of the interaction between maturation and training. The same pair of identical twin girls at different ages was studied by several investigators (Gesell and Thompson, 1929; Strayer, 1930; and Hilgard, 1933). They trained one co-twin and then the other in the same skills, with the major emphasis on the age at which each co-twin was trained. Another co-twin control study of this variety involved reading ability (Fowler, 1965).

A second category compares the effectiveness of different training programs. For example, Vandenberg (1968) reported a study in Stockholm that separated identical twins during their first years in elementary school to test two different instructional programs. The third, and most general, category includes studies of the extent to which environmental influences can produce differences in genetically identical individuals. Two Russian studies (Mirenva, 1935; Levit, reported by Newman, Freeman, and Holzinger, 1937) applied the co-twin method in this way to the study of motoric training. Other studies have applied the method to reading and numerical concepts (Vandenberg, Stafford, and Brown, 1968) and to personality (Plomin and Willerman, 1975).

The co-twin method provides a unique opportunity to control genetic variability in certain kinds of research in order to study environmental influence. The usual experimental method randomizes genetic differences, tests large numbers of subjects, and reports mean differences between the groups. In instances where the intensity and the time requirements of an experimental procedure preclude large numbers of subjects, the co-twin method may be a most useful alternative. A variant of the co-twin method is to study identical twins who are very different for a trait in order to identify nonshared environmental factors responsible for the difference. Because identical twins do not differ genetically, any behavioral differences between identical twin partners must be due to nonshared environmental influences. A study of identical twins discordant for schizophrenia is described later.

Studies of Genetic Similarity Within Pairs of Fraternal Twins

Even though fraternal twins average half of their segregating genes in common, genetic segregation assures that some fraternal twin pairs are more genetically similar than others, as is the case with nontwin siblings. Attempts have been made to use this fact to test the relationship between genetic similarity, determined by blood group analyses, and phenotypic similarity. If genes influence a particular behavior, then fraternal twins who are more similar genetically should be more similar phenotypically. The method has been applied to cognitive and personality traits (Carter-Saltzman and Scarr, 1975). However, the usefulness of the method is doubtful because the vast majority of fraternal twins fall between 45 percent and 55 percent genetic resemblance and it is difficult to detect associations between behavioral resemblance and genetic resemblance that varies by only a few percent (Loehlin, Willerman, and Vandenberg, 1974).

TWIN STUDIES AND BEHAVIOR

The twin method has been applied to a wide range of behaviors. Many of these studies have been reviewed by Fuller and Thompson (1978), Mittler (1971), and Vandenberg (1976). In the following sections, we shall continue with our examples of IQ and schizophrenia. Twin research on other behaviors is presented in Chapter 14.

We should note in passing that twins are not as difficult to obtain for research as one might imagine. For example, there are several large registers of twins. In the United States, there is a register of 16,000 pairs maintained by the National Research Council of the National Academy of Sciences (Hrubec and Neel, 1978), and another register, called the Kaiser-Permanente Twin Registry, begun in California in 1974 (Friedman and Lewis, 1978). There are also large registers in Finland (Kaprio et al., 1978), Sweden (Cederlöf and Lorich, 1978), and Norway (Kringlen, 1978). In addition, there are many mothers of twins clubs in the United States, and many of the clubs are interested in participating in research on twins.

Intelligence Quotient (IQ)

The first behavioral genetics twin study focused on IQ (Merriman, 1924). Since then, IQ tests have been administered to thousands of twins in studies of the heritability of general intelligence. Merriman also first noted the slightly lower average IQ for twins—about 96, compared to the population average of 100. Although the reason for this is not clear, it does not greatly affect the results of twin studies because the variance of twins is the same as that of the general population.

Table 12.7 lists 19 twin studies of general cognitive ability (Loehlin and Nichols, 1976). These are studies in which more than 25 pairs of each twin type were tested. They include 3,454 pairs of identical and 2,885 pairs of fraternal twins. The correlations in all studies except one indicate substantially greater similarity for identical twins.

The median correlation for all identical twins in these studies is 0.86. The median correlation for all fraternal twins is 0.62. A summary of the world's literature on IQ yielded very similar correlations: The median IQ correlation for 4,672 pairs of identical twins is 0.86 and the correlation for 5,546 pairs of fraternal twins is 0.60 (Bouchard and McGue, 1981). Falconer's formula for estimating heritability by doubling the difference between these correlations suggests a broad-sense heritability of about 0.50 for general cognitive ability. Although this estimate is not adjusted for assortative mating or nonadditive genetic variance, such adjustments would not alter the conclusion that roughly half of observed variation in general cognitive ability is due to genetic differences.

If about half of the phenotypic variation in IQ is due to genetic differences, then about half must be caused by all other sources of variance. Figure 12.5 shows these median correlations for general cognitive ability in terms of the diagram with which we described the twin method. It shows that about half of the phenotypic variance is genetic in origin. Also, most of the environmental variance operates between families. In other words, twin studies suggest that shared (common) familial influences are the major environmental sources of individual differences in general cognitive ability. However, the correlation for nontwin siblings is lower than the correlation for fraternal twins. (See Chapter 11.) This means that twin studies overestimate the role of

Table 12.7

Identical and fraternal twin correlations for measures of general cognitive ability

Test	Correlation		Number of pairs		Source
	Identical	Fraternal	Identical	Fraternal	
National and Multi-Mental	0.85	0.26	45	57	Wingfield and Sandiford (1928)
Otis	0.84	0.47	65	96	Herrman and Hogben (1933)
Binet	0.88	0.90	34	28	Stocks (1933)
Binet and Otis	0.92	0.63	50	50	Newman, Freeman, and Holzinger (1937)
I-Test	0.87	0.55	36	71	Husén (1947)
Simplex and C-Test	0.88	0.72	128	141	Wictorin (1952)
Intelligence factor	0.76	0.44	26	26	Blewett (1954)
JPQ-12	0.62	0.28	52	32	Cattell, Blewett, and Beloff (1955)
I-Test	0.90	0.70	215	416	Husén (1959)
Otis	0.83	0.59	34	34	Gottesman (1963)
Various group tests	0.94	0.55	95	127	Burt (1966)
PMA IQ	0.79	0.45	33	30	Koch (1966)
Vocabulary composite	0.83	0.66	85	135	Huntley (1966)
PMA total score	0.88	0.67	123	75	Loehlin and Vandenberg (1968)
General ability factor	0.80	0.48	337	156	Schoenfeldt (1968)
ITPA total	0.90	0.62	28	33	Mittler (1969)
Tanaka B	0.81	0.66	81	32	Kamitake (1971)
NMSQT total score					
1962	0.87	0.63	687	482	Nichols (1965)
1965	0.86	0.62	1,300	864	Loehlin and Nichols (1976)

SOURCE: Adapted from Loehlin and Nichols, 1976.

the familial environment because twins share more common familial influences than do nontwin siblings. A recent analysis suggests that disadvantageous environmental factors unique to identical twins lower their resemblance for IQ, thus lowering twin estimates of the heritability of IQ scores (Bailey and Horn, 1986).

A recent Norwegian twin study of general cognitive ability has been reported for 757 identical and 1,093 fraternal male twin pairs tested upon induction into the Norwegian Armed Forces (Sundet et al., 1988). The correlations for identical and fraternal twins are, respectively, 0.83 and 0.51. Cohort changes were examined for twins born from 1930 through 1960, but no simple changes in heritability were observed. For example, the identical and fraternal twin correlations for births from 1931 to 1935 (before World War II) were 0.84 and 0.51, respectively; for births after the war, the correlations were 0.83 and 0.51, despite the more egalitarian social and educational policies implemented in Norway after the war. The heritability estimates for this recent report are higher than in other recent studies—as noted in Chapter 11, recent studies tend to show lower IQ heritability than older studies. Although this difference may be due to factors unique to Norway, it will no doubt receive further attention in research.

Behavioral genetic analyses can also be applied to developmental data. A common mistake is to think that genes are somehow turned on at the moment of conception and continue to run at full throttle for the rest of our lives. As explained in Chapter 4, genes can be turned on and off. Furthermore, environmental circumstances change as the organism develops. Thus,

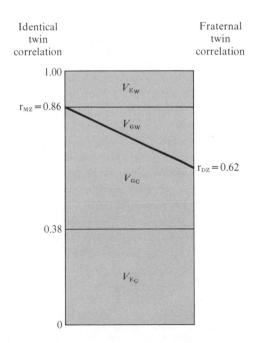

Figure 12.5
Pictorial representation of the average results of twin studies of general cognitive ability.

we might expect to find different patterns of genetic and environmental influences at different stages of development. Although it is reasonable to expect greater similarity as the twins grow up in similar environments, the early twin studies by Galton, Thorndike, and Merriman found little change in twin similarity as a function of age.

These earlier studies were cross-sectional, meaning that they obtained data on twins of different ages at the same time. A longitudinal study follows the same individuals over time. An exemplary longitudinal study of cognitive development in nearly 500 pairs of twins was conducted in Louisville, Kentucky, during the past two decades (Wilson, 1983). Twins were tested on the Bayley scales of mental and motor development at 3, 6, 9, 12, 18, and 24 months of age. IQ tests were administered at 2½ and 3 years (the Stanford-Binet), at 4, 5, and 6 years (the Wechsler Preschool and Primary Scale of Intelligence), and at 7, 8, 9, and 15 years (Wechsler Intelligence Scale for Children). The twin correlations and age-to-age stability correlations are shown in Figure 12.6 which summarizes 20 years of research. These data suggest little genetic influence in infancy. After the first six months, during

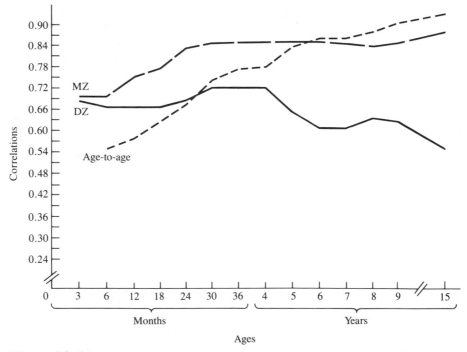

Figure 12.6
Twin and age-to-age IQ correlations from the longitudinal Louisville Twin Study. Data from Wilson (1983) for over 100 pairs each of identical and fraternal twins. Age-to-age correlations involve from 242 to 424 individuals and represent the correlation from that age to the next; for example, the age-to-age correlation for three months is the correlation from three to six months.

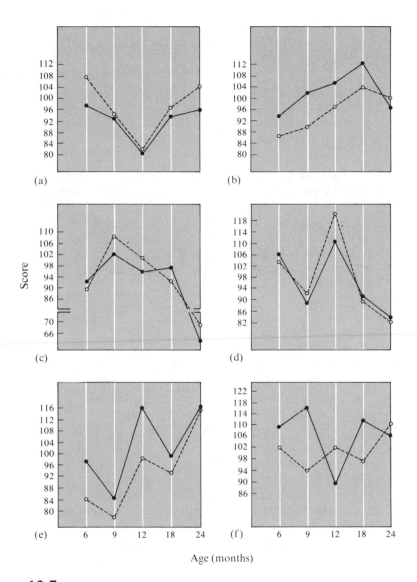

Figure 12.7

Profiles of mental development scores for MZ twins at ages 6 through 24 months. The pairs in (a) through (e) exhibit moderate to high profile congruence; the pair in (f) is obviously noncongruent. (From "Twins: early mental development" by R. S. Wilson. *Science*, 175, 914–917. Copyright © 1972 by the American Association for the Advancement of Science.)

which there is no difference between identical and fraternal twin correlations, identical twin correlations are about 0.10 greater than those for fraternal twins. The average twin correlations from 9 to 36 months are 0.77 for identical twins and 0.67 for fraternal twins. Although the differences in correlations are small, they are consistent in suggesting genetic influence;

heritabilities during infancy are about 0.20. Other twin studies of infant mental development yield similar results (Griffiths and Phillips, 1976; Nichols and Broman, 1974).

An interesting aspect of these data in infancy is that the twin correlations are so high, higher than the age-to-age stability. In other words, the score of one member of a twin pair at a given age is a better predictor of the co-twin's score than an earlier score of the co-twin him- or herself. Part of the reason for the high twin correlations in infancy is that twins have substantially more perinatal problems than singletons, and twins share these environmental factors to a greater extent than do nontwin siblings (Plomin, 1986a). In addition, longitudinal analysis suggests that twin pairs follow a similar pattern of mental development across age. Sample profiles for six pairs of identical twins in infancy presented in Figure 12.7 indicate substantial similarities within twin pairs in "spurts and lags" in developmental patterns. Heredity appears to play a role in these developmental changes in that these developmental patterns are somewhat more similar for identical than for fraternal twins (Wilson, 1986).

Figure 12.6 shows that heritability increases during childhood. Identical twin correlations do not increase during childhood; however, fraternal twin correlations decline from above 0.7 at 3 to below 0.60 at 6 and 7 years. For this reason, the twin results show progressively increasing heritability up to middle childhood. Heritability appears to decline in early adolescence. The heritability estimates at each year from 3 to 9 years are 0.18, 0.24, 0.38, 0.54, 0.50, 0.34, and 0.36. For a small sample at 15 years, the heritability estimate is higher, as we would expect from other twin studies of adolescents and young adults. The increase in heritability during childhood is interesting because you might guess that as children develop and experience more diverse environments, environmental variance increasingly accounts for phenotypic variance. This would imply that heritability will wane, not wax, during childhood. It has been suggested that heritability continues to increase during adulthood, although there are suprisingly few studies of older adults (Plomin, 1986a). For example, a recent study of 80 pairs of twins 40 years old on average yielded identical and fraternal twin correlations of 0.88 and 0.47, respectively, which suggests an IQ heritability of about 0.80 (Tambs, Sundet, and Magnus, 1984).

Schizophrenia

Twin studies of psychopathology usually report *concordance* values. When a sample of twins has been ascertained for psychopathology, concordance is the percentage of co-twins with the same diagnosis. However, there are several ways of computing concordance (Gottesman and Shields, 1982). The most common method is called *pairwise* concordance. In a sample of twin pairs in which at least one twin is affected, pairwise concordance is the number of concordant pairs divided by the total number of pairs. For example, in Irving Gottesman and James Shields' (1972) study, both members of the pair were schizophrenic in 10 of 24 pairs of identical twins; so the pairwise concordance would be $10/24 = 0.42$. The other major method considers index cases rather than pairs. In this *proband* method, the number of affected twins in concordant pairs is divided by the number of index cases (probands). Gottesman and Shields began with 34 identical twin probands and found that 20 of these individuals were in concordant pairs. The probandwise concordance would thus be $20/34 = 0.59$. Although there are merits to both approaches, the proband method is more appropriate from a sampling point of view if each affected member of a twin pair is identified independently. The probandwise concordance can be compared to risk figures for other family groups and to general population rates because it expresses concordance in terms of the risk of the co-twin of an affected twin.

Family studies of schizophrenia suggested a risk of about 8 percent for first-degree relatives of schizophrenics. Such familial influence could have either genetic or environmental origins, and twin studies have been useful in investigating these possibilities. Because twins show no more psychopathology than singletons, the results of twin studies can probably be generalized to the rest of the population. Eleven twin studies in seven countries reported concordances for identical and fraternal twins (Kringlen, 1966). The median pairwise concordance for identical twins was 42 percent, while that for fraternal twins was 9 percent. Table 12.8 presents the results of five twin studies of schizophrenia since 1966 that included 210 identical twin pairs and 309 pairs of fraternal twins (Gottesman and Shields, 1982). The weighted average probandwise concordance is 0.46 for the identical twins and 0.14 for the fraternal twins. The most recent twin study involves all male twins who are U.S. veterans of World War II (Kendler and Robinette, 1983). Twin probandwise concordances are 30.9 percent for 164 pairs of identical twins and 6.5 percent for 268 pairs of fraternal twins. Although these concordances are

Table 12.8

Schizophrenia: probandwise concordance in recent twin studies

Investigator	Year	Country	Identical		Fraternal	
			Concordance %	Number of pairs	Concordance %	Number of pairs
Kringlen	1967	Norway	45	55	15	90
Fischer	1973	Denmark	56	21	27	41
Gottesman and Shields	1972	U.K.	58	22	12	33
Tienari	1971	Finland	35	17	13	20
Pollin et al.	1972	U.S.	43	95	9	125

SOURCE: After Gottesman and Shields, 1982.

lower than in earlier studies, it is noteworthy that genetic influence on schizophrenia exceeds that for common medical conditions, as shown in Table 12.9.

How heritable is schizophrenia? One problem with the concordance statistic is that it is difficult to interpret concordance in terms of variance explained, in part because concordance needs to be compared to population prevalences of the disorder to determine the extent of resemblance. Nonetheless, if co-twins of a schizophrenic identical twin have a thirty-fold greater risk of becoming schizophrenic, compared to individuals in the general population, and if the risk for fraternal twins is only six times greater than the general population's, genetic factors appear to be significantly and substantially implicated in schizophrenia. However, the concordance of 30 percent for identical twins means that 70 percent of the time genetically identical individuals are discordant for schizophrenia, which implies that nongenetic factors primarily determine why one person is diagnosed as schizophrenic and another is not.

As discussed in Chapter 11, researchers often convert concordances to a special type of correlation that assumes that a normal distribution of liability (risk) underlies the either–or diagnosis (Falconer, 1965). For example, if a disorder yields a concordance of 10 percent and the base rate in the population is 1 percent, the liability correlation is 0.45. Correlations for the construct of liability towards schizophrenia are about 0.85 for identical twins and about 0.50 for fraternal twins. These results suggest that the heritability of the assumed construct of liability towards schizophrenia is 0.70. However, it should be emphasized that this heritability refers to a hypothetical construct of an underlying liability towards schizophrenia, not to the actual diagnosis of schizophrenia.

Extensive analyses of differences in diagnostic criteria, the biases introduced by different types of sampling procedures (resident hospital population, consecutive admission, or twin registry), and the possibility that schizophrenia is really a heterogeneous complex of psychotic conditions have been

Table 12.9
Twin concordances for schizophrenia and common medical disorders

Disorder	Pairwise concordance	
	Identical	Fraternal
Schizophrenia	30.9	6.5
Diabetes mellitus	18.8	7.9
Ulcers	23.8	14.8
Obstructive pulmonary disease	11.8	8.2
Hypertension	25.9	10.8
Ischemic heart disease	29.1	18.3

SOURCE: After Kendler and Robinette, 1983.

examined (Gottesman and Shields, 1982). The extent to which the data might be influenced by these factors can be illustrated by results of analyses of two of the twin studies. Gottesman and Shields (1977) analyzed separately the concordance of identical co-twins of mild and severe proband cases from their twin study of about 45,000 consecutive admissions in a large London hospital. As indicated in Table 12.10, the concordance for identical twins is only 27 percent for probands in the hospital less than two years. In contrast, the concordance was 77 percent when probands were hospitalized over two years and 75 percent when probands were not able to stay outside the hospital for longer than six months. The relationship between severity and concordance has been documented for most twin studies (Gottesman and Shields, 1977). Clearly, the severity of the cases chosen for study can affect the results obtained. It has been suggested, for example, that the largest twin study of schizophrenia, conducted by F. J. Kallmann (1946), consisted mostly of severe or chronic cases; this might account for the relatively higher identical twin concordance (69 percent) obtained by Kallmann as compared to subsequent investigators.

Another diagnostic matter of considerable importance concerns the strictness of the definition of schizophrenia. Three twin studies can be analyzed using both a strict definition of schizophrenia and a broader definition that includes borderline cases (Fischer, Harvald, and Hauge, 1969). The results are shown in Table 12.11, which shows that concordances for both identical and fraternal twins are higher when the broader definition of schizophrenia is employed.

The issue of diagnostic criteria spills over into the larger question of the heterogeneity of schizophrenia. Although many attempts have been made to break down schizophrenia into more discrete genetic entities, current interest focuses on the possibility that there is a spectrum of schizophrenia that extends as far as antisocial behavior. Generally, researchers talk about a "hard" spectrum, which includes borderline and questionable schizophrenia as well as chronic schizophrenia, as distinct from a "soft" spectrum, which includes personality disorders. Table 12.12 lists the concordances for schizophrenic identical twins when we expand the concept of schizophrenia to include the "hard" and "soft" spectra (Shields, Heston, and Gottesman,

Table 12.10

Schizophrenia: relationship between severity and concordance for identical twins

Degree of severity of proband	Concordance
Less than two years in hospital	27% (4/15)
More than two years in hospital	77% (10/13)
Inability to stay out of hospital for at least six months	75% (12/16)

SOURCE: Gottesman and Shields, 1977.

Table 12.11

Concordance for schizophrenia as a function of strictness of diagnostic criteria

Study	MZ		DZ	
	Strict schizophrenia	Including borderline cases	Strict schizophrenia	Including borderline cases
Kringlen (1966)	28% (14/50)	38% (19/50)	6% (6/94)	14% (13/94)
Gottesman (1966) and Shields	42% (10/24)	54% (13/24)	9% (3/33)	18% (6/33)
Fischer, Harvald, and Hauge (1968)	24% (5/21)	48% (10/21)	10% (4/41)	19% (8/41)

SOURCE: After Fischer, Harvald, and Hauge, 1969.

1975). On the average, only about 25 percent of the identical co-twins showed no psychopathology. The median identical twin concordance for the "hard" spectrum was 61 percent. For the combined categories of "hard" and "soft," the concordance was 77 percent. These twin data suggest that the genetic propensity toward schizophrenia may be broader than we once suspected. See Gottesman and Shields (1982) for a discussion of this issue. Another issue that needs to be considered is age correction. None of the concordances that we

Table 12.12

Schizophrenic spectrum: pairwise concordance for schizophrenic identical twins for hard and soft schizophrenic spectra

Study	Number of pairs	Concordance for hard spectrum	Concordance for hard or soft spectrum	Percentage of pairs with one co-twin normal
Luxenburger (1928)	14	72	86	14
Rosanoff et al. (1934)	41	61	68	32
Kallmann (1946)	174	69	95	5
Slater (1953)	37	64	78	22
Kringlen (1967)	45	38	67	33
Fischer et al. (1969)	21	48	57	43
Gottesman and Shields (1972)	22	50	77	23

SOURCE: After Shields, Heston, and Gottesman, 1975.

have mentioned so far has been corrected for age. There is no satisfactory age correction for twin concordances, but it is an important issue because age of onset can vary. One study of Finnish twins (Tienari, 1963) differed from all other twin studies in finding *no* concordance for identical twins. The twins were 30 to 40 years of age at the time of that report. In a continuation of the study, one of the identical twin pairs was concordant by 1968. By 1971, 7 of the 20 affected identical twins were in concordant pairs, so that the proband-wise concordance was 7/20, or 35 percent (Gottesman and Shields, 1976).

In summary, although twin data do not conform to a simple genetic model, they clearly support a genetic hypothesis. The risk of schizophrenia for fraternal twins is about the same as for other siblings (although opposite-sex twins have a consistently lower concordance). Identical twins are much more concordant than fraternal twins. Thus, these studies suggest the influence of genetic differences. Like the family studies, they also suggest that within-family environmental influences play a major role.

Because differences within pairs of identical twins are environmental in origin, the co-twin control method can be used to study identical twins discordant for schizophrenia. This method is particularly important for studying psychopathology, because differences within pairs of identical twins must be caused by within-family environmental influences. These, as we have seen, are important in the etiology of psychopathology. In one study, in which 17 discordant identical twin pairs were studied for a decade (Belmaker et al., 1974), the salient within-family environmental factors have remained elusive. A difference in birthweight was suggested as a possible factor, but other studies have not supported this finding. Another possibility is submissiveness on the part of one twin, but it is difficult to determine the cause and effect of this possible factor.

Gottesman and Shields (1982, p. 120) conclude:

> Despite high hopes, the study of discordant MZ pairs has not yet led to a big payoff in the identification of crucial environmental factors in schizophrenia. The problem is simply more difficult than we can cope with: Environmental variation within twin pairs is limited to a relatively narrow range, sample sizes are small, the data needed are subject to retrospective distortions, and the culprits may be nonspecific, time-limited in their effectiveness, and idiosyncratic.

They suggest that the co-twin control method may also be useful in identifying biochemical "endophenotypes" ("inside" phenotypes) that contribute to the genetic etiology of schizophrenia. The hope is to identify markers that occur for both members of discordant identical twin pairs.

SUMMARY

Twin studies take advantage of the natural experimental situation resulting from the fact that identical twins are twice as similar genetically as fraternal twins. If genes make a difference for a particular behavior, identical twins should be more similar than fraternals. Zygosity determination and the reasonableness of the equal environments assumption were discussed, as well as the estimation of broad-sense heritability and between- and within-family environmental influences.

For general cognitive ability (IQ), the average correlations for identical and fraternal twins are 0.86 and 0.62, respectively. This pattern of correlations suggests substantial genetic influence (broad-sense heritability of about 0.50). It also points to a substantial role for between-family environmental influences (E_c) on the behavior of twins.

Genetic influences also play an important role in schizophrenia. The average probandwise concordances for schizophrenia in five studies are 0.46 and 0.14 for identical and fraternal twins, respectively.

CHAPTER·13

Adoption Studies

The first adoption study was conducted in the same year (1924) as the first twin study. However, far more twin studies have been reported since then, no doubt because of the greater ease of conducting twin studies. Nonetheless, adoption studies of complex human behaviors provide the most convincing demonstration of genetic influence. These studies untangle genetic and environmental factors common to members of natural families by studying genetically unrelated individuals living together (to assess environmental influences common to family members), and genetically related individuals living apart (to test genetic influences). In this way, adoption studies can determine the extent to which familial resemblances are due to genetic or environmental similarity.

Adoption studies include several types of designs. Most important from the genetic point of view are those designs that provide direct estimates of genetic influence by studying adopted-apart relatives, typically biological parents and their adopted-away offspring, and twins and nontwin siblings separated early in life. Most important from the environmental perspective are those designs that provide direct estimates of shared environmental influence by studying genetically unrelated individuals adopted together, such as adoptive parents and their adopted children, and genetically unrelated children adopted into the same family. Designs that include both adopted-apart relatives and adoptive relatives are called full designs; designs that include just one are partial. The full adoption design can be seen as a simple quasi-experimental design in which family environment is randomized in comparisons, for example, between biological parents and their adopted-away offspring.

Correspondingly, heredity is randomized in comparisons between adoptive parents and their adopted children.

In adoption studies it is useful to include families in which both heredity and family environment are shared to provide a comparison group for the direct estimates of hereditary influence and shared environment from the full adoption design. For example, in the case of parents and offspring comparisons, there are "genetic" parents (biological parents and their adopted-away children), "environmental" parents (adoptive parents and their adopted children), and "genetic-plus-environmental" parents (nonadoptive or control parents and their children). Comparisons among the three family types can disentangle the causes of familial resemblance. That is, in nonadoptive families, parents share both heredity and environment and thus parent–offspring resemblance could be due to either factor. Resemblance between biological parents and their adopted-away offspring can be due only to shared heredity; resemblance between adoptive parents and their adopted children can be due only to shared environment. A frequently used type of partial design estimates genetic influence by comparing environmentally based resemblance in adoptive relatives to resemblance in nonadoptive families in which family members share both heredity and family environment.

Because of the relative rarity of most psychopathology and the qualitative nature of diagnoses of psychopathology, adoption studies in this area have not been able to take advantage of quantitative genetic methods that are used to analyze normal variation in a population. Instead, they study the frequency of psychopathology in adopted-away relatives of affected individuals. For example, in the classic adoption study of schizophrenia (Heston, 1966), 5 of 47 adopted-away offspring of schizophrenic biological mothers were found to be schizophrenic. This incidence of 10 percent could be compared to the population incidence of about 1 percent, a comparison which indicates genetic influence. Because it is possible that biological parents and adoptees are not representative of the general population, a control group is obtained that consists of adoptees whose biological parents have no known psychopathology. If heredity influences a disorder, its incidence will be greater in the adopted-away offspring of the affected biological parents than in the adoptees whose biological parents were not affected. If heredity is unimportant, the incidence should not differ for the two groups of adoptees.

This is a partial adoption design that uses the genetic prong of the two-pronged full adoption design. Table 13.1 illustrates possible adoption designs for qualitative disorders. Most studies have tested the influence of genetic relatedness by making comparisons between adoptees with affected biological parents reared by normal adoptive parents (Ba/An) and adoptees with normal biological relatives reared by normal adoptive parents (Bn/An). A few studies, called *cross-fostering* studies, have tested the influence of the rearing environment by examining the incidence of the disorder in adoptees with normal biological relatives reared by affected adoptive parents (Bn/Aa) and in a control group of adoptees (Bn/An). Rarely do studies of psychopathology include the Ba/Aa cell—adoptees with affected biological relatives

Table 13.1
Paradigm for adoption studies of psychopathology

Biological relatives	Adoptive relatives	
	Normal (An)	Affected (Aa)
Normal (Bn)	Bn/An	Bn/Aa
Affected (Ba)	Ba/An	Ba/Aa

reared with affected adoptive relatives. This is not surprising because of the obvious difficulty of obtaining such a sample.

The partial adoption design described above (Bn/An versus Ba/An) is called the *adoptees' study method* because the adopted-away offspring are studied to determine the incidence of the disorder in the adoptees. The incidence of the disorder in the adopted-away offspring is compared to the incidence in a control group that consists of adoptees whose biological parents and adoptive parents have no known psychopathology. A second major strategy is a full adoption design called the *adoptees' family method*. Rather than beginning with parents, this method begins with adoptees who are affected (probands) and adoptees who are unaffected. The incidence of the disorder in the biological and adoptive families of the adoptees is assessed. Genetic influence is suggested if the incidence of the disorder is greater for the biological relatives of the affected adoptees than for the biological relatives of unaffected control adoptees. Environmental influence is indicated if the incidence is greater for the adoptive relatives of the affected adoptees than for the adoptive relatives of the control adoptees.

Adoption studies also provide a powerful tool for evaluating environmental influences as distinct from genetic variables. As discussed in Chapter 14, adoption studies provide direct estimates of shared environmental influence, they indicate the extent to which associations between specific measures of family environment and children's development in control families are mediated by hereditary similarity between parents and children, and they facilitate analyses of genotype–environment correlation and interaction.

ISSUES IN ADOPTION STUDIES

Because the logic of the adoption design is so straightforward, few criticisms have been leveled at it. If you want to know the extent to which observed resemblance between parents and their offspring is genetic in origin for a particular behavior, you can assess the resemblance between adopted-away children and their biological parents (called "birth parents" by adoption agencies), who gave their children genes but no familial environment. Conversely, you could also study the resemblance between adopted children and their adoptive parents, who provide a familial environment, but no genes. As in any experiment, certain biases must be avoided, or if that is impossible, they must be assessed. In the case of adoption studies, the representativeness of the sample and the possibility of selective placement need to be considered.

Representativeness

Although many stereotypes of birth parents, adoptive parents, and adoptees have been commonly accepted, such preconceptions tend to fade in the face of data. For example, many people believed that the average IQ of birth parents is lower than that of the general population. However, in the state of Minnesota during the period from 1948 to 1952, when IQ tests were required for all women giving up children for adoption, the average IQ score of 3,600 women was 100, with a standard deviation of 15.4. These values are the same as those for the general population (Scarr, 1977). One of the largest adoption studies is the Colorado Adoption Project, a full adoption design that includes 245 biological parents and their adopted-away offspring, the adoptive parents of these children, 245 matched nonadoptive families, and adoptive and nonadoptive siblings in these families (Plomin and DeFries, 1985; Plomin, DeFries, and Fulker, 1988). Extensive analyses indicate that biological and adoptive parents in the Colorado Adoption Project are quite representative of the general population not only for cognitive abilities but also for demographic characteristics such as educational and socioeconomic level, personality, and family environment.

The important point is that the representativeness of these groups in terms of variances and means can be measured. If some degree of unrepresentativeness is found, it does not invalidate the results of an adoption study. Rather, it can be taken into account in the interpretation of data.

Selective Placement

An issue that is more specific to adoption studies is *selective placement*, which means that adoptees are placed with adoptive parents who resemble the birth parents in some ways. For example, adoption agencies tend to place children whose birth parents are tall with tall adoptive parents. However, in terms of behavior, adoption agencies have only limited information (usually just education and occupation). Thus, they could not accurately match children for behavioral characteristics even if they wanted to do so. In fact, many adoption agencies now avoid selective placement altogether (even for physical characteristics) because they feel that it causes adoptive parents to have the false expectation of receiving a child similar in many ways to themselves.

Selective placement may increase the resemblance between adoptive parents and their adopted children (if there is genetic influence on the trait being studied). It also may increase the resemblance between birth parents and their adopted-away children (if there is environmental influence). Figure 13.1 presents a simplified path diagram illustrating the influence of selective placement on parent–child resemblance. Path analysis was briefly described in Chapter 9. For now, you need only remember that a path (such as e or g) represents the effect of one variable on another, independent of other influences. In Figure 13.1, e is the path that represents the influence of the adoptive parents (A) on the adopted child (C), independent of the birth parents (B). Similarly, g represents the influence of the birth parents on the adopted child, independent of the adoptive parents. The double-headed path (s) represents the selective placement correlation between adoptive and birth parents. It is possible to solve for these paths to determine parental influences independent of selective placement effects. In this way, we can assess the environmental influence of adoptive parents independent of selective placement (path e), and the genetic influence of birth parents independent of selective placement (path g).

As indicated in Chapter 9, we can use a path model to specify relationships. For example, the correlation between adoptive parents and their adopted children includes not only path e, but also the g and s chain of influences. Thus,

$$r_{AC} = e + gs$$

This means that the correlation between adoptive parents and their adopted children will be inflated if there is selective placement for a characteristic as well as genetic influence on that trait. Similarly, the correlation between birth parents and their adopted-away children will be inflated if there is both selective placement and environmental influence:

$$r_{BC} = g + es$$

Thus, in partial adoption studies, selective placement can result in overestimates of both genetic and shared family environmental influences.

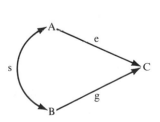

Figure 13.1

Path diagram illustrating the influence of selective placement on parent–child resemblance in an adoption study. The letter *A* symbolizes the adoptive parents, *B* birth parents, and *C* the adopted child. (From "Behavioral genetics" by J. C. DeFries and R. Plomin. Reproduced, with permission, from *Annual Review of Psychology*, 29, 473–515. Copyright © 1978 by Annual Reviews, Inc. All rights reserved.)

Other Issues

DeFries and Plomin (1978) have outlined several criteria for an ideal adoption study. These include such general requirements as the inclusion of environmental assessments, studying more than one type of family relationship, measuring many variables, and studying developmental phenomena using a longitudinal design. Although we suggested that an ideal study should meet all these criteria, that is not really essential for the success of an adoption study. One other criterion, the analysis of assortative mating, is critical. We have mentioned several times that the resemblance between parents and offspring is inflated if assortative mating occurs. Assortative marriage is known to occur for cognitive abilities (Johnson et al., 1976) and for psychopathology (Dunner et al., 1976). Unwed birth parents of adopted children also are known to mate assortatively for physical and behavioral characters (Plomin, DeFries, and Roberts, 1977). In short, we can expect that about the same degree of assortative mating occurs for unwed birth parents as for married couples. We need to take this into account when interpreting the resemblance between birth parents and their offspring in adoption studies.

Another important issue involves the effect of prenatal environment. Because birth mothers provide the prenatal environment for their adopted-away children, the phenotypic resemblance between them in an adoption study may reflect prenatal environmental influences. This possibility can be tested by comparing correlations between birth mothers and their adopted-away children to correlations between the birth fathers and the children. This test has been incorporated in recent adoption studies of psychopathology, as we shall see later in this chapter. A particularly valuable asset of adoption studies is their ability to test the influence of prenatal maternal environment, independent of postnatal environment.

HERITABILITY AND ENVIRONMENTALITY

Like the twin studies described in Chapter 9, adoption studies can be used to estimate the extent to which phenotypic variance is due to genetic variance (heritability) and environmental variance (environmentality). The simplest adoption design to understand is the rare but dramatic situation in which identical twins are adopted separately at birth and reared apart in uncorrelated environments. The resemblance of these pairs, expressed as a correlation, can be attributed to heredity. For example, the correlation for reared-apart identical twins for height is about 0.90. This implies that about 90 percent of the variation in height is shared by reared-apart identical twins. This resemblance—unlike the resemblance for identical twins reared together—cannot be due to shared environment, because the twins were adopted apart and lived in different families that are uncorrelated in terms of height. As described in previous chapters, heritability is a statistic that quantifies the extent of genetic influence. In the case of identical twins reared apart, their correlation directly represents heritability; in other words, the correlation of 0.90 suggests that 90 percent of the variation in height is due to genetic variation. (See path diagrams in Figure 9.12 and related text.)

Identical twins reared apart are rare; all the world's literature adds up to fewer than 100 pairs. However, one ongoing study in the United States (Bouchard, 1984) and two studies using nationwide records of twins in Scandinavia (Pedersen et al., 1984; Langinvainio et al., 1984) are tripling that number.

Other adoption designs involving first-degree relatives are also useful. Most of these studies investigate resemblance between biological parents and their adopted-away offspring who were relinquished for adoption early in life. Adopted-apart siblings can also be used to assess the influence of heredity; two ongoing studies of identical twins reared apart also include reared-apart fraternal twins, which will provide important comparisons with data from adopted-apart identical twins. Because first-degree relatives resemble each other only 50 percent genetically, the observed correlation for adopted-apart first-degree relatives for a measured trait such as height includes only half of the genetic variance—the other half of the genetic variance that affects the trait makes first-degree relatives different. That is, if a trait were perfectly heritable, the observed correlation for adopted-apart first-degree relatives would be 0.50, not 1.0 as in the case of identical twins. Thus, if a correlation of 0.50 is observed for adopted-apart first-degree relatives, we would estimate heritability as 1.0. Using height as an example, the observed correlation for adopted-apart parents and offspring is about 0.45, suggesting again that about 90 percent of the variation among individuals in height is genetic in origin.

The design is illustrated in a path diagram of the relationship between parents and offspring in natural families (Figure 13.2) that shows that parents share both genetic and environmental influences with their children. The point of an adoption study is to separate these two sets of influence. Figure 13.3 illustrates this separation for parents and offspring. The genetic side of the adoption design involves birth parents who give their adopted-away children genes but not environment; the environmental side involves adoptive parents who give their children a familial environment but not genes.

Table 10.1 described the relationship between familial correlations and heritability. The genetic side of an adoption study, in the absence of selective

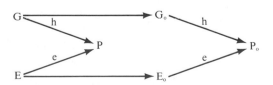

Figure 13.2
Path diagram of the relationship between parents and offspring in natural families. The subscript *o* refers to offspring. See text for explanation.

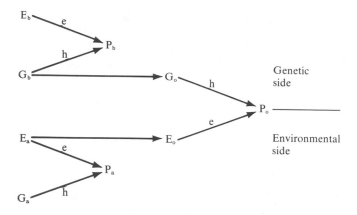

Figure 13.3
Path diagram showing the genetic and the environmental sides of the adoption design. The subscript *a* refers to the adoptive parent, *b* to the birth parent, and *o* to the adopted offspring. See text for explanation.

placement, estimates genetic influence independent of environment. The regression of adopted-away offspring on the score of one birth parent estimates half of the narrow-sense heritability ($\frac{1}{2}V_A/V_P$). The regression of adopted children on the midparent score of their birth parents directly estimates narrow-sense heritability. The correlation between full siblings (or fraternal twins) adopted into different families estimates roughly half of the broad-sense heritability, which may also be directly estimated from the correlation between identical twins reared in different families.

Phenotypic variance that cannot be explained by genetic differences is ascribed to the environment. Even better, the environmental side of adoption studies permits the direct assessment of shared-family (between-family) environmental influences. The between-family environmental components described in Table 9.9 can each be assessed by a particular adoptive relationship. For example, common environmental influences shared by parents and their offspring can be measured by the resemblance between adoptive parents and their adopted children. Between-family environmental influences shared by siblings are revealed by the correlation between genetically unrelated children adopted into the same family.

As is true of the results of other behavioral genetics methods, the results of adoption studies can be affected by genotype–environment interaction and correlation. In fact, adoption studies are the only practical tool for isolating interactions and correlations between genetic and environmental influences on human behavior.

Until recently, the adoption design has been applied primarily to the study of IQ and schizophrenia. Adoption studies for other domains of behavior are described in Chapter 14.

INTELLIGENCE QUOTIENT (IQ)

The first adoption study on the topic of general cognitive ability was conducted in 1924 by S. Theis, and every adoption study for about the next 40 years continued to focus on this same trait. A summary of the world's IQ adoption research as of 1980 is presented in Table 13.2 (Bouchard and McGue, 1981). These results confirm the conclusion from twin studies (Chapter 12) that heredity is importantly involved in IQ scores. The correlation for 814 biological parents and their adopted-away offspring is 0.22, and the correlation for 203 adopted-apart siblings is 0.24. These "genetic" results fit with the "environmental" results of adoptive relatives in adding up to the correlations for "genetic-plus-environmental" relatives. For example, the sum of the correlations between children and their "genetic" parents (0.22) and between children and their "environmental" parents (0.19) is similar to the correlation between children and their "genetic-plus-environmental" parents (0.42). In contrast, the correlation for 65 pairs of identical twins adopted apart is 0.72, a dramatic result that is being confirmed in an ongoing study of twins reared apart, which reports a correlation of 0.69 for 48 pairs (Bouchard, 1987). The correlation between identical twins reared apart may exceed twice that of first-degree relatives because of ascertainment bias or nonadditive genetic variance.

Most adoption studies utilize a partial design in which correlations for adoptive ("environmental") relatives are compared to correlations for biological relatives reared together ("genetic plus environmental"); these studies are reviewed elsewhere (DeFries and Plomin, 1978). The most powerful estimate of genetic influence comes from the comparison of genetically related individ-

Table 13.2

IQ correlations from adoption studies prior to 1980

	Number of pairs	Average weighted correlation
Parents and offspring		
Genetic (biological parents and adopted-away offspring)	814	0.22
Environmental (adoptive parents and adopted children)	1,397	0.19
Genetic plus environmental (parents and offspring)	8,433	0.42
Siblings		
Genetic (siblings adopted apart)	203	0.24
Environmental (genetically unrelated adopted together)	714	0.32
Genetic plus environmental (siblings together)	26,473	0.47
Identical twins		
Genetic (adopted apart)	65	0.72
Genetic plus environmental (identical twins together)	4,672	0.86

SOURCE: Bouchard and McGue, 1981.

uals who are reared apart in uncorrelated environments. The original study by Theis in 1924 had several severe problems. For example, mental ability was simply rated on a three-point scale, and only 35 percent of the 910 children were adopted before they were five years old. These problems make it difficult to interpret the finding that the adopted children's rated mental ability was affected by the social status of their biological parents more than by that of their adoptive parents.

More than 20 years later, a full adoption study of IQ was reported by Marie Skodak and Harold Skeels (1949). A group of 100 illegitimate children adopted before six months of age was administered IQ tests on at least four different occasions at about two, four, seven, and 13 years of age. As indicated in Table 13.3, the mean IQ of the adopted children was found to be above average at all ages, even when they were only about two years old. In contrast, the mean IQ of the 63 birth mothers who were tested was 86 (with a normal standard deviation of 15.8). However, the lower-than-average IQ of the birth mothers may be partially attributable to the fact that the test was administered shortly after the birth of the baby, and usually after the mother had decided to relinquish the child for adoption, conditions unconducive to optimal performance. Although studies of average differences between adopted children and their birth parents often suggest possible environmental influences (e.g., Schiff et al., 1978; Willerman, 1979a), such studies are particularly prone to problems such as the one just mentioned, as well as to others, such as assortative mating (Munsinger, 1975).

Both the birth mothers and the adopted children were similar in variance to the general population. However, there was some selective placement. The correlation between education level of the birth mothers and the adoptive midparent level was 0.30. Correlations between the IQs of the adopted children at various ages and those of their birth mothers are also reported in Table 13.2. The correlation is significant by four years of age and

Table 13.3

Mean IQs of adopted children and correlations between IQs of the children and their birth mothers

| Test | Mean Age of Adopted children | | Mean | Standard deviation | Correlation with birth mothers |
	Years	Months			
Kuhlman revision of Binet	2	2	117	13.6	0.00
1916 Stanford-Binet	4	3	112	13.8	0.28
1916 Stanford-Binet	7	0	115	13.2	0.35
1916 Stanford-Binet	13	6	107	14.4	0.38
1937 revision of Stanford-Binet	13	6	117	15.5	0.44

SOURCE: After Skodak and Skeels, 1949.

increases somewhat thereafter, reaching a level of 0.44 for the 13-year-olds' test scores on the 1937 revision of the Stanford-Binet. A similar pattern of correlation was observed between the children's IQ and the educational level of the birth mothers. It is interesting to note that the same pattern has also been observed between children's IQ and the educational level of the birth fathers (Honzik, 1957), which suggests that the role of prenatal maternal influences is minimal.

Marjorie Honzik (1957) compared these developmental data to those obtained from her study of parents rearing their own children, as illustrated in Figure 13.4. Clearly, the similarity between the birth mothers and their adopted-away children is much like the similarity between natural mothers and their own children whom they reared. Similar results have been obtained in comparisons between birth fathers and natural fathers.

Thus, the genetic side of the Skodak and Skeels study strongly suggests hereditary influence on IQ. What about the environmental side of the full adoption design? Figure 13.4 also addresses that issue by showing the correlations between the adoptive mothers' education and the IQ of their adopted children as a function of age. The fact that the correlations hover around the

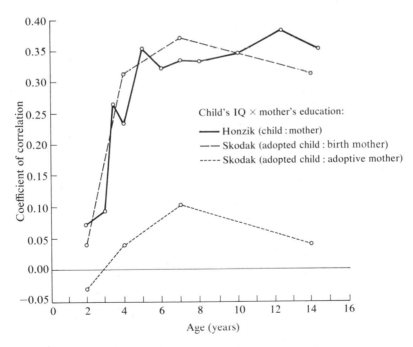

Figure 13.4
Coefficients of correlation at different ages between the child's IQ and the educational level of the birth mother or of the adoptive mother. (After "Developmental studies of parent–child resemblance in intelligence" by M. P. Honzik. *Child Development*, 28, 215–228. Copyright © 1957 by the Society for Research in Child Development, Inc.)

0.05 level suggests little between-family environmental influence of mothers' educational levels on the children's IQs.

Although there are six other full adoption studies (reviewed by DeFries and Plomin, 1978), only two obtained IQ scores for both biological and adoptive parents. The results of the Texas Adoption Project are described in Table 13.4 (Horn, Loehlin, and Willerman, 1979, 1982). The significant IQ correlation between biological mothers and their adopted-away offspring again suggests genetic influence. The second study is the longitudinal Colorado Adoption Project, which is described later.

Taken together with the twin data for IQ described in the previous chapter, these data make it difficult to escape the conclusion that heredity influences individual differences in IQ scores. Not only is genetic influence significant, it is also substantial, although the various estimates of heritability that can be obtained from Table 13.2 vary considerably. The correlation for identical twins reared apart suggests a heritability of about 70 percent; doubling the correlation for parents and offspring adopted apart yields an estimate of 0.44; doubling the correlation for siblings adopted apart is 0.48; doubling the difference between the correlation for biological parents and offspring living together (0.42) and the correlation for adoptive parents and their adopted children (0.19) leads to an estimate of 0.46; doubling the difference between the correlation for biological siblings reared together (0.47) and the correlation for adoptive siblings (0.32) provides an estimate of 0.30.

As indicated in Chapter 9, many refinements need to be considered in providing more precise estimates of heritability, such as nonadditive genetic variance, assortative mating, and genotype–environment correlation and interaction. These refinements, which have different effects on the various estimates of heritability, are discussed in Chapter 14. Nonetheless, the data in Table 13.2 converge on the conclusion that the heritability of IQ scores is about 0.50, meaning that genetic differences among individuals account for about half of the differences among them in their performance on IQ tests. The error surrounding this estimate may be as high as 20 percent, so we can only say with confidence that the heritability of IQ scores is at least at the

Table 13.4

IQ correlations between adopted children and their adoptive parents and birth mothers from the Texas Adoption Project

	Correlation with adopted children's IQs	Number
Adoptive mothers' IQ	0.17	459
Adoptive fathers' IQ	0.14	462
Birth mothers' IQ	0.31	345

SOURCE: After Horn, Loehlin, and Willerman, 1979.

bottom of this range. Nonetheless, it is a remarkable finding: To explain 30 percent of the variance of anything as complex as IQ scores is an important achievement.

If half of the variance of IQ scores is due to heredity, the other half is due to environment. Much of the environmental variance appears to be of the type shared by family members. Genetically unrelated children adopted together yield a correlation of 0.32, suggesting that about a third of the total variance of IQ scores is due to this class of environmental influence. The correlation between adoptive parents and their adopted children (0.19) suggests less shared environmental influence, although it seems reasonable that parents and their children share less similar environments than do siblings.

Recent evidence suggests that shared environmental influence that affects IQ scores may be much less after adolescence (Plomin, 1988). Earlier studies involving adoptive siblings happened to study them in childhood. In 1978 the first study of older adoptive siblings yielded a strikingly different result from the usual correlation of about 0.30: The IQ correlation was -0.03 for 84 pairs of adoptive siblings from 16 to 22 years of age (Scarr and Weinberg, 1978). Other studies of older adoptive siblings have also found similarly low IQ correlations (Kent, 1985; Teasdale and Owen, 1984). The most impressive evidence comes from an ongoing ten-year longitudinal follow-up of the Texas Adoption Project, which finds an IQ correlation of 0.02 for 223 pairs of postadolescent adoptive siblings (Willerman, 1987). These same adoptive siblings yielded an IQ correlation of 0.26 ten years earlier at the average age of eight years. These data suggest that shared environment is important for IQ during childhood when children are living at home and then fades in importance after childhood as extrafamilial influences become more influential.

Twin data described in Chapter 12 indicate that the heritability of IQ scores increases sharply during childhood. Adoption data add to this finding. As mentioned earlier, the Colorado Adoption Project is a longitudinal, prospective adoption study of 245 adoptive families and 245 matched nonadoptive families in which parents have been tested (including both the biological and adoptive parents of the adoptees) and the children have been studied at one, two, three, and four years of age (Plomin, DeFries, and Fulker, 1988). The results of the parent–offspring comparisons confirm the twin data in suggesting increasing IQ heritability during childhood. In addition, this adoption study has explored the extent to which genetic effects in childhood are related to genetic effects in adulthood. The approach uses the parent–offspring design as an "instant" longitudinal study from childhood to adulthood. In brief, if genetic changes occur during development, relatives of different ages, such as parents and offspring, should not be as similar as relatives of the same age, such as twins. Model-fitting analyses suggest that genetic effects on IQ scores in early childhood are highly correlated with genetic effects on IQ scores in adulthood (DeFries, Plomin, and LaBuda, 1987). The concept of the parent–offspring design as an instant longitudinal study from childhood to adulthood is being put to practical use in screening infant behaviors for those that best predict adult IQ (Fulker et al., 1988).

SCHIZOPHRENIA

A single adoption study in 1966 by Leonard Heston turned the tide towards acceptance of genetic influence on schizophrenia. The first adoption study of schizophrenia (Heston, 1966) used the adoptees' study method. The study identified hospitalized chronic schizophrenic women, who had been hospitalized while pregnant and whose children had been placed in foundling homes or foster homes during the first two weeks of life. The children of 47 such women were interviewed at the average age of 36, and were compared to 50 adoptees whose birth parents had no known psychopathology. The well-known results, summarized in Table 13.5, indicate significant genetic influence. All five of the affected adoptees had been reared by normal adoptive parents (the Ba/An condition). All these individuals had been hospitalized, and three were chronic schizophrenics hospitalized for several years. Four other Ba/An individuals were regarded as schizophrenic or borderline schizophrenic by one or two of the three psychiatric raters, so a broader definition of schizophrenia would indicate occurrence of the disorder in 20 percent (9/47) of the offspring of schizophrenic mothers. Because half of the Ba/An individuals demonstrated some degree of psychosocial disability, Heston suggested that the definition of schizophrenia should be broadened to include schizoid dimensions (an inclusive definition, referred to as "soft schizophrenic spectrum").

The major findings of Leonard Heston's study—that schizophrenia is heritable, and that the heritable complex may include schizoid dimensions as well as borderline schizophrenias—have been confirmed by David Rosenthal and his associates in their adoption studies in Denmark. These investigators used the *Folkeregister* of Copenhagen to find approximately 5,500 individ-

Table 13.5

Chronic schizophrenia in adopted offspring of schizophrenic and nonschizophrenic birth mothers

Birth mothers	Foundling and foster homes	
	Normal	Affected
Normal	0% (0/50)	—
Affected	11% (5/47)	—

SOURCE: After Heston, 1966.

uals who had been adopted between 1924 and 1947. Rosenthal and co-workers (1968, 1971, 1972) employed the same design as that used by Heston, but with the added experimental control provided by systematic assessment of adoptees, the use of blind psychiatric interviews, and the fact that schizophrenic mothers were not in a mental hospital while pregnant. After the birth parents of the adoptees were identified through the *Folkeregister*, their names were traced in a psychiatric register, hospital records were obtained, and a consensus diagnosis was used to select schizophrenic mothers or fathers whose children had been placed in adoptive homes. Forty-four birth parents (32 mothers and 12 fathers) who were diagnosed as certain or uncertain chronic schizophrenics were in this group, and their adopted-away children (Ba/An) were matched to controls whose birth parents had no psychiatric history (Bn/An). The adoptees, 33 years old on the average and thus still at substantial risk, were interviewed for three to five hours by an interviewer blind to the status of their birth parents.

Table 13.6 presents the incidence of schizophrenia (chronic), "hard schizophrenic spectrum," and "soft schizophrenic spectrum" for the two groups of adoptees. For all three classifications, the results suggest substantial genetic influence. There is a lower incidence of schizophrenia in the adoptees than in Heston's study, but it is likely that the birth parents in this study were less severely affected.

The unusually high rates of psychopathology in the control data (Bn/An) shown in Table 13.6 occurs in the early Danish studies because these studies relied on hospital records to assess psychiatric status and may have

Table 13.6
Schizophrenia and hard and soft schizophrenic spectrum in adopted offspring of schizophrenic and nonschizophrenic birth parents

Birth parents	Adoptive parents	
	Normal	Affected
Normal	0% (0/67)* 4% (3/67)† 18% (12/67)‡	— — —
Affected	7% (3/44)* 14% (6/44)† 27% (12/44)‡	— — —

*Chronic schizophrenia.
†Chronic, acute, borderline, and uncertain schizophrenia (hard spectrum).
‡Hard spectrum plus paranoid and schizoid personality (soft spectrum).
SOURCE: Rosenthal et al., 1968.

overlooked significant psychopathology in the control birth parents. For example, psychiatric interviews in a later study (Wender et al., 1974) revealed 36 persons who were schizophrenic but had not been hospitalized. The birth parents of controls had been interviewed, and it appears that one-third of them fall in the schizophrenic spectrum. Thus, with respect to the early studies, they concluded that "our controls are a poor control group and . . . our technique of selection has minimized the differences between the control and index groups" (Wender et al., 1974, p. 127). This bias is conservative in terms of demonstrating genetic influence.

In the early Rosenthal study described above, adoptees were included in the Ba/An group if they had only one schizophrenic parent, and data were obtained only on that parent. In a follow-up study (Rosenthal, 1975a), the other birth parent of each adoptee was interviewed. The results of interviewing these "co-parents" suggest considerable assortative mating in terms of soft schizophrenic spectrum. The effects of assortative mating have been discussed previously. In the present context, this co-parents study provides additional support for a genetic hypothesis: The adopted-away offspring of chronic schizophrenic birth mothers or fathers were significantly more often diagnosed as in the schizophrenic spectrum when the co-parent also exhibited some type of schizophrenic disorder.

Other studies in Denmark (Kety et al., 1968, 1971, 1975, 1976; Kety, 1987) have used the adoptees' family method rather than the adoptees' study method used in the studies described above. In other words, the investigators first identified schizophrenic adoptees and then assessed the psychiatric status of their biological and adoptive relatives. Thus, they focused on the extent of psychopathology among the relatives of schizophrenic and nonschizophrenic adoptees. The design of these studies is illustrated in Table 13.7.

The pool of adoptees in the studies reported by Rosenthal and associates included 507 individuals who had been admitted to a psychiatric facility. Of these, 33 were diagnosed as falling within the hard schizophrenic spectrum (16 were chronic schizophrenics), and these index adoptees (probands) were matched to controls having no history of psychiatric problems. The *Folkeregister* was searched for the names of parents, siblings, and half-siblings from the biological and adoptive families of the index and control adoptees. Nearly all of the birth parents (N = 126) and adoptive parents (N = 129) were

Table 13.7

Design of adoption studies reported by Kety and associates

	Nonschizophrenic adoptees	Schizophrenic adoptees
Biological relatives	Control group for genetic test	Test of genetic influence
Adoptive relatives	Control group for environmental test	Test of environmental (rearing) influence

identified. Although the search revealed few full siblings of the index or control adoptees (N = 32), a large number of biological half-siblings (N = 176) was found.

The results of an early study relying on hospital diagnoses were mixed (Kety et al., 1968, 1971), but results based on extensive psychiatric interviews (Kety et al., 1975, 1976) are less ambiguous. Approximately 90 percent of the relatives were interviewed, and psychiatric diagnoses were obtained by consensus. Table 13.8 shows the frequency of hard schizophrenic spectrum in first-degree biological relatives, biological half-siblings, and adoptive relatives of schizophrenic and nonschizophrenic adoptees. The data for first-degree biological relatives support a genetic hypothesis, as do the results for the biological half-siblings.

The comparison of biological half-siblings who have the same father (paternal half-siblings) with those who have the same mother (maternal half-siblings) is particularly useful for examining the possibility that results of adoption studies may be affected by prenatal or early maternal care, rather than by genetic transmission. Data on paternal half-siblings are not influenced by these environmental variables. Using results of both hospital diagnoses and psychiatric interviews, the frequency of hard schizophrenic spectrum in paternal half-siblings of schizophrenic and nonschizophrenic adoptees was determined (Table 13.8). These important data again confirm a genetic hypothesis, suggesting that prenatal maternal environmental variables do not play an important role in the results of these adoption studies.

Jon L. Karlsson (1966, 1970) conducted several small adoption studies in Iceland. Different designs were used, but three studies yield 52 first-degree relatives separated from schizophrenic probands. Of these relatives, 12 (23 percent) were schizophrenic, providing strong support for a genetic hypothesis. This percentage is considerably higher than the usual familial incidence in first-degree relatives, possibly due to greater severity of schizophrenia in the probands in these studies.

Table 13.8

Hard schizophrenic spectrum (based on psychiatric interviews) in biological and adoptive relatives of schizophrenic and nonschizophrenic adoptees

	Nonschizophrenic adoptees	Schizophrenic adoptees
First-degree biological relatives	4% (3/68)	12% (8/68)
Adoptive parents and adoptive siblings	4% (4/90)	3% (2/73)
Biological half-siblings (total sample)	3% (3/104)	16% (16/101)
Biological half-siblings (paternal only)	3% (2/64)*	18% (11/61)

*Both psychiatric interviews and hospital diagnoses.
SOURCE: Kety et al., 1976.

Investigators in Denmark have also tested the effect of a schizophrenic rearing environment by studying adoptees whose birth parents were not affected but whose adoptive parents were schizophrenic. Using the same pool of adoptees as in the studies reported by Rosenthal and Kety, hospital records were searched for those whose adoptive parents had been diagnosed as schizophrenic (Wender et al., 1974). A total of 28 such cases were found. Of these schizophrenic adoptive parents, nine were chronic, nine were acute, four were borderline, and six were diagnosed as falling within the soft schizophrenic spectrum. The adoptees in this group (the Bn/Aa cell in Table 13.1) were compared to those in the two cells of the design that were discussed earlier (Bn/An and Ba/An). Diagnosis of the adoptees was by means of a three- to five-hour psychiatric interview and 1½ days of psychological testing (providing a rich data source that will continue to be mined in the future). The essential aspects of the cases were typed on cards and sorted by four raters into 20 categories of severity. Adoptees with scores higher than 15 were considered to be in the hard schizophrenic spectrum.

The 24 adoptees classified in the hard schizophrenic spectrum were found to be distributed in the three groups as indicated in Table 13.9. These findings suggest that being reared by a parent in the schizophrenic spectrum is not sufficient to produce schizophrenia. The influence of genetic factors is again supported by the comparison between adoptees with no schizophrenic birth parents (Bn/An and Bn/Aa) and those with a birth parent in the schizophrenic spectrum (Ba/An).

These results tend to confirm the findings of an earlier study by the same investigators (Wender, Rosenthal, and Kety, 1968) that was conducted in the United States using a different design. Natural parents rearing their own schizophrenic offspring were compared to adoptive parents of schizophrenics, and adoptive parents of unaffected children. Significantly higher psychopathology was found among the natural parents of schizophrenics than among the adoptive parents of schizophrenics. This finding supports a genetic hypothesis. However, the adoptive parents of schizophrenics received a significantly higher psychopathology rating than adoptive parents of normal children, suggesting that either these parents contributed to schizophrenia in their

Table 13.9

Hard schizophrenic spectrum as a function of rearing and schizophrenic status of birth parents

Birth parents	Adoptive parents	
	Normal	Affected
Normal	10% (8/79)	11% (3/28)
Affected	19% (13/69)	—

SOURCE: Wender et al, 1974.

children or they responded to the worry caused by having a chronically ill child.

Because this has been the only study to suggest a possible role for schizophrenic rearing, a similar design was used in a subsequent investigation (Wender et al., 1977). The study involved the adoptive status of all patients in 14 New York metropolitan hospitals. Individuals who had been adopted before one year of age and who were currently 15 to 30 years old were diagnosed by consensus for disorders falling within the hard schizophrenic spectrum. The 33 parents of 19 such adoptees were compared to 33 natural parents who had reared schizophrenic children. The parents were diagnosed by psychiatric interviews for hard and soft schizophrenic spectrum, and the results are summarized in Table 13.10. Data for the soft schizophrenic spectrum strongly support a genetic hypothesis and deny the influences of rearing, although results for the hard spectrum are not as clearcut. Where the diagnoses were based on computer analysis rather than psychiatric interviews, the results support a genetic hypothesis for both types of schizophrenic spectrum.

In closing this section on adoption studies of schizophrenia, it is appropriate to mention another type of adoption design involving identical twins reared separately. Gottesman and Shields (1982) have reviewed studies including a total of 12 pairs of identical twins, separated by the age of two years, in which at least one co-twin was schizophrenic. Of the 12 pairs, 7 were concordant for schizophrenia, a rate that is even higher than the incidence in identical twins reared together. Although this approach is inherently limited by small sample sizes, its results corroborate other adoption data in suggesting a major role for genetic influences.

With respect to schizophrenia, results of the adoption studies clearly point to genetic influence. Including the study by Heston, the Danish studies, and the study by Karlsson (with the results given in Tables 13.5, 13.6, and 13.7, and in the summary of Karlsson's data), there are 211 first-degree biological relatives of schizophrenic adoptees and 185 control individuals.

Table 13.10

Schizophrenic spectrum in the birth and adoptive parents of schizophrenics

	Diagnosis by psychiatric consensus		Diagnosis by computer analysis	
	Hard spectrum	Soft spectrum	Hard spectrum	Soft spectrum
Birth parents who reared schizophrenics	18% (6/33)	45% (15/33)	18% (6/33)	36% (12/33)
Adoptive parents of schizophrenics	15% (5/33)	15% (5/33)	3% (1/33)	6% (2/33)

SOURCE: After Wender et al., 1977.

The incidence of schizophrenia among the biological relatives of schizophrenics was 13 percent (28/211), while that in the control group was 1.6
percent (3/185). This summary result is consistent with the findings of each
individual study. Thus, an overall view of the adoption study data on schizophrenia clearly allows us to reject the hypothesis of no genetic influence. This
conclusion is in line with a recent reanalysis of the Denmark adoption data
(Kendler and Gruenberg, 1984) and with an excellent summary of genetic
studies of schizophrenia (Gottesman and Shields, 1982).

The finding of schizophrenia in 13 percent of the biological relatives in
these studies deserves more discussion. This is the incidence in biological
relatives who share no environment with the schizophrenic individual. It is
actually higher than the incidence in biological relatives of schizophrenics
sharing the same family environment, as mentioned in our discussion of
family studies in Chapter 11. This suggests that all of the familial resemblance
for schizophrenia is due to heredity, and that none is due to between-family
environmental influences, whose hypothesized role in the etiology of schizophrenia has seemed so reasonable. This conclusion is supported by the cross-
fostering research that found no increase in the incidence of schizophrenia in
adoptees reared by schizoid adoptive parents.

This does not mean that the environment is unimportant in triggering
schizophrenia. However, it does indicate that the environmental culprit that
we have traditionally blamed for schizophrenia (between-family influences) is
actually blameless. As was true in the family and twin studies, the adoption
studies suggest that the environment plays a very substantial role but operates
within families, making members of the same family distinct from one another. Unfortunately, we are a long way from tracking down these within-
family environmental influences. From the point of view of prevention, we
can only hope that Gottesman and Shields are wrong when they suggest that
"the 'culprits' may be nonspecific, time-limited in their effectiveness, and
idiosyncratic" (1976, p. 379).

SUMMARY

Adoption studies provide the most convincing demonstration of the genetic
influence on complex human behaviors. Issues of representativeness, selective
placement, and assortative mating were discussed. Adoption studies, like the
twin studies reviewed in Chapter 12, suggest that about half of the observed
variation in IQ scores is due to genetic differences. Though the majority of the
environmental variance occurs between families for young siblings, such
between-family variance has less influence on the resemblance between parents and offspring and between older siblings. For schizophrenia, adoption
data suggest that the familial resemblance long known to occur for psychosis
is due to heredity and not to between-family environmental influences. The
same data, however, provide strong evidence for the role of nonshared environmental factors.

CHAPTER·14

Human Behavioral Genetics Research

Chapters 11, 12, and 13 described family, twin, and adoption methods and used IQ and schizophrenia as examples of the application of these methods. The aim of the present chapter is to illustrate the diverse areas in the social and behavioral sciences to which behavioral genetic methods have been applied. Consistent with this goal, our overview of human behavioral genetics research emphasizes breadth, not depth. That is, what follows is not an exhaustive review of human behavioral genetic studies—such a review requires a book in itself. Instead, our overview considers some of the major areas of behavioral genetic applications to cognitive abilities, psychopathology, and personality.

For each of these domains, diverse topics are covered. In the section on cognitive abilities, we pull together the family, twin, and adoption research presented in previous chapters. We also discuss related topics: occupational status, race and class differences in IQ, specific cognitive abilities, academic performance, reading disability, mental retardation, and dementia. The section on psychopathology includes a review of our previous discussion of schizophrenia, affective disorders, other psychopathology such as anorexia, delinquent and criminal behavior, and alcoholism. The personality section focuses on the two major "superfactors" of extraversion and neuroticism and also discusses research on other personality traits, attitudes and beliefs, and vocational interests.

We hope that this litany of topics is not overwhelming but rather entices students to want to learn more. For each topic we present results from one of the best studies in the area, as well as a summary of other studies and references that will allow students to pursue these topics in greater depth. In an attempt to keep this overview to a manageable length, we have focused on behavioral genetics results themselves, rather than, for example, discussing issues related to the assessment of each domain.

In addition to introducing the diverse applications of behavioral genetic research and illustrating the widespread influence of heredity on behavior, we also hope that this overview indicates how much remains to be learned. By no means have all the questions about nature and nurture been answered for any domain. Indeed, the ratio of what is known to what needs to be learned is small. The excitement and the future of behavioral genetics lies in improving this ratio and in considering other areas of the social and behavioral sciences from the behavioral genetics perspective that recognizes the possibility of genetic as well as environmental influences.

The second section of the chapter illustrates this point. A new area for research in human behavioral genetics focuses on the environment. There is much to be learned about the interface between the environment and behavior when it is explored from a behavioral genetics perspective. One example of research in this area is the topic of nonshared environment, which, although mentioned in previous chapters, is highlighted in the first part of this section. A second example is research on "the nature of nurture," which shows that heredity can affect measures of the environment.

The third section of the chapter considers five future directions for research in behavioral genetics. The final section discusses the relationship between science and politics and returns to the opening theme of this text concerning genetics, individual differences, and social equality.

COGNITIVE ABILITIES

As mentioned, this section provides an overview of diverse topics related to cognitive abilities. We begin with a model-fitting analysis that pulls together the data on general cognitive ability, IQ.

Intelligence Quotient (IQ)

As indicated in the three previous chapters, family, twin, and adoption data for IQ converge on the conclusion that individual differences in IQ scores are substantially influenced by heredity. Figure 14.1 summarizes behavioral genetics IQ research prior to 1980 (Bouchard and McGue, 1981). Each dot represents one study; the number of studies, number of individuals, and the average weighted correlations are also listed. These are the IQ correlations referred to in earlier chapters and they suggest that genetic influence is

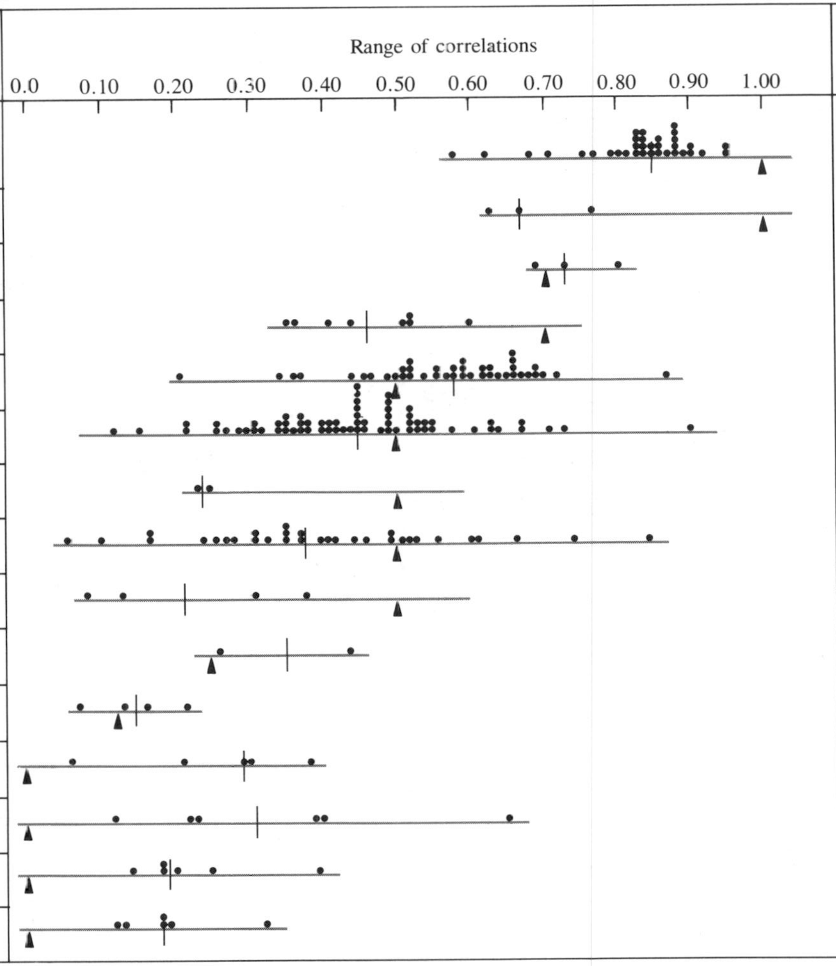

	Number of correlations	Number of pairings	Weighted average
Monozygotic twins reared together	34	4,672	0.86
Monozygotic twins reared apart	3	65	0.72
Midparent–midoffspring reared together	3	418	0.72
Midparent–offspring reared together	8	992	0.50
Dizygotic twins reared together	41	5,546	0.60
Siblings reared together	69	26,473	0.47
Siblings reared apart	2	203	0.24
Single parent–offspring reared together	32	8,433	0.42
Single parent–offspring reared apart	4	814	0.22
Half-siblings	2	200	0.31
Cousins	4	1,176	0.15
Nonbiological sibling pairs (adopted/natural pairings)	5	345	0.29
Nonbiological sibling pairs (adopted/adopted pairings)	6	369	0.34
Adopting midparent–offspring	6	758	0.24
Adopting parent–offspring	6	1,397	0.19

Figure 14.1

A summary of IQ correlation coefficients for family, twin, and adoption studies compiled by Bouchard and McGue from various sources. The horizontal lines show the median correlation coefficients, and the arrows show the correlation expected if IQ were entirely due to additive genetic variance. (From "Familial studies of intelligence: A review" by T. J. Bouchard, Jr., and M. McGue. Science, 212, 1055–1059. Copyright © 1981 by the American Association for the Advancement of Science.)

substantial on IQ scores, that genetic influence is mostly additive, and that environmental influence is in part due to shared environmental factors.

As discussed in Chapter 9, model-fitting techniques are useful for analyzing data from different groups simultaneously, rather than examining family, twin, and adoption results separately. The data in Figure 14.1 were fitted to a simple model similar to the model described in Chapter 9 (Chipuer, Rovine, and Plomin, 1989). (See Figure 9.14 and accompanying text.) The data from the 12 groups in Figure 14.1, excluding the midparent comparisons, were fitted to the model, which includes additive and nonadditive genetic parameters and shared and nonshared environmental parameters. As discussed in Chapter 9, this model keeps open the possibility that shared environmental influence differs in magnitude for twins, siblings, and parents and their offspring. Because cousin data are available for IQ, the possibility of a different shared environmental effect for cousins was added to the model described in Chapter 9. In addition, the analysis adjusted for assortative mating, which, as indicated in Chapter 8, is important for IQ.

Four models were analyzed using model-fitting procedures, with results shown in Table 14.1. The "null model," which tests whether all familial correlations are equal, can confidently be rejected. Although the tremendous power gained by including so many groups in this analysis detects significant differences between the data and the other three models, these models clearly fit better than the null model. Two findings in Table 14.1 are of particular interest. First, the fit of model 4, which includes a nonadditive genetic parameter, is significantly better than model 3, which drops that parameter, indicating a significant effect of nonadditive genetic variance. Second, model 4, with separate shared environmental parameters for twins, siblings, parents and their offspring, and cousins, fits the data significantly better than model 2, which equates these parameters. This finding indicates that the magnitude of shared environment differs for the four groups.

Table 14.1
Model-fitting results for IQ data in Figure 14.1

Model	χ^2	df	p
1. All rs equal (null)	2,478.76	11	0.000
2. $h^2 + d^2 + c^2$	363.47	9	0.000
3. $h^2 + c_1^2 + c_2^2 + c_3^2 + c_4^2$	101.58	7	0.000
4. $h^2 + d^2 + c_1^2 + c_2^2 + c_3^2 + c_4^2$	67.98	6	0.000
Model 4 vs. model 3 (testing nonadditive genetic variance)	33.60	1	<0.05
Model 4 vs. model 2 (testing equality of shared environment)	325.61	3	<0.001

SOURCE: Adapted from Chipuer, Rovine, and Plomin, 1989.

Table 14.2 lists parameter estimates for the full model. The standard errors indicate that both additive and nonadditive genetic variance is significant, as are the shared environmental components for all four groups. Translating the parameter estimates to percentage of variance explained by squaring the parameter estimates indicates that genetic variance accounts for 45 percent of the phenotypic variance in IQ scores, 32 percent for additive, and 13 percent for nonadditive genetic variance. As suggested by the results of model fitting, the magnitude of shared environment differs for the four groups. The differences seem reasonable: The magnitude of shared environment from most to least is twins (37 percent), siblings (24 percent), parents and offspring (20 percent), and cousins (11 percent). However, these genetic and shared environmental parameters differ developmentally, as discussed in Box 14.1.

Although the issue of genetic influence on IQ scores has traditionally been one of the most controversial areas in the behavioral sciences, a recent survey of over a thousand scientists and educators indicates that most now believe that individual differences in IQ scores are at least partially inherited (Synderman and Rothman, 1987).

An area relevant to IQ is occupational status, which correlates greater than 0.50 with IQ (Jensen, 1980). Controversy in this area was fueled by the suggestion that our society is moving toward a hereditary meritocracy, stratified according to hereditary differences in ability (Herrnstein, 1973). The reasoning behind this argument is that social class is likely to show heritable influence, because measures of social class show substantial correlations with IQ, which is known to be substantially heritable. Studies since then have shown that occupational status is significantly influenced by heredity. For example, a study of 1,900 pairs of 50-year-old male twins yielded identical and fraternal twin correlations of 0.42 and 0.21 for current occupation and 0.54 and 0.30 for earnings (Taubman, 1976). An adoption study of occupational status yielded correlations of 0.20 and 0.14, respectively, between biological fathers and their adult adopted-away sons (2,467 pairs) and daugh-

Table 14.2

Parameter estimates for model 4 solutions from Table 14.1

Parameter	Parameter estimates $\pm$ standard error	Percentage variance
h^2	0.58 ± 0.01	34
d^2	0.3 ± 0.04	12
c^2_{twin}	0.6 ± 0.01	38
$c^2_{sibling}$	0.50 ± 0.01	25
$c^2_{parent-offspring}$	0.44 ± 0.01	19
$c^2_{cousins}$	0.33 ± 0.01	11

Box 14.1

Developmental Behavioral Genetics and IQ

Behavioral genetics results for IQ differ as a function of age. A new subdiscipline, developmental behavioral genetics, focuses on developmental change rather than assuming that genetics is only involved in continuity (Plomin, 1986). Two types of developmental changes are considered: changes in heritability and age-to-age genetic changes. Heritability can change as different genetic and environmental systems come into play during development. Age-to-age genetic change refers to the extent to which the genes that affect a trait at one age overlap with genetic effects at another age.

In the previous discussion of behavioral genetic results for IQ, no mention was made of the age of the subjects in the various studies, even though the ages vary widely. However, as the twin studies discussed in Chapter 12 indicate, the results for IQ differ dramatically during development. Heritability of IQ scores increases substantially during childhood and may increase further during adolescence. Genetic effects account for about 15 percent of the variance in infant mental test scores and increase in importance to about 40 percent of the variance by the early school years. Increasing heritability means that the phenotypic variance of IQ scores is increasingly due to genetic differences. This may occur because more genes come to affect IQ scores, because early genetic effects produce increasingly larger IQ differences, or— less likely— because environmental differences come to be less important. You might think that the increase in heritability is due to increasing reliability of IQ tests, but that is not the case. Infant and childhood tests are highly reliable in the short term, even though they do not predict later IQ very well.

This increase in heritability is particularly interesting. You would guess that as children develop and experience more diverse environments, environmental variance will increasingly account for phenotypic variance, which implies that heritability will thus decrease. However, this is not the case for IQ.

Heritability is not the only component of variance that changes developmentally. In childhood, shared environment accounts for about 30 percent of the variance of IQ, but its importance wanes during adolescence. As mentioned in Chapter 13, the IQ

ters (1,519 pairs) (Teasdale, 1979). Correlations between biological mothers' occupational status and that of their adopted-away offspring were 0.26 for sons (505 pairs) and 0.16 for daughters (326 pairs). A study of 99 pairs of adopted-apart siblings yielded a correlation of 0.22 (Teasdale and Owen, 1981). All of these results are consistent with a heritability of about 40 percent for occupational status. However, it is not yet clear to what extent genetic influence on occupational status is due to its correlation with IQ.

An even more controversial area involves racial/ethnic differences in IQ scores. Average IQ differences between blacks and whites have most often been studied, and blacks, on the average, score about one standard deviation (about 15 IQ points) lower than whites on conventional tests of general intelligence. The etiology of this average difference is very much disputed.

correlation for unrelated children adopted together into the same family provides the most direct test of the importance of shared environment. Earlier studies involving such adoptive siblings happened to study them in childhood. These studies of young adoptive siblings still living at home yield an IQ correlation of about 0.30. However, four recent studies of older adoptive siblings yield IQ correlations of zero on average (Plomin, 1988). This finding suggests that although shared environmental influences are important for IQ in childhood, their importance wanes to negligible levels during adolescence.

In addition to developmental changes in heritability, the second type of genetic change in development, age-to-age change, is also a topic of importance for IQ. That is, to what extent are genetic effects at one age related to genetic effects at a later age? Genetic change means that genetic effects at one age differ from genetic effects at another age. DNA does not change—different genes are expressed or the products of genes change in their effects during development. Even though heritability is substantial for a trait in childhood and in adolescence, different genetic effects might operate at the two ages. Longitudinal data in which subjects are tested repeatedly are particularly valuable in developmental behavioral genetics research, because they permit the analysis of age-to-age genetic change and continuity.

The distinction between the two types of genetic change during development is important in terms of IQ. Even though heritability increases from childhood to adulthood, evidence is beginning to mount that genetic effects on IQ scores in early childhood are highly correlated with genetic effects on IQ scores in adulthood (De-Fries, Plomin, and LaBuda, 1987). In other words, even though the heritability of IQ is relatively low in early childhood, genetic effects at that age continue to have an effect on individual differences in IQ in adulthood. You might wonder why phenotypic stability from childhood to adulthood is not also high for IQ. The answer is that heritability of IQ is much less than 1.0 at each age. That is, although genetic effects on IQ in childhood correlate very highly with genetic effects on IQ in adulthood, environmental effects are also important, and these environmental effects are apparently not stable from childhood to adulthood.

Some argue that it is simply caused by cultural bias of IQ tests, although research does not generally support this hypothesis (Jensen, 1980). A controversy of unprecedented proportions in the behavioral sciences began in 1969 with the publication of a monograph by Arthur Jensen, in which he reviewed the evidence for the heritability of IQ scores and hypothesized that genetic factors may be implicated in the observed difference between blacks and whites. However, the fact that IQ scores are heritable within black and white groups does not mean that the difference between the groups is necessarily heritable. Even if heritability within each of the two groups were 100 percent, the difference between the groups could be completely environmental in origin. For example, suppose that two groups were equal in means and variance. Then suppose that some environmental variables (those related to prejudice, for example) led to lower scores for everyone in one group. If these variables had equal effects on all members of that group, variance within the group would not be changed, even though the mean would be lower. Thus, heritability within both groups would still be 1.0, and the difference between the groups would be entirely environmental. Though this example is extreme, it makes the point that high within-group heritability does not necessarily imply that an observed difference between group means is also highly heritable.

As discussed in Chapter 9, it is difficult to determine whether an average difference between groups is due to genetic or environmental factors. An admirably balanced review and evaluation of studies in the area of racial/ethnic IQ differences concludes that the available data are essentially equivocal (Loehlin, Lindzey, and Spuhler, 1975; see also Vernon, 1979; Willerman, 1979). It is especially important to distinguish differences *between* groups from differences *within* groups in discussions of social class and ethnic differences in IQ. For both blacks and whites in the United States, social class (independent of race) accounts for about 8 percent of the variance of IQ. Race (independent of social class) accounts for about 14 percent. However, individual differences, independent of race and class, are responsible for 78 percent of the variance. In fact, the average IQ difference between full siblings in a family is twice as great as the average differences between social classes and as great as the average difference between blacks and whites (Jensen, 1976). Thus, although average group differences are deceptively easy to grasp, individual differences within groups make a far more important contribution to total variance.

Specific Cognitive Abilities

There is certainly more to characterizing cognitive ability than can be achieved by the IQ score alone. Although specific abilities tend to be moderately correlated with one another, lending support to the concept of "general intelligence," they are sufficiently different to permit a more fine-grained analysis of cognitive functioning. Depending on their level of analysis, researchers have focused on two general abilities (such as verbal and nonverbal performance), the six to twelve group factors measured by L. L. Thurstone, and the 120 postulated by J. P. Guilford (1967) in his model of intelligence. The specific cognitive abilities that have been studied most in behavioral genetics are verbal ability, spatial ability, memory, and perceptual speed; each is a complex ability that needs to be studied at a more fine-grained level of analysis.

An important family study of specific cognitive abilities is the Hawaii Family Study of Cognition, described in Chapter 11 (DeFries et al., 1979). Fifteen tests of cognitive abilities were administered to over 6,000 individuals in nearly 2,000 families. Application of a technique called *factor analysis* to correlations among scores on the 15 tests yielded four group factors: verbal (including vocabulary and fluency), spatial (visualizing and rotating objects in two- and three-dimensional space), perceptual speed (simple arithmetic and number comparisons), and visual memory (short-term and longer-term recognition of line drawings). A multivariate analysis (see Chapter 9) of cross-correlations between parents and offspring yielded similar factors (DeFries, Kuse, and Vandenberg, 1979). This suggests that the genetic and environmental influences salient to specific cognitive abilities are neither very broad nor idiosyncratic in their effect. Rather, there are several sets of genetic and environmental influences that correspond to the phenotypic factors.

Regressions of offspring on midparent (the upper limit of heritability) for the four factors and the 15 cognitive tests are plotted separately for two ethnic groups in Figure 14.2. The most obvious fact is that familiality differs for the various tests and factors. These data were corrected for unreliability of

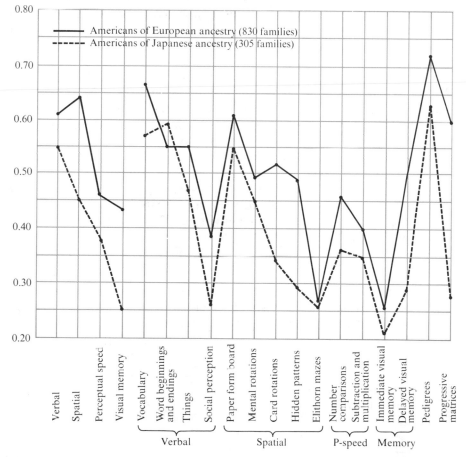

Figure 14.2

Family study of specific cognitive abilities. Regression of midchild on midparent for four group factors and 15 cognitive tests in two ethnic groups. (Data from DeFries et al., 1979.)

the tests to make sure that the differences in familial resemblance were not caused by reliability differences among the tests. For both ethnic groups, the verbal and spatial factors show more familial resemblance than the perceptual speed and memory factors. Other family studies (reviewed in DeFries, Vandenberg, and McClearn, 1976) also indicate that the greatest familial similarity occurs for verbal ability.

Figure 14.2 makes another important point: Tests within each factor show dramatic differences in familial resemblance. For example, one spatial test, "Paper Form Board," shows high familiality. The test involves showing how to cut a figure to yield a certain pattern — for example, to cut a circle to yield a triangle and three crescents. Another spatial test shows the lowest familial resemblance. This test involves drawing one line that connects as many dots as possible in a maze of dots. For this reason, future research on specific cognitive abilities is likely to consider more finely differentiated abilities, especially information processing and other new approaches that are emerging from experimental psychology. In general, the pattern of resemblance is quite similar in the two ethnic groups, although, as discussed in Chapter 11 for IQ, familial resemblance tends to be slightly lower in the families of Japanese ancestry.

The results of dozens of twin studies are summarized in Table 14.3 (Nichols, 1978). Specific cognitive abilities show slightly less genetic influence than IQ. Memory and verbal fluency show lower heritability (about 30 percent); the other abilities suggest heritabilities from about 40 to 50 percent. The largest twin studies do not consistently find greater heritability for particular cognitive abilities (Bruun, Markkanen, and Partanen, 1966; Loehlin and Nichols, 1976; Schoenfeldt, 1968), although it has been argued that verbal and spatial abilities in general show greater heritability than do perceptual speed and especially memory abilities (Plomin, 1986a). From 20 to 40 percent of the variance is due to shared environment, although it is likely that twins share environmental influences to a greater extent than do nontwin

Table 14.3
Twin correlations for tests of specific cognitive abilities

Ability	Number of studies	Twin correlations	
		Identical	Fraternal
Verbal comprehension	27	0.78	0.59
Verbal fluency	12	0.67	0.52
Reasoning	16	0.74	0.50
Spatial visualization	31	0.64	0.41
Perceptual speed	15	0.70	0.47
Memory	16	0.52	0.36

SOURCE: Nichols, 1978.

siblings (Plomin, 1988). Multivariate analyses suggest that intercorrelations among cognitive abilities are largely due to genetic covariance (LaBuda, DeFries, and Fulker, 1987; Tambs, Sundet, and Magnus, 1986).

Few adoption studies have included batteries of tests of specific cognitive abilities. In one study, nonadoptive parent–offspring and sibling correlations were greater than adoptive parent–adopted child and adoptive sibling correlations for subtests of an intelligence test (Scarr and Weinberg, 1978a). The only significant correlations for the adoptive relationships emerged for vocabulary, suggesting that vocabulary scores are influenced by shared environmental factors. Genetic influence plus substantial shared environmental influence may explain the high familial resemblance for vocabulary in the Hawaii Family Study described above.

Specific cognitive abilities is one focus of the ongoing Colorado Adoption Project (Plomin, DeFries, and Fulker, 1988). Although the children in this study are still young, significant genetic influence is found in parent–offspring model-fitting analyses of verbal, spatial, and perceptual speed abilities, but not for memory (Plomin et al., 1988). Multivariate analyses suggest that genetic effects on the various cognitive abilities are highly correlated — that is, the genes that affect one cognitive ability also affect other cognitive abilities (Rice, Fulker, and DeFries, 1986).

In summary, diverse cognitive tests show significant and often substantial genetic influence throughout the life span. Not enough work has been done developmentally to warrant conclusions concerning developmental changes in components of variance for specific cognitive abilities. Memory abilities may be less influenced than other specific cognitive abilities by heredity. One dimension in the cognitive realm that appears to show little genetic influence is creativity. A review of ten twin studies of creativity yields an average correlation of 0.61 for identical twins and 0.50 for fraternal twins (Nichols, 1978). Some work indicates that this genetic influence of about 20 percent is primarily due to the correlation between creativity and IQ. When IQ is controlled, identical and fraternal twin correlations for creativity tests are scarcely different (Canter, 1973).

Academic Performance

If heredity plays an important role in most specific cognitive abilities, it seems likely that heredity also affects school achievement. Several large studies of twins have employed measures of academic achievement and scholastic ability rather than measures of specific cognitive abilities; however, the distinction between these domains is not at all clear.

Report card grades show substantial genetic influence. For example, in one study, school grades were obtained for 352 pairs of identical and 668 pairs of fraternal 13-year-old twins in Sweden (Husén, 1959). Correlations for identical and fraternal twins, respectively, were 0.72 and 0.57 for reading grades, 0.76 and 0.50 for grades in writing, 0.81 and 0.48 for arithmetic, and 0.80 and 0.51 for history.

Twin studies of academic achievement test scores also show substantial genetic influence. For example, the largest twin study in the United States utilized data from the National Merit Scholarship Qualifying Test for 1,300 identical and 864 fraternal twin pairs (Loehlin and Nichols, 1976). The twin correlations, listed in Table 14.4, are quite similar to those for specific cognitive abilities. The average difference between the correlations for the two types of twins is about 0.20, suggesting heritabilities of about 40 percent for scholastic achievement tests. The consistency of results across tests is not so surprising, since the tests intercorrelate highly, about 0.60. Multivariate analysis of these data indicates that genetic correlations among the tests are substantial, suggesting that a general genetic factor affects the various tests, although each test is also influenced by independent genetic factors (Plomin and DeFries, 1979; Martin, Jardine, and Eaves, 1984). In other words, these results suggest that there are genetic reasons why some students who do well in English, for example, also do well in other scholastic areas.

Reading Disability

As many as 25 percent of children have difficulty learning to read. For some, specific causes can be identified, such as mental retardation, brain damage, sensory problems, and cultural or educational deprivation. However, many children without such problems find it difficult to read.

Family studies have shown that reading disability runs in families. For example, a study of 1,044 individuals in 125 families with a reading-disabled child and 125 matched control families shows familial resemblance in that the siblings and parents of the reading-disabled children performed significantly worse on reading tests than did siblings and parents of control children (DeFries, Vogler, and LaBuda, 1986). In addition, this study shows that

Table 14.4
Twin correlations for tests of scholastic achievement

	Twin correlations	
Test	Identical	Fraternal
English usage	0.72	0.52
Mathematics	0.71	0.51
Social studies	0.69	0.52
Natural sciences	0.64	0.45

SOURCE: Loehlin and Nichols, 1976.

reading-disabled children whose parents also have difficulty reading are less likely to improve after a five-year period.

Twin studies suggest that genetic factors may play an important role in familial resemblance for reading disability. In one study of 97 twin pairs, concordance for reading disability was 84 percent for identical twins and 29 percent for fraternal twins (Bakwin, 1973). In another study of 40 pairs, identical and fraternal twin concordances were 80 percent and 45 percent, respectively (Decker and Vandenberg, 1985). A third small study found evidence for genetic influence on spelling disability but not on other aspects of reading disability (Stevenson et al., 1987).

In the largest twin study of reading disability conducted to date (De-Fries, Fulker, and LaBuda, 1987), extensive psychometric test data were obtained from a sample of 64 pairs of identical twins and 55 pairs of fraternal twins in which at least one member of each pair had been diagnosed as being reading disabled. In order to test the hypothesis that reading disability is due at least in part to heritable influences, these data were subjected to a variant of the multiple regression analysis described in Chapter 12 (Box 12.1). This analysis, which is of general importance as a means to compare the etiology of the extremes of a distribution with the etiology of individual differences, is described in Box 14.2. This analysis was first applied to twin data on reading disability, and results suggested that reading disability may just be the lower end of the distribution of reading disability (DeFries and Fulker, in press).

No adoption data have been reported as yet for reading disability.

Mental Retardation

As indicated earlier, individual differences in IQ scores show substantial genetic influence. Although one might expect that very low IQ scores are due to genetic factors as well, the sources of extreme scores can differ from those of individual differences in the middle of the distribution. For example, very low IQ scores can be caused by chromosomal abnormalities such as Down's syndrome (see Chapter 6) and single-gene disorders such as PKU (see Chapter 4), as well as by environmental trauma, such as birth complications, nutritional deficiencies, and head injuries. Indeed, severe mental retardation (IQs less than 50) appears to be due largely to nonheritable factors: Severely retarded individuals are likely to have siblings of normal IQ (Nichols, 1984). In one study of over 17,000 white children, 0.5 percent were severely retarded (Nichols, 1984). As shown in Figure 14.4, none of the 20 siblings of the severely retarded children was retarded and the siblings' average IQ was 103.

In contrast, siblings of mildly retarded individuals tend to be mildly retarded. In the study just mentioned, 1.2 percent of the children were mildly retarded. Of the 58 siblings of the mildly retarded children, 12 were retarded and their average IQ was only 85. Findings for a black population also showed familial resemblance for mild retardation.

Box 14.2

Multiple Regression Analysis of Twin Data: Etiology of the Normal and the Abnormal

As mentioned in Box 12.1, the multiple regression approach to the analysis of twin data is especially useful in addressing one of the core issues in psychopathology: the relationship between the normal and the abnormal. When quantitative data, such as reading scores, are obtained, rather than qualitative data such as a diagnosis of reading disability, the multiple regression procedure can be used to assess the extent to which the magnitude of genetic and environmental factors that affect the disorder are similar to those that affect normal variability. In other words, the procedure tests whether the disorder is merely the extreme of a normal distribution of variability (DeFries and Fulker, in press).

To the extent that the performance deficits of probands are heritable, the scores of DZ probands should regress more than those of MZ co-twins toward the mean of the unselected population. For example when probands have been identified by low test scores (see Figure 14.3), the scores of both MZ and DZ co-twins will regress toward the mean of the unselected population. But to the extent that reading disability is heritable, this regression toward the mean should differ for MZ and DZ co-twins—DZ co-twins should regress more than MZ co-twins toward the mean of the unselected population.

When the regression model (without the interaction term) discussed in Box 12.1 is fitted to quantitative data from twins selected on the basis of a disorder in one twin, one of the regression coefficients estimate between-group heritability. In other words, it indicates the extent to which the difference between the probands and the unselected population is heritable. If this group heritability is comparable to the usual individual heritability, the disorder may represent the lower end of the normal continuum of variability.

This approach was first applied to the study of reading disability (DeFries, Fulker, and LaBuda). The results indicate that group heritability of reading disability—the extent to which the difference in reading scores between the reading disabled and the unselected population is genetically influenced—is about 30 percent. This group heritability is comparable to individual heritability of reading scores, which led the authors to conclude that reading disability may represent the lower end of the normal continuum of reading ability.

Results such as these have suggested that mild mental retardation is familial, even though severe retardation is not. An important family study demonstrating the familial nature of mild retardation involved 80,000 relatives of 289 mentally retarded individuals (Reed and Reed, 1965). If one parent is retarded, the risk for retardation in their children is about 20 percent; if both parents are retarded, the risk is nearly 50 percent; if a sibling is also retarded, the risk is about 70 percent. This data set also suggests that mild retardation is strongly familial but severe retardation is not (Johnson, Ahern,

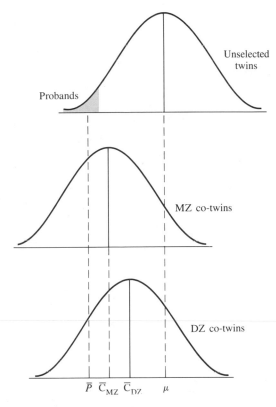

Figure 14.3
Hypothetical distributions for reading performance of an unselected sample of twins, and of the MZ and DZ co-twins of probands (P) with reading disability. The differential regression of the MZ and DZ co-twin means toward the mean of the unselected population (μ) provides a test of genetic influence. (From DeFries, Fulker, and LaBuda, 1987)

and Johnson, 1976). No twin or adoption data are as yet available, although one twin study is in progress (Detterman, 1987).

Dementias

Two diseases, Huntington's chorea and Alzheimer's disease, involve mental and motor deterioration in middle age or later. Huntington's disease is a single-gene dominant trait (described in Chapter 3). A gene on chromosome 21 has been implicated in some cases of familially transmitted Alzheimer's (Goldgaber et al., 1987), though, unlike Huntington's disease, Alzheimer's is not a simple single-gene characteristic. For example, many identical twin pairs are discordant for Alzheimer's disease (Nee et al., 1987; Renvoize et al., 1986). Familial resemblance was found in a study of 30 well-documented

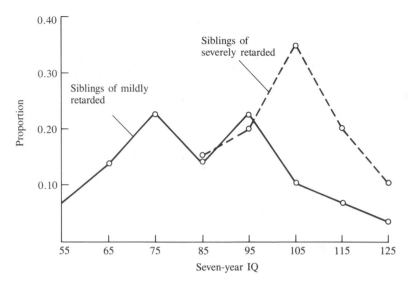

Figure 14.4
Siblings of mildly retarded children tend to be mildly retarded, and siblings of severely retarded children tend to be of normal IQ, suggesting that mild retardation is familial but severe retardation is not. (From Nichols, 1984.)

cases of Alzheimer's disease: The risk was 23 percent for parents and 10 percent for siblings of the Alzheimer's patients (Heston and Mastri, 1977). A twin study of general senility yielded concordances of 43 percent for identical twins and 8 percent for fraternal twins, suggesting substantial genetic influence (Kallman, 1955). No adoption data have been reported for the dementias.

PSYCHOPATHOLOGY

Psychopathology has received at least as much attention as cognitive abilities from behavioral geneticists. The field is intrinsically more complicated because, unlike general cognitive ability, there is no general factor that pervades psychopathology. What follows is only an overview of the highlights of research in psychopathology. Details can be found in a recent book-length review by Vandenberg, Singer, and Pauls (1986).

Schizophrenia

Schizophrenia — characterized by long-term thought disorders, hallucinations, and disorganized speech — has shown genetic influence in numerous family, twin, and adoption studies (see Chapters 11, 12, and 13). Family studies clearly indicate that first-degree relatives of schizophrenics are at considerably greater risk for schizophrenia than are individuals in the general population (risk of about 1 percent). The risk for parents of schizophrenics is greater than 5 percent, the risk for siblings exceeds 10 percent, and the risk for children of schizophrenics is about 13 percent. The average weighted risk for first-degree relatives of schizophrenics is 8.4 percent, more than eight times the risk for individuals chosen randomly from the population.

Twin studies suggest that this familial resemblance is substantially genetic in origin. Identical twin concordance is about 40 percent, and fraternal twin concordance is similar to nontwin sibling concordance, about 10 percent. The much greater similarity of identical twins as compared to fraternal twins suggests the possibility of nonadditive genetic factors that are shared by identical twins but not by fraternal twins or other first-degree relatives. This hypothesis is supported by adoption data for first-degree relatives. The average incidence of schizophrenia among the biological first-degree relatives of schizophrenic adoptees is 13 percent, whereas that in the control group is 1.5 percent. Although these adoption data indicate genetic influence, the magnitude of genetic influence appears to be less than suggested by twin studies, which would be the case if nonadditive genetic variance affected schizophrenia.

An excellent and very readable account of behavioral genetic research on schizophrenia is available (Gottesman and Shields, 1982); it also discusses diagnosis and other general issues concerning schizophrenia. As discussed in Chapter 7, schizophrenia has been linked to a gene on chromosome 5 in two Icelandic families with a high incidence of schizophrenia (Sherrington et al., 1988), although other studies have excluded the possibility of linkage to chromosome 5 in other families (e.g., Kennedy et al., 1988).

A final note concerning schizophrenia involves infantile autism, which shows schizophrenic-like symptoms beginning before the age of three years. However, behavioral genetic research suggests that autism is largely distinct from adult schizophrenia. For 936 parents of autistic children, about 2 percent were hospitalized for schizophrenia, compared to the population risk of about 1 percent for schizophrenia. For 743 siblings of autistic children, only 1.7 percent were schizophrenic. Autism is a rare disorder, with a prevalence of two to five per 10,000, a male–female ratio of 3:1, and an association with mental retardation (Smalley, Asarnow, and Spence, 1988). Two twin studies suggest genetic influence on autism (Folstein and Rutter, 1977; Ritvo et al., 1985). Twin concordances pooled across studies are 64 percent for MZ twins and 9 percent for DZ twins (Smalley et al., 1988). It has been suggested that some basic genetic abnormality of language or sociability is involved (Folstein and Rutter, 1988).

Affective Disorders

Behavioral genetics research during the past decade has turned to the study of affective disorders of mood, which are of two major types; depression (unipolar, major depressive disorder) and depression alternating with manic elevations of mood (bipolar depression).

Depression, the most common form of mental illness, is marked by feelings of worthlessness, sadness, disturbances of sleep and appetite, loss of energy, and suicidal ideation. Mania is characterized by hyperactivity, reduced need for sleep, and euphoria. Difficulties in diagnosing affective disorders create ambiguity in ascertaining base rates in the population; one recent attempt suggests that the lifetime risk for a major depressive disorder is about 5 percent and that about 1 percent of the population has experienced both a manic and depressive psychotic episode (Robins et al., 1984). As is the case for schizophrenia, manic-depressive disorder occurs with equal frequency for males and females; depression, however, occurs twice as frequently for females.

Affective psychoses are genetically distinct from schizophrenia: There is no more schizophrenia in relatives of affective psychotics than in the general population, and affective psychosis occurs no more frequently in relatives of schizophrenics than in the general population. An in-between disorder, called *schizoaffective* disorder, is generally agreed to be related to the affective disorders, because family members of these individuals usually have affective disorders rather than schizophrenia.

Familial resemblance for the affective disorders is as great as for schizophrenia: In nine studies of nearly 6,000 first-degree relatives of affected individuals, the risk is 9 percent (Rosenthal, 1970). Eight family studies published since 1975 also consistently show familial resemblance (Nurnberger and Gershon, 1981). There is evidence from several studies that the prevalence of depression is increasing and that the onset of depression is occurring earlier in younger cohorts, a change that has occurred too quickly to be accounted for by genetic factors (Weissman, Kidd, and Prusoff, 1987). The most recent study consists of 235 probands with major depressive disorder and their 826 first-degree relatives (Reich et al., 1987). Major depression was diagnosed in 13 percent of the male relatives and in 30 percent of the female relatives. For males especially, a cohort effect is seen, inasmuch as the frequency of depressive disorder is higher in younger cohorts (9 percent for fathers, 12 percent for brothers, and 19 percent for sons).

The familial risk for bipolar illness is lower. In seven studies of 2,500 first-degree relatives of bipolar probands, the average risk of bipolar illness is 5.8 percent (Rice et al., 1987). The most recent study involved 187 families of bipolar probands and yielded a frequency of 5.7 percent bipolar illness in 557 first-degree relatives as compared to a risk of 1.1 percent in a control sample (Rice et al., 1987).

Twin studies indicate that familial resemblance for the affective disorders is largely genetic in origin. The overall concordance for identical twins is 65 percent and that for fraternal twins is 14 percent in seven studies involving a total of 146 pairs of identical twins and 278 pairs of fraternal twins (Nurnberger and Gershon, 1981). For example, the largest twin study of affective disorders yields concordances of 67 percent and 18 percent, respectively, for 55 pairs of identical twins and 52 pairs of fraternal twins in Denmark (Bertelsen, Harvald, and Hauge, 1977).

These twin results suggest even greater genetic influence than for schizophrenia. However, unlike schizophrenia, adoption studies of affective disorders indicate far less genetic influence for affective disorders. Four adoption studies of affective disorders have been reported, and they yield mixed results (Loehlin, Willerman, and Horn, 1988). One of the best studies, however, indicates genetic influence (Wender et al., 1986). The biological and adoptive relatives of 71 affectively ill adoptees and 71 control adoptees were studied. Unipolar and bipolar disorders occurred in 5.2 percent of the 387 biological relatives of the probands and in 2.4 percent of the 344 biological relatives of the controls. The biological relatives of affected adoptees also showed greater rates of alcoholism (5.4 percent versus 2.0 percent) and attempted or actual suicide (7.3 percent versus 1.5 percent). The sharp differences between the twin and adoption data suggest caution in reaching conclusions about the relative importance of genetic and environmental factors.

An interesting recent study involves the offspring of identical twins discordant for manic-depressive illness (Bertelsen, 1985). The same 10 percent risk of affective disorder was found in the offspring of identical twins, regardless of whether the identical twin was affected. This suggests that the identical twin who does not evidence manic-depressive illness nonetheless transmits the illness to his or her offspring to the same extent as does the ill co-twin.

Family studies suggest that depressive and manic-depressive disorders may be distinct. For example, relatives of manic-depressives have higher rates of manic-depressive disorders than do relatives of depressive probands (Rice et al., 1987; Vandenberg, Singer, and Pauls, 1986). Many other subtypes of affective disorders have been proposed, primarily on the basis of differential response to therapy. As discussed in Chapter 7, manic-depressive illness has been linked to a gene on chromosome 11 in the Older Order Amish (Egeland et al., 1987), and another gene on the X chromosome may also be involved (Baron et al., 1987).

Other Psychopathology

Although the vast majority of research on psychopathology has focused on the psychoses, attention has begun to turn to milder disorders. Brief descriptions of several areas of recent interest follow.

Anxiety neurosis, sometimes known as panic disorder, showed about 20 percent risk in first-degree relatives of affected individuals, compared to only about 3 percent risk in relatives of controls (Crowe et al., 1983). Family studies also indicate that anxiety disorders are heterogeneous and that some are related to depression (Weissman, 1988).

Family studies of *anorexia nervosa* indicated some familial resemblance, as well as familial links with other forms of eating disorders (Strober, et al., 1985) and affective disorders (Gershon et al., 1984). The first twin study of anorexic twins yields concordance rates of 55 percent for 16 identical twin pairs and 7 percent for 14 fraternal twin pairs, suggesting substantial genetic influence (Holland et al., 1984).

Attention deficit disorder, previously called hyperactivity, runs in families. The most recent study found a risk of 31 percent for first-degree relatives of children with diagnoses of attention deficit disorder, compared to a risk of 6 percent in a control group (Biederman et al., 1986). No reasonably sized twin or adoption studies have as yet been reported for this disorder. As in the case of many other disorders of childhood, transmission of psychopathology from parents to offspring does not appear to be syndrome specific (Earls, 1987). That is, parents of hyperactive children show an increased incidence of several other disorders.

Tourette's syndrome, characterized by chronic motor and phonic tics, suggests evidence of substantial genetic influence in a recent twin study (Price et al., 1985). Twin concordances were 53 percent and 8 percent, respectively, for 30 MZ pairs and 13 DZ pairs. Family studies suggest that Tourette's syndrome is related to obsessive-compulsive disorder (Pauls et al., 1986).

Somatization disorder, formerly known as Briquet's syndrome, involves multiple and chronic physical complaints of unknown origin; it occurs primarily in females and may be the female counterpart of antisocial behavior in males (Cloninger et al., 1986). The familial nature of somatization disorder has been demonstrated (Guze et al., 1986), and an adoption study indicates that adopted-away daughters of criminal and alcoholic biological fathers more often complain of multiple medical problems than does a control group of female adoptees (Bohman et al., 1984).

Twin studies suggest genetic influence on *antisocial personality disorder* (previously called psychopathy), although this area is most often studied in terms of criminal behavior (discussed in the following section).

Delinquent and Criminal Behavior

Delinquent or criminal behavior seems so much a matter of choice for which one is held accountable that most people probably never consider the possibility of genetic influence. Controversy has arisen with the publication of a recent book that argues that heredity importantly affects criminal behavior (Wilson and Herrnstein, 1985). Of course, genetic influence by no means implies that crime is destined by heredity. Criminal behavior is multiply determined, and possible effects of genes on behavior that we call criminal could include genetic influence on diverse characteristics, such as body build, neurological factors, mental ability, personality, and psychopathology.

Six twin studies of juvenile delinquency yield 87 percent concordance for identical twins and 72 percent concordance for fraternal twins, suggesting slight genetic influence and substantial environmental sources of resemblance (Gottesman, Carey, and Hanson, 1983). A recent quantitative study of delinquent acts indicates greater genetic influence than do the earlier studies that attempt to diagnose delinquency: Correlations of 0.71 and 0.47 were found for identical and fraternal twins, respectively (Rowe, 1983a). This study also provides evidence that environmental resemblance for delinquency in twins in part reflects the direct influence of one twin on the other, inasmuch as twins are often partners in crime.

The relationship between juvenile delinquency and adult criminality is not clear. Although most adult criminals were delinquents, most delinquents do not become criminals. An important addition to this area is an adoption study of aggressive conduct disorder in children as related to antisocial personality in parents (Jary and Stewart, 1985). For 37 adoptees diagnosed for aggressive conduct disorder, 30 percent of the biological fathers and 59 percent of the biological mothers of the adoptees were diagnosed as antisocial, whereas none of the adoptive parents was similarly diagnosed. These results suggest genetic links between aggressive conduct disorder in childhood and adult antisocial personality.

Evidence for genetic influence is stronger for adult criminality than for delinquency. In eight twin studies of adult criminality, the average concordance for identical and fraternal twins was 51 percent and 30 percent, respectively (Cloninger and Gottesman, 1987). The best twin study involved all male twins born on the Danish Islands from 1881 to 1910 (Christiansen, 1977). Significant and substantial genetic influence is suggested both for serious crimes against persons and for crimes against property. These two categories of crime appear to be uncorrelated genetically.

Adoption studies are consistent with the hypothesis of significant genetic influence on adult criminality. For example, one of the best studies again comes from Denmark, based on 14,427 adoptees and their biological and adoptive parents (Mednick, Gabrielli, and Hutchings, 1984). For 2,492 adopted sons who had neither adoptive nor biological criminal parents, 14 percent had at least one criminal conviction. For 204 adopted sons whose adoptive (but not biological) parents are criminals, 15 percent had at least one conviction. If biological (but not adoptive) parents are criminal, 20 percent (of 1,226) adopted sons have criminal records. If both biological and adoptive parents are criminal, 25 percent (of 143) adopted sons are criminals. In addition, the Danish adoption study obtained data for siblings raised apart (20 percent concordance), half siblings raised apart (13 percent concordance), and pairs of unrelated children reared together in the same adoptive families (9 percent concordance). Other adoption studies in Sweden and in the United States yield similar results. Although parent–offspring adoption data suggest somewhat less genetic influence than do the twin data, all of the data are consistent with a hypothesis of some genetic influence.

Alcoholism

Alcoholism runs in families: About 25 percent of the male relatives of alcoholics are themselves alcoholics, as compared with less than 5 percent of the males in the general population (Cotton, 1979). Alcoholism in a first-degree relative is by far the single best predictor of alcoholism.

Twin and adoption studies have found evidence for genetic influence on the quantity of alcohol consumed among normal drinkers. For example, in a Swedish study, correlations for identical and fraternal twins were 0.64 and 0.27, respectively; correlations for identical and fraternal twins reared apart were similar—0.71 and 0.31, respectively (Pedersen et al., 1984). These data suggest substantial genetic influence and negligible influence of shared environment.

No twin studies have focused on alcoholism per se; because the relationship between alcohol use and abuse is not understood, studies of normal drinkers may not be relevant to alcoholism. One twin study of liver cirrhosis is noteworthy because advanced alcoholism is the major cause of liver cirrhosis. In a study of nearly 16,000 middle-age male pairs of twins, concordances for liver cirrhosis were 15 percent for identical twins and 5 percent for fraternal twins (Hrubec and Omenn, 1981).

A Swedish adoption study also provides evidence for genetic influence on alcoholism, at least in males (Bohman, Sigvardsson, and Cloninger, 1982; Cloninger, Bohman, and Sigvardsson, 1981). Twenty-two percent of the adopted-away sons of biological fathers who abused alcohol were alcoholic, suggesting appreciable genetic influence. Hereditary influence from biological fathers to their adopted-away sons is especially strong when the adopted-away sons are reared in lower-class adoptive families (Sigvardsson, Cloninger, and Bohman, 1985). Several other smaller adoption studies are consistent with the hypothesis of genetic influence on alcoholism in males. Alcoholism appears to be much less frequent in females than in males, and genetic influence also appears to be less for females: Only 4 percent of the adopted-away daughters of alcohol-abusing biological fathers were alcoholic (Sigvardsson et al.).

The heritability of alcohol use and abuse is a good example of what we mean when we say that genes influence behavior. No matter how strong the hereditary propensity towards alcohol might be, few people will become alcoholic unless large quantities of alcohol are consumed over long periods of time. Furthermore, it is unlikely that genes drive us to drink; what is more likely to be inherited is an absence of brakes—physiological and psychological factors that make most people want to stop drinking after a certain point of intoxication.

Finding subtypes of alcoholism would increase the likelihood that interventions can be found to prevent alcoholism before it irrevocably devastates the lives of affected individuals and their families. A recent attempt along these lines suggests three subtypes of alcoholics (Bohman et al., 1987). The milieu-limited type of alcoholic tends to have parents with only mild alcohol abuse; such an individual is affected by environmental factors, such as low socioeconomic status. The male-limited form involves alcoholics whose fathers severely abuse alcohol; this condition appears to be unaffected by environmental circumstances. A third type is associated with antisocial behavior, including violent criminality and recurrent alcohol abuse in males and somatic disorders in females.

PERSONALITY

Personality includes dozens of diverse domains of behavior that are not primarily cognitive in nature: overt behaviors, such as activity level; feelings, such as emotionality; preferences, such as sociability; and attitudes, such as traditionalism. Most research on personality, especially research in behavioral genetics, involves self-report questionnaires.

Extraversion and Neuroticism

Although personality traits are incredibly diverse, two "superfactors," extraversion and neuroticism, cut across many traits. Extraversion encompasses dimensions like sociability, impulsivity, and liveliness. Neuroticism includes traits like moodiness, anxiousness, and toughness; it is a broad dimension of emotional stability–instability, not just a matter of neurotic tendencies. A review of twin research involving over 25,000 pairs of twins suggests substantial heritability for both extraversion and neuroticism (Henderson, 1982). The review also pointed out that extraversion generally suggests evidence for nonadditive genetic variance: Fraternal twin correlations are less than half the size of identical twin correlations (see also Lykken, 1982). These phenomena can be seen in two recent studies. The largest twin study involves a Swedish sample of 4,987 identical twin pairs and 7,790 fraternal twin pairs from 17 to 49 years of age (Floderus-Myrhed, Pedersen, and Rasmuson, 1980). For the entire sample, the identical and fraternal twin correlations were, respectively, 0.51 and 0.21 for extraversion and 0.50 and 0.23 for neuroticism. As another example, a study in Australia involving 2,903 twin pairs found identical and fraternal twin correlations of 0.52 and 0.17 for extraversion and 0.50 and 0.23 for neuroticism, respectively (Martin and Jardine, 1986).

A study of adopted-apart twins and matched twins reared together whose average age was 59 years also finds evidence for nonadditive genetic variance for extraversion (Pedersen et al., 1988). In a sample of 220 identical twins reared together (MZT) and 204 fraternal twins reared together (DZT), twin correlations for extraversion are 0.54 and 0.06, respectively; the fraternal twin correlation is much less than half the size of the identical twin correlation. The correlation for 95 pairs of identical twins reared apart (MZA) is 0.30; the correlation for 220 pairs of fraternal twins reared apart (DZA) is 0.04. Model-fitting analyses suggest that broad heritability is about 40 percent, considering data from all four twin groups, and that nonadditive genetic variance is significant. Correlations for neuroticism are as follows: 0.41 for MZT, 0.24 for DZT, 0.24 for MZA, and 0.28 for DZA. Results obtained from model-fitting analyses of data for all four groups suggest that heritability of neuroticism is about 30 percent and that genetic variance is additive.

Studies of first-degree relatives, however, suggest less genetic influence. The average parent–offspring and sibling correlation for extraversion and neuroticism is 0.13 (Henderson, 1982). Even if this familial resemblance were entirely due to heredity, heritability estimates would only be about half that suggested by the twin studies. One factor contributing to the discrepancy may be nonadditive genetic variance, which makes identical twins, but not first-degree relatives, similar (see Chapter 9). The correlations are even lower in the largest family study conducted as part of the Hawaii Family Study of Cognition, which includes more than a thousand families (Ahern et al., 1982). The parent–offspring and sibling correlations are, respectively, −0.05 and 0.25 for extraversion, and 0.17 and 0.07 for neuroticism.

Adoption studies suggest that familial resemblance seen in family studies is nearly altogether genetic in origin. In a review of extraversion and neuroticism data from three adoption studies, the average correlation for nonadoptive relatives was about 0.15, and the average correlation for adoptive relatives was nearly zero (Henderson, 1982). These data suggest a heritability estimate of about 0.30 for extraversion and neuroticism, although it should be remembered that estimates based on first-degree relatives involve only additive genetic variance.

In summary, two conclusions can be drawn concerning genetic influence on extraversion and neuroticism. First, both traits show appreciable heritability. Second, the classical twin design yields heritability estimates that are greater than estimates from family and adoption designs. This discrepancy may be due in part to nonadditive genetic variance, especially for extraversion.

EAS Traits

As mentioned earlier, extraversion and neuroticism are global traits that encompass many dimensions of personality. The core of extraversion, however, is sociability, the extent to which individuals prefer to do things with others rather than alone. The key component of neuroticism is emotionality, the tendency to become aroused easily. From infancy to adulthood, emotionality and sociability, along with another trait, activity level, have been proposed as key heritable components of personality, a theory referred to by the acronym EAS (Buss and Plomin, 1984). Activity level has not been studied nearly as much as sociability and emotionality simply because it tends not to be included in personality questionnaires for adults, although it is nearly always assessed by rating instruments for children, perhaps because of the conspicuousness of this trait in childhood. A review of behavioral genetic data for these three traits in infancy, childhood, adolescence, and adulthood lends support to the EAS theory (Plomin, 1986a); however, it should be noted that many personality traits display genetic influence, and it is difficult to prove that some traits are more heritable than others. The issue of differential heritability of personality traits is a topic of current interest in the field (Loehlin, 1987).

A recent study of EAS traits uniquely focuses on personality in the last half of the life span—the average age of the twins studied was 59 years (Plomin et al., 1988). This fact is important because heritability can change during the life course; in fact, the results of the Swedish study suggest that heritability of the EAS traits is somewhat lower later in life. For example, the correlations for 90 pairs of identical twins reared apart are 0.30 for emotionality, 0.27 for activity level, and 0.20 for sociability. Model-fitting analyses that consider results simultaneously from all four groups of twins yield heritability estimates of about 40 percent for emotionality, 25 percent for activity level, and 25 for sociability. Some evidence for nonadditive genetic variance was also found (Plomin et al., 1988).

Other Personality Traits

Personality includes a bewildering array of traits such as rebelliousness, empathy, suspiciousness, anomie, and sensation seeking. Unlike extraversion and neuroticism or the EAS traits, traits like these have rarely been examined in more than one behavioral genetic study, and for this reason the dozens of relevant studies are not easily summarized. However, most personality traits show some genetic influence, and evidence for nonadditive genetic variance often appears.

A landmark study involving nearly 800 pairs of adolescent twins utilized several personality questionnaires and concluded that identical and fraternal twin correlations center around 0.50 and 0.30, respectively (Loehlin and Nichols, 1976). A more recent example, again using the powerful combination of twin and adoption designs, is a study conducted in Minnesota of 44 pairs of identical twins reared apart, a small sample of fraternal twins reared apart, and large samples of identical and fraternal twins reared together (Tellegen et al., 1988). Correlations for identical twins reared apart are as follows:

Sense of well-being	0.49
Social potency (leader who likes to be the center of attention)	0.57
Achievement (works hard, strives for mastery)	0.38
Social closeness (intimacy)	0.15
Stress reaction (neuroticism)	0.70
Alienation	0.59
Aggression	0.67
Control (cautious, sensible)	0.56
Harm avoidance (low risk taking)	0.45
Traditionalism (follows rules and authority)	0.59
Absorption (imagination)	0.74

These correlations (the average is 0.54) are as high as those found for identical twins reared together. Model-fitting analyses incorporating data from all four groups of twins yield an average heritability of about 0.50 for these diverse personality traits; nonadditive genetic variance appears important for about half of the traits.

Social closeness (intimacy) shows lower heritability than the other personality traits. Although this may simply be a chance result due to the relatively small sample size, other studies provide some support for the hypothesis that need for intimacy may be less heritable than other personality traits. However, it is surprising that nearly all personality traits, at least as assessed by self-report questionnaire, show moderate genetic influence. It is generally agreed that part of the reason for this outcome involves the pervasive influence of extraversion and neuroticism in personality: Because these two traits are substantially heritable, other traits related to them will also show genetic influence. Some traits that are independent of extraversion and neuroticism are less influenced by heredity. For example, traditional masculinity–femininity and tolerance of ambiguity appear to show little genetic influence in adolescence and substantial shared environmental influence: Identical and fraternal twin correlations are, respectively, 0.46 and 0.40 for masculinity and 0.49 and 0.38 for tolerance of ambiguity (Loehlin, 1982).

Attitudes and Beliefs

Some traits involve attitudes and beliefs to a greater extent than others. For example, in the study just described, traditionalism assesses conformity and conservativeness—the extent to which one follows rules and authority and endorses high moral standards and strict discipline. It has been a surprise to find as much genetic influence for such traits as for other personality traits. For example, heritability was estimated as 0.63 for traditionalism in the Minnesota study by Tellegen and associates (1988), although other studies find lower, but nonetheless significant, heritability. For example, a review of three English twin studies yields an average identical twin correlation of 0.67 for 894 pairs and an average fraternal twin correlation of 0.52 for 523 pairs (Eaves and Young, 1981); a recent Australian study of 1,797 identical and 1,101 fraternal twin pairs yielded correlations of 0.63 and 0.46, respectively (Martin et al., 1986). These twin correlations suggest a heritability of about 30 percent. What is most striking about this pattern of correlations is the high fraternal twin correlation, which could suggest substantial shared environmental effects. However, assortative mating is extremely high for this trait (spouses correlate about 0.50), in contrast to other personality traits, which seldom yield significant assortative mating. When assortative mating is taken into account, heritability is estimated to be about 50 percent.

Although some attitudes and beliefs, such as conservatism, appear to be influenced by heredity, others are, as we might expect, due primarily to shared environmental influences. In Loehlin and Nichols' (1976) large study of high school twins, identical and fraternal twin correlations were, respectively, 0.56 and 0.67 for belief in God and 0.60 and 0.58 for involvement in religious affairs. Similarly, attitudes toward racial integration showed no genetic influence; identical and fraternal twin correlations were 0.37 and 0.40, respectively. It is noteworthy that some traits show shared environmental influence and no genetic influence.

Vocational Interests

Vocational interests are related both to personality and to cognitive abilities. One of the most widely used measures is the Strong Vocational Interest Blank. In two twin studies—one in 1932 and the other in 1974—the average identical twin correlation was about 0.50 and the average fraternal twin correlation was about 0.25 (Carter, 1932; Roberts and Johansson, 1974). The 1974 study included over 1,500 twin pairs and indicated that all of the vocational types—called realistic, intellectual, social, enterprising, conventional, and artistic—show approximately the same level of genetic influence. An adoption study has also found evidence for genetic influence in comparisons between nonadoptive and adoptive parents and their children as adults (Scarr and Weinberg, 1978b).

Observational Data

It should be noted that all of the preceding studies of personality employed self-report questionnaires. There have also been several twin studies of children using parental ratings of personality, and these also tend to yield results suggesting substantial genetic influence (Buss and Plomin, 1984). However, there are very few studies in which behavior is objectively observed, no doubt because of the much greater expense of conducting such studies. One observational study of infant twins suggested genetic influence on infants' social responding to a stranger in their home (that is, shyness), whereas the same social behaviors directed toward the mother showed no genetic influence (Plomin and Rowe, 1979). Another observational study of school-aged twins involved modeled aggression in which children were videotaped hitting an inflated clownlike plastic figure, a measure that has been shown to be valid and to relate to teacher and peer ratings of aggressiveness (Plomin, Foch, and Rowe, 1981). No genetic influence was found: Twin correlations were about 0.45 for both types of twins. Observational studies of infant twins in the laboratory also show less evidence of genetic influence than do questionnaire studies (Goldsmith and Campos, 1986; Wilson and Matheny, 1986). In sum, the few observational studies that have been conducted suggest much more diverse results—and less pervasive genetic influence—than do questionnaire studies. This is not simply because observational measures tend to be less reliable (Plomin, 1981); it may be because observations provide a more precise snapshot of behavior that is not averaged over many situations and across much time.

ENVIRONMENTAL ANALYSES

Behavioral genetics, with its balanced view that recognizes genetic as well as environmental influences, is making important contributions to our understanding of nurture, as well as nature. That is, explorations of environmental influence can profit from considering heredity. Two examples are described in this section: nonshared environment and genetic effects on environmental measures.

Nonshared Environmental Influence

The perceptive reader may have noticed something very striking in the behavioral genetic results reviewed above: Children growing up in the same family do not resemble each other unless they are genetically related. As explained in Chapter 9, environmental variance can be divided into two components, one shared by family members and another that is not shared. Behavioral genetic results imply that environmental influences important to behavioral development operate in such a way as to make children in the same family different from one another. That is, environmental influences do not operate on a family-by-family basis but rather on an individual-by-individual basis. They are specific to each child rather than general for an entire family.

One of several results that converge on the conclusion that shared family environment is unimportant is the correlation for adoptive "siblings"— genetically unrelated children adopted at an early age into the same family. Their resemblance cannot be due to shared heredity, and thus their resemblance directly assesses the importance of environmental influence shared by children growing up in the same family. For personality, adoptive sibling correlations are about 0.05 on average. Genetically unrelated individuals reared together also show no greater than chance resemblance for psychopathology. For cognitive abilities, although adoptive siblings are similar in childhood (correlations of about 0.25), by adolescence their correlations are near zero, suggesting that the long-term impact of shared family environment is slight (Plomin, 1988). Possible exceptions to the rule that nonshared environment accounts for nearly all environmental influence are a few personality traits that involve attitudes, such as traditional masculinity–femininity, measures of beliefs per se, such as religious and political beliefs, delinquency, and—in the cognitive realm—vocabulary. These results are reviewed in an article with numerous commentaries and a response to the commentaries (Plomin and Daniels, 1987).

The importance of nonshared environmental factors suggests the need for a reconceptualization of environmental influences that focuses on experiential differences between children in the same family. That is, many environmental factors differ across families, such as socioeconomic status, parental education, and child-rearing practices. However, environmental factors that do not differ between children growing up in the same family are not likely to influence behavioral development. The critical question becomes: Why are children in the same family so different from one another? The key to unlock this riddle is to study more than one child per family. This permits the study of experiential differences within a family and their association with differences in behavioral outcome. Because heredity contributes to differences between siblings, sibling differences in experiences might reflect rather than affect differences in their behavior. Behavioral genetics methods are useful in addressing this issue. For example, because members of identical twin pairs do not differ genetically, one approach is to relate behavioral differences within pairs of identical twins to their experiential differences.

So far, research on this new topic indicates that siblings in the same family experience considerably different environments in terms of parental treatment and in their interaction with each other and with their peers. There is growing evidence that these differential experiences are systematically related to sibling differences in developmental outcomes. Family structure variables such as birth order, sibling age spacing, and gender differences account for a small portion of variance. In addition to such systematic nonshared factors, it is possible that nonsystematic factors, such as accidents, illnesses, or other experiences, initiate differences between siblings that, when compounded over time, make children in the same family different in unpredictable ways.

Although the first few steps have been taken toward identifying specific sources of nonshared environment, much remains to be learned. Answers to the question "Why are children in the same family so different from one another?" pertain not only to sibling differences. Their importance is far more general: They help us to understand the environmental origins of individual differences in behavioral development.

The Nature of Nurture

In addition to demonstrating the importance of environmental variance and, more specifically, of nonshared environment, behavioral genetics research has made important advances in understanding the interface between nature and nurture. Recent research has shown that heredity can affect environmental measures and also be involved in associations between measures of the environment and behavioral outcomes.

The possibility of genetic influence on measures of the environment is not as paradoxical as it seems once you realize that measures of the environment, especially measures of the family environment, are indirectly measures of behavior. For example, two frequently studied dimensions are parental love and control, obviously measures of parental behavior. Parental behavior is also ultimately responsible for the physical features of the family environment. For example, the most widely used environmental item in studies of mental development is the number of books in the home—parents put the books on the shelves. Because behavioral genetics research has shown that variability for most behaviors shows some genetic influence, it should come as no surprise that behavioral genetic studies of environmental measures also show genetic influence.

Genetic influence can affect measures of family environment in two ways. First, genetically influenced characteristics of parents, such as parents' cognitive abilities and personality, can affect ratings on measures of the family environment. Number of books in the home, for example, could show genetic influence, because brighter people read more. Second, parents can respond to genetically influenced characteristics of their children, such as the children's personality, abilities, and physique. For example, genetic influence could be observed for a measure such as the amount of time that parents read to their children if brighter children get their parents to read to them often.

The first studies of genetic influence on environmental measures were twin studies that analyzed adolescent twins' perceptions of their family environment (Rowe, 1981; 1983b). Twins were asked to rate the affectionateness and control of their parents. In two studies using different measures of the family environment, identical twin correlations were significantly greater than fraternal twin correlations for measures of parental affection but not for measures of parental control. In other words, genetic factors affect adolescents' perceptions of their parents' affection but not of their control. This could occur because adolescents' perceptions of their parents' affection are a function of genetic differences in the adolescents' personality and adjustment. For example, more anxious adolescents might be sensitive to imagined slights or rebuffs from their parents. Another possibility is that adolescents' perceptions are accurate—their parents' affection may be related to genetically influenced characteristics of the adolescents, such as their sociability.

Similar results have been found in a study of elderly twins in Sweden who were asked about their perceptions of their childhood family environment, viewed retrospectively half a century later (Plomin et al., 1988). This study, as mentioned earlier, combines the classical study of twins reared together with the adoption design of twins reared apart. Despite the procedural differences between this study and Rowe's two studies of adolescents, the results confirm those of Rowe's studies in that perceptions of control show little genetic influence, whereas other aspects of the family environment show genetic influence. For example, an affection-related dimension labeled Cohesiveness yielded the highest correlation of eight scales for identical twins reared apart (0.41), and the lowest correlations (−0.03) were observed for a scale called Control. Model-fitting analyses using the data from reared-apart and reared-together identical and fraternal twins confirm the conclusion of genetic influence on all dimensions of the family environment except control. The correlations for identical twins reared apart from early in life are particularly impressive, because these individuals were reared in different families! This could mean that genetic influence is in the eye of the beholder—that is, heredity may be involved in subjective characteristics that affect perceptions of the family environment. However, it is also possible that members of the two families responded similarly to genetically influenced characteristics of the separated identical twins.

Recent research using objective measures of the family environment and the adoption design has also found evidence of genetic influence. For example, one of the most widely used objective measures of the home environment relevant to cognitive development, called the HOME (Caldwell and Bradley, 1978), yielded correlations for nonadoptive siblings that are greater than for adoptive siblings. When each sibling was 12 months, the nonadoptive and adoptive sibling correlations for the HOME are 0.50 and 0.36, respectively; at two years of age, the correlations are 0.50 and 0.32, respectively (Plomin, DeFries, and Fulker, 1988). Ratings made from videotapes of mothers interacting with each of two adoptive or nonadoptive siblings when each child was one, two, and three years old (Dunn and Plomin, 1986) also indicate that mothers treat their children more similarly if they are biological siblings than if they are adoptive siblings. Moreover, this evidence for genetic influence again emerged for affection, not for control.

The most general implication of this research is that labeling a measure environmental does not make it an environmental measure. Heredity can play a role in such measures via genetically influenced characteristics of parents and children. As interesting as it is to consider genetic influence on environmental measures, the real importance of this topic lies in the possibility that genetic influences may mediate on the relationship between environmental measures and children's development.

Genetic influence on such environment-development associations could come about because the measure of family environment is related to genetically influenced characteristics of the parents, which are then inherited by their children. In nonadoptive families, parents share heredity as well as environment with their children, and it is possible that hereditary similarity between parents and children mediates environment-development associations. In adoptive families, however, parents and their children do not share heredity, so the environmental measure is associated with children's behavior solely for environmental reasons. This suggests a way to determine the extent of genetic involvement in environment–behavior associations: Compare nonadoptive and adoptive families. If heredity is involved in such associations, correlations in nonadoptive families will exceed those in adoptive families. Moreover, the greater the difference between the correlations in the nonadoptive and adoptive families, the greater the extent to which the environment–behavior association is mediated genetically.

Three earlier adoption studies that included measures of the home environment suggest substantial genetic mediation of the association between environmental measures and children's IQ. Across these three studies, the average correlation between measures of the home environment and children's IQ was 0.45 in nonadoptive families and only 0.18 in adoptive families. In the Colorado Adoption Project, similar results were found in infancy. For example, the HOME measure of the family environment correlated 0.44 with mental development scores at two years of age in nonadoptive homes; in adoptive homes, the correlation was 0.29. The HOME also correlated 0.50 with a measure of language development at two years in nonadoptive homes; in adoptive homes, the correlation was 0.32.

Genetic mediation of environment–development associations is not limited to cognitive development. For example, genetic mediation was found between several measures involving familial affection and measures of infant easiness (Plomin, Loehlin, and DeFries, 1985). In the Colorado Adoption Project, environment-development associations are generally weaker in early childhood than in infancy; nonetheless, correlations continue to be greater in nonadoptive homes than in adoptive homes. Moreover, longitudinal associations between environmental measures in infancy and developmental outcomes in childhood also show genetic mediation. For example, various environmental measures at 12 months of age correlate 0.42 with IQ at three years in nonadoptive families; in adoptive families, the correlation is 0.27 (Plomin, DeFries, and Fulker, 1988).

Again, the most general implication of these results is that, in nonadoptive families, relationships between environmental measures and measures of development cannot be assumed to be environmental in origin. Behavioral genetic analyses indicate that it is safer to assume that fully half of most environment-development associations is due to genetic mediation.

Other examples of the research at the interface between nature and nurture include the topics of genotype–environment interaction and correlation, described briefly in Chapter 9. Behavioral genetic methods provide useful tools for studying these important issues as well (e.g., Plomin, DeFries, and Fulker, 1988). However, research has not yet made sufficient progress to warrant discussing these topics in this introduction to behavioral genetics.

REVIEW OF HUMAN BEHAVIORAL GENETICS RESEARCH

Table 14.5 provides an overview of the major findings of human behavioral genetics, including research beyond that reviewed in this chapter. In addition to illustrating the widespread influence of heredity on behavior, this overview shows how much remains to be learned. For each entry in the table, we would like to know the following: precise estimates of genetic and environmental parameters, the extent to which genetic variance is additive and nonadditive, the extent to which environmental variance is shared and nonshared, the developmental course of genetic and environmental parameters, age-to-age genetic change and continuity, and whether major-gene effects can be found. As indicated in Table 14.5, we are a long way from these objectives for most domains. The table also shows that most evidence for genetic influence involves twin comparisons; relatively few adoption studies have been conducted. Linkage has been studied primarily in relation to specific syndromes of mental retardation. As discussed in Chapter 7, reports of such effects for

Table 14.5
Summary of human behavioral genetic research

Study method				Evidence to date
Family	Twin	Adoption	Linkage	

IQ

Family	Twin	Adoption	Linkage	Evidence to date
++	++	++	+	$h_B^2 = 40\%-60\%$ $(r_{MZ} = 0.85; r_{DZ} = 0.60)$ h_B^2 increases during childhood Substantial genetic continuity from childhood to adulthood Some evidence for V_d V_{Ec} in childhood V_{Ec} diminishes after childhood V_{Ec} twin > sibling > parent–offspring

Occupational Status

+	+	+	0	$h_B^2 \simeq 40\%$ $(r_{MZ} = 0.40; r_{DZ} = 0.20)$

Specific Cognitive Abilities

+	+	+	0	$h_B^2 = 30\%-50\%$ $(r_{MZ} = 0.70; r_{DZ} = 0.50)$ h_B^2 verbal and spatial > memory? Genetic correlations among abilities Some evidence for V_{Ec}

Creativity

+	++	0	0	$h_B^2 \simeq 20\%$ $(r_{MZ} = 0.60; r_{DZ} = 0.50)$ No genetic influence when IQ is controlled?

Academic Performance

+	+	0	0	$h_B^2 \simeq 50\%$ $(r_{MZ} = 0.75; r_{DZ} = 0.50)$ No evidence that some tests are more heritable than others Genetic correlations among tests Some evidence for V_{Ec}

Reading Disability

+	+	0	+	Familial resemblance Twin studies suggest genetic influence Reading disability may be the lower end of the distribution of reading ability

Mental Retardation

+	0	0	++	No familial influence for severe retardation? Mild retardation is familial Mild retardation may be the lower end of IQ distribution h_B^2 and V_{Ec} unknown Linkage for specific syndromes, e.g., PKU

(continued)

Table 14.5
Summary of human behavioral genetic research (*cont.*)

Study method				Evidence to date
Family	Twin	Adoption	Linkage	
Dementias				
+	+	0	+	Some evidence for genetic influence
Schizophrenia				
++	++	+	+	Risk for first-degree relatives $\approx$10% Risk for MZ twins $\approx$40% Adoption data suggest less genetic influence than twin data No evidence for V_{Ec}
Major Depressive Disorder				
++	++	+	+	Genetically distinct from schizophrenia Risk for first-degree relatives $\approx$10% Risk for MZ twins $\approx$65% Adoption data suggest less genetic influence than twin data No evidence for V_{Ec}
Bipolar Depression				
++	0	0	+	Genetically distinct from major depressive disorder Risk for first-degree relatives $\approx$5% No evidence for V_{Ec}
Anxiety Neurosis				
+	0	0	0	Risk for first-degree relatives $\approx$20%
Anorexia Nervosa				
+	+	0	0	Some familial resemblance Risk for MZ twins $\approx$55%
Attention Deficit Disorder				
+	0	0	0	Risk for first-degree relatives $\approx$30%
Tourette's Syndrome				
+	+	0	+	Risk for first-degree relatives $\approx$10% Risk for MZ twins $\approx$50%
Somatization Disorders				
+	0	+	0	Some familial resemblance
Delinquent Behavior				
+	++	+	0	Only slight genetic influence (85% MZ; 70% DZ) Substantial V_{Ec}
Criminal Behavior				
+	+	+	0	Substantial genetic influence (70% MZ; 30% DZ) Adoption data suggest less genetic influence than twin data No evidence for V_{Ec}

Table 14.5
Summary of human behavioral genetic research (*cont.*)

Study method				Evidence to date
Family	Twin	Adoption	Linkage	
Alcoholism				
++	0	+	0	Some adoption-study evidence for genetic influence Some twin evidence for genetic influence on quantity of alcohol consumed No evidence for V_{Ec}
Personality				
++	++	++	0	$h_B^2 \simeq 40\%$ ($r_{MZ} = 0.50$; $r_{DZ} = 0.30$) Adoption data suggest less genetic influence than twin data Possible V_d for extraversion Little evidence that any traits are more heritable than others No evidence for V_{Ec}
Attitudes and Beliefs				
+	++	+	0	Substantial h_B^2 for traditionalism Negligible h_B^2 for some dimensions
Vocational Interests				
+	+	+	0	$h_B^2 \simeq 0.50$ ($r_{MZ} = 0.50$; $r_{DZ} = 0.25$)
Observational Measures				
+	+	+	0	More diverse results than self-report questionnaires
Environmental Measures				
+	+	+	0	Genetic influence for most family environment measures except control Half of most environment–behavior associations is due to genetic mediation

NOTE: $0 =$ none; $+ =$ few; $++ =$ many.
$h_B^2 =$ broad heritability;
$V_{Ec} =$ shared environmental variance; $V_d =$ nonadditive genetic variance (dominance and epistasis).

other domains (Alzheimer's dementia, reading disability, schizophrenia, and bipolar depression) have not been replicated.

Much of the future of behavioral genetics research will lie in refining Table 14.5 and adding other domains of behavior, as discussed in the following section.

DIRECTIONS IN BEHAVIORAL GENETICS

Predicting the near future of behavioral genetics research is not a matter of crystal ball gazing, because the momentum of current developments makes them certain to continue. The future will no doubt witness the application of behavioral genetics strategies to the study of many behavioral domains. It is fair to say that most behavior is uncharted in terms of behavioral genetics issues. Even within the major domains of cognitive abilities, psychopathology, and personality, behavioral genetics has just scratched the surface of possible applications. For cognition, most research has focused on IQ and group factors of cognitive abilities. The future of behavioral genetics research in this area lies in more fine-grained analyses of cognitive abilities and in the use of information processing and other experimental approaches to cognition. Behavioral geneticists have just begun to consider psychopathology other than schizophrenia and affective disorders. Personality is so complex that it can easily keep behavioral geneticists busy for decades, especially if researchers begin to use observations and other more objective data in addition to self-report questionnaires.

Although much more basic behavioral genetics research is needed even within the major behavioral domains reviewed in this chapter, other domains represent rich territory for future behavioral genetics exploration. For example, vulnerability and invulnerability to stress and coping with stress (Garmezy and Rutter, 1983) are important topics relevant to adjustment that have not been addressed by behavioral geneticists. Next to nothing is known about life satisfaction and other aspects of adjustment that are broader than traditional diagnostic categories of psychopathology. Health-related behaviors represent an especially exciting and important area for future research. Interest in emotion and motivation has increased in recent years in the behavioral sciences, but behavioral genetics research has scarcely considered the diverse topics within these domains. Only a smattering of behavioral genetics research has addressed issues relevant to the major area of social cognition, such as attributional biases. Behavioral genetics research has just begun to move beyond the individual to consider genetic and environmental influences on social relationships in the family and beyond the family. Areas of the behavioral sciences most ripe for behavioral genetics are those that have traditionally ignored individual differences. Prime among these are the fields of perception, learning, and language. Entire disciplines within the social and behavioral sciences, such as sociology and economics, have essentially been untouched by behavioral genetics research.

In addition to behavioral domains that are relevant throughout the life span, behaviors important in the context of life events and their transitions represent outstanding opportunities for future research. The possibilities are limitless, including neonatal behavior, children's relationships with siblings and parents, the stresses of beginning school, adjustment to the educational system, the physical and social transitions of adolescence, entrance into the adult world of work, marriage, and child-rearing, and adjustment to the changes of later life.

Thus, one direction for research is to apply behavioral genetics theory and methods to behaviors that have not previously been studied from this perspective. However, the future of behavioral genetics holds much more than answers to the "how much" question for one behavior after another. Five directions for research are particularly important.

The first is multivariate analysis, described in Chapter 9 and applied in later chapters. Multivariate analysis will be particularly important in breaking down the heterogeneity that exists within psychopathology and in exploring the nexus of associations within cognition and personality. It will also be useful in studying the etiology of associations between biology and behavior —for example, assessing the genetic and environmental underpinnings of behavior, as it relates to the phenomena of neuroscience.

A second methodological direction for research is already substantially underway: model fitting, described in Chapter 9 and discussed in later chapters. Testing the fit between an explicit model and data from a combination of behavioral genetic designs adds considerable sophistication and precision to behavioral genetics analyses.

The third and fourth directions for future research have also been launched. One direction is the analysis of genetic change and continuity during development. Two types of developmental change in gene expression are of interest: change in heritability and age-to-age genetic change. Another direction is the analysis of subjective and objective measures of the environment and of associations between these environmental measures and behavioral outcomes, as discussed earlier in this chapter.

Signs of a fifth direction for behavioral genetics research are just now appearing: the application of the techniques of molecular genetics, described in Chapter 7.

Each of these trends is as applicable for animal research as for human research. Behavioral research on mice and *Drosophila*, as well as simpler organisms, will continue to provide more controlled studies of genetic influence on behavior and its developmental transaction with environmental influences.

SCIENCE AND POLITICS

The controversy that swirled around behavioral genetics research during the 1970s has largely faded. Increasing acceptance of hereditary influence on individual differences in behavior represents one of the most remarkable changes during the modern history of the social and behavioral sciences. Indeed, the wave of acceptance of genetic influence on behavior is growing into a tidal wave that threatens to engulf the second message of behavioral genetics research reviewed in this chapter. The first message is that heredity influences many aspects of behavior. The second message is just as important: Variability in complex behaviors of interest to social and behavioral scientists and to society is due at least as much to environmental influences as it is to genetic influences. It is critically important to recognize that behavioral genetic methods assess environmental as well as genetic influences on behavior, rather than assuming that either nature or nurture is altogether sufficient.

It is good for the social and behavioral sciences that they have moved away from environmentalism and now recognize genetic influences on behavior. As the pendulum swings from environmentalism, it is important that the pendulum be caught midswing before its momentum carries it over to biological determinism. Behavioral genetics research clearly demonstrates that both nature and nurture are critical to individual differences in behavior.

Behavioral genetics must strive to remain apolitical:

> . . . human behavior genetics, of all disciplines in science, is always in danger of edging into the political domain. To the extent this happens, it becomes less of a science. Consequently, there is a real onus on all workers in the area to avoid, as far as possible, any political biases that may erode their impartiality as scientists and to follow closely the dictates of facts and logic. (Fuller and Thompson, 1978, p. 225)

As discussed in Chapter 1, some of the concerns about finding genetic influence on behavior are misplaced—for example, thinking that genes determine one's destiny or deny democratic equality. Behavioral genetics findings are compatible with a wide range of social action, including no action at all. We believe that better decisions can be made with knowledge than without. But knowledge alone by no means accounts for societal and political decisions. Values are just as important in the decision-making process, and decisions (bad or good) can be made with or without knowledge. The relationship between knowledge and values is complex. Most scientists view themselves as fact seekers, ferreting out facts using the objective and verifiable methods of science. Values and politics, in this view, are what other people do with scientific facts. At the opposite extreme is the view that the scientific method is not at all objective, but rather riddled with values and politics. Some even argue that this should be so—that science is politics and should be used as a political tool.

Values undoubtedly do enter the scientific process from the beginning, when we decide what problems to study and when we interpret the findings and their implications. Scientists need to worry about how their findings will be used, although this concern often leads to frustration, since scientists have little control over the use of their published results. Some research is so esoteric that no one ever gets excited about it or its implications. However, powerful findings with important implications for society can create problems as well as solutions. For example, consider the new techniques of prenatal screening. Amniocentesis has obvious benefits in terms of detecting chromosomal and genetic disorders before birth; combined with therapeutic abortion, it can relieve parents and society of the tremendous burden of birth defects. However, it raises ethical problems concerning abortion and creates the possibility of abuses, such as compulsory screening and mandatory abortion if defects are found. Of course, despite the problems created by scientific findings, we would not want to cut off the flow of knowledge in order to cut down on the problems.

We equate knowledge with the scientific method. When scientists face a problem, they must be objective if they want to arrive at the answer, rather than merely provide support for their preconceived notions. Of course, we are not so naive as to argue that scientists can or should totally eschew politics. But there are some telling examples of what happens when politics and science become mixed. Time and again, science has been interpreted—and misinterpreted—to advance political causes. Two notorious examples related to genetics include the eugenics movement in the early part of the twentieth century and the Nazi carnage during World War II. A less well-known example occurred in the Soviet Union from the 1930s until the 1960s.

A political takeover of biology in the USSR was accomplished by T. D. Lysenko during the Stalin era (Medvedev, 1969). Like Lamarck, Lysenko believed that acquired characteristics could be inherited. He attempted to apply this belief to agriculture, with little success; eventually, his belief in his theory overcame his belief in the scientific method. With the help of a Marxist philosopher in the 1930s, Lysenko mounted a political campaign to brand Mendelian scientists as idealist and obstructionist. It has been suggested that the real reasons for the adverse reaction to Mendelism were matters of values:

> One is the dislike of Mendelism because Mendelian heredity, with its self-copying genes and its random undirected mutations, seems to offer too much resistance of man's desire to change nature, and to elude the control we would like to impose. Lamarckism, on the other hand, holds out a promise of speedy control. . . . This is relevant not only in agriculture, but also in human affairs, for it would be politically very convenient and agreeable if a few generations of life under improved communist conditions would level up the genetic quality of the population of the U.S.S.R. . . . The second ideological reason . . . is a dislike of Mendelism because it implies human inequality, and because it can be taken to imply human helplessness in the face of genetic predestination. (Huxley, 1949, pp. 182–184)

By 1939, Lysenko had a firm grip on the reins of science, and some of the Stalinist purges were used to eliminate the leading geneticists. In a conference in 1948 Lysenko announced that his views were endorsed by the Central Committee of the Party. It became impossible to challenge Lysenko's views; this signaled the end of Mendelian genetics in Russia for two decades. Geneticists who refused to acquiesce left Russia, and geneticists who remained were forced to recant. Many lost their positions, and some lost their lives. By 1964 the crop failures that resulted from Lysenko's policies were too much to be ignored, and they may even have contributed to the downfall of Nikita Khrushchev, who had supported Lysenko during the post-Stalinist political upheaval. Lysenko was disgraced and Mendelian genetics returned to Russia.

Lysenko's belief in the inheritance of acquired characteristics and many of his experiments have no scientific merit, as Soviet scientists now recognize. However, the scientific issues are less important in the present context than understanding the process of politicization and its disastrous results. The methods used by Lysenko to force his ideas on science deserve special attention to sensitize us to such politicization. First, the scientific controversy quickly became personal. Attacks were focused on men and their motives, rather than on their ideas (Brill, 1975). Second, jingoism prevailed. Simplistic slogans with popular appeal took the place of honest discussions of complex issues. A third and related tactic was Lysenko's greater use of the popular press than of scientific journals to promulgate his views. Fourth, he disavowed the scientific method. For example, Lysenko told his staff that "to obtain a certain result, one must wish to obtain such a particular result; if you want a particular result you will obtain it" (Lerner and Libby, 1976, p. 396). Fifth, no compromise or middle ground was allowed. The issue was treated as a class struggle in science, and the revolutionary slogan "Who is not with us is against us" was applied.

The example of Lysenko suggests the problems that can occur when science is controlled by politics. In the 1970s similar tactics of politicization were used in relation to human behavioral genetics research, for example, in stopping a newborn chromosomal screening program (Culliton, 1975b) and in attacking Arthur Jensen (Jensen, 1972) and E. O. Wilson (Wade, 1976; Wilson, 1978a). Although scientific criticism is healthy, attempts to politicize science are ill considered even when well motivated because of the bizarre course that politicization can take once it has begun. For example, Lysenko began with the good intention of advancing science by making science more responsive to socialist needs. If we were able to distinguish between concern with values and scientific issues, there would perhaps be less rancor in discussions of both values and scientific issues.

We suggest that genetic and environmental effects on individual differences are scientific issues. Social equality, on the other hand, is a value—and a very important one. If examined closely, there need be no contradiction between the value of equality and the possibility of genetic differences among individuals. Surely, we do not treat people equally because there are no differences among us. People differ in morophological characteristics—such as height, strength, age, gender, and so on—and they also differ behaviorally. However, none of these differences implies that discrimination is justified. Equality of opportunity and equality before the law are independent of the scientific question of individual differences, regardless of their genetic or environmental etiology.

The basic scientific message of behavioral genetics is that each of us is an individual. Recognition of, and respect for, individual differences is essential to the ethic of individual worth. Proper attention to individual needs, including the provision of the environmental circumstances that will optimize the development of each person, is a utopian ideal and no more attainable than other utopias. Nevertheless, we can approach this ideal more closely if we recognize, rather than ignore, individuality. Although much research effort will be needed, acquiring the requisite knowledge would seem to warrant a high priority. Human individuality is the fundamental natural resource of our species.

SUMMARY

The first part of this chapter emphasizes the first message of behavioral genetics research: For most behaviors studied so far, genetic influence is significant and substantial. Genetic influence has been found for IQ, specific cognitive abilities, school achievement, reading disability, mental retardation, psychoses, delinquent and criminal behavior, alcoholism, diverse personality traits, some attitudes and beliefs, and vocational interests. Genetic influence is so ubiquitous and pervasive in behavior that a shift in emphasis is warranted: Ask not what is heritable; ask what is not heritable. So far, few domains show no genetic influence; these include some beliefs (religiosity and political values) and perhaps creativity independent of IQ.

However, behavioral genetics has two messages. The second message of behavioral genetics research is just as important as the first: The same data provide the best available evidence for the importance of environmental influences. Two examples of the usefulness of behavioral genetics strategies for studying the environment are identification of the importance of non-shared environment and exploration of genetic influence on environmental measures and on environment–behavior associations.

Future directions for behavioral genetics research include studying new domains of behavior and applying new multivariate, model-fitting, developmental, environmental, and molecular genetic techniques. Although values enter the scientific process, science must remain as independent as possible from politics. Lysenko's politicization of genetics in the USSR serves as a reminder of the disastrous consequences that can occur when science becomes controlled by politics. The value of social equality is not compromised by the finding of genetic differences among individuals.

References

Aceves-Pina, E. O.; Booker, R.; Duerr, J. S.; Livingstone, M. S.; Quinn, W. G.; et al. 1983. Learning and memory in *Drosophila*, studied with mutants. *Cold Spring Harbor Symposia in Quantitative Biology*, 48, 831–840.

Adelman, J. P.; Bond, C. T.; Douglas, J.; and Herbert, E. 1987. Two mammalian genes transcribed from opposite strands of the same DNA locus. *Science* 235, 1514–1517.

Adler, J. 1976. The sensing of chemicals by bacteria. *Scientific American*, 234, 40–47.

Agrawal, N.; Sinha, S. N.; and Jensen, A. R. 1984. Effects of inbreeding on Raven Matrices. *Behavior Genetics*, 14, 579–585.

Ahern, F. M.; Johnson, R. C.; Wilson, J. R.; McClearn, G. E.; and Vandenberg, S. G. 1982. Family resemblances in personality. *Behavior Genetics*, 12, 261–280.

Allen, G. 1975. *Life science in the twentieth century*. New York: John Wiley.

Ambros, V., and Horvitz, H. R. 1984. Heterochronic mutants of the nematode Caenorhabditis elegans. *Science*, 226, 409–416.

Ambrus, C. M.; Ambrus, J. L.; Horvath, C.; Pedersen, H.; Sharma, S.; Kant, C.; Mirand, E.; Guthric, R.; and Paul, T. 1978. Phenylalanine depletion for the management of phenylketonuria: Use of enzyme reactors with immobilized enzymes. *Science*, 201, 837–839.

Anastasi, A. 1958. Heredity, environment, and the question "How?" *Psychological Review*, 65, 197–208.

Anisman, H. 1975. Task complexity as a factor in eliciting heterosis in mice: Aversively motivated behaviors. *Journal of Comparative and Physiological Psychology*, 89, 976–984.

Arnold, S. J., and Wade, M. J. 1984. On the measurement of natural and sexual selection: Theory. *Evolution*, 38, 709–718.

Arveiler, B.; Oberle, I.; Vincent, A.; Hofker, M. H.; Pearson, P. L.; and Mandel, J. L. 1988. Genetic mapping of the Xq27q28 region: New RFLP markers useful for diagnostic applications in fragileX and hemophiliaB families. *American Journal of Human Genetics*, 42, 380–389.

Ashton, G. C. 1986. Blood polymorphisms and cognitive abilities. *Behavior Genetics*, 16, 517–529.

Ayala, F. J. 1978. The mechanism of evolution. *Scientific American*, 239, 56–69.

Bailey, D. W. 1971. Recombinant inbred strains. *Transplantation*, 11, 325–327.

––––––. 1981. Strategic uses of recombinant inbred and congenic strains in behavior genetics research. In *Genetic research strategies in psychobiology and psychiatry*, eds. E. S. Gershon, S. Matthysee, X. O. Breakefield, and R. D. Ciaranello, pp. 189–198. Pacific Grove, Calif.: Boxwood.

Bailey, J. M., and Horn, J. M. 1986. A source of variance in IQ unique to the lower-scoring monozygotic (MZ) cotwin. *Behavior Genetics*, 16, 509–516.

Baker, W. K. 1978. A genetic framework for *Drosophila* development. *Annual Review of Genetics*, 12, 451–470.

Bakwin, H. 1973. Reading disability in twins. *Developmental Medicine and Child Neurology*, 15, 184–187.

Bammer, G. 1983. The Australian high and low avoidance rat strains: Differential effects of ethanol and [alpha]-methyl-p-tyrosine. *Behavior and Brain Research*, 8, 317–333.

Barash, D. P. 1975. Ecology of paternal behavior in the hoary marmot (*Marmota caligata*): An evolutionary interpretation. *Journal of Mammalogy*, 56, 612–615.

––––––. 1977. *Sociobiology and behavior.* New York: Elsevier.

Barnicot, N. A. 1950. Taste deficiency for phenylthiourea in African Negroes and Chinese. *Annals of Eugenics*, 15, 248–254.

Baron, M.; Gruen, R.; Rainer, J. D.; Kane, J.; Asnis, L.; et al. 1985. A family study of schizophrenic and normal control probands: Implications for the spectrum concept of schizophrenia. *American Journal of Psychiatry*, 142, 447–455.

Baron, M., and Risch, N. 1987. The spectrum concept of schizophrenia: Evidence for a genetic–environmental continuum. *Journal of Psychiatric Research*, 21, 257–267.

Baron, M.; Risch, N.; Hamburger, R.; Mandel, B.; and Kushner, S. 1987. Genetic linkage between X-chromosome markers and bipolar affective illness. *Nature*, 326, 289–292.

Barr, M. L., and Bertram, E. G. 1949. A morphological distinction between neurones of the male and female, and the behaviour of the nucleolar satellite during accelerated nucleoprotein synthesis. *Nature*, 163, 676.

Bashi, J. 1977. Effects of inbreeding on cognitive performance. *Nature*, 266, 440–442.

Baskin, Y. 1984. Doctoring the genes. *Science 84*, Dec., 52–60.

Bateson, W. 1909. *Mendel's principles of heredity.* Cambridge, England: Cambridge University Press.

Beadle, G. W., and Tatum, E. L. 1941. Experimental control of developmental reaction. *American Naturalist*, 75, 107–116.

Bell, A. G., and Corey, P. N. 1974. A sex chromatin and Y body survey of Toronto newborns. *Canadian Journal of Genetics and Cytogenetics*, 16, 239–250.

Belmaker, R.; Pollin, W.; Wyatt, R. J.; and Cohen, S. 1974. A follow-up of monozygotic twins discordant for schizophrenia. *Archives of General Psychiatry*, 30, 219–222.

Bender, B.; Linden, M.; and Robinson, A. (In press.) Environment and developmental risk in children with sex chromosome abnormalities. *Journal of the American Academy of Child Psychiatry*.

Bender, B.; Puck, M.; Salbenblatt, J.; and Robinson, A. 1984. The development of four unselected 47,XYY boys. *Clinical Genetics*, 25, 435–445.

Benzer, S. 1967. Behavioral mutants of *Drosophila* isolated by countercurrent distribution. *Proceedings of the National Academy of Sciences*, 58, 1112–1119.

———. 1973. Genetic dissection of behavior. *Scientific American*, 229, 24–37.

Bertelsen, A. 1985. Controversies and consistencies in psychiatric genetics. *Acta Psychiatrica Scandinavica*, 71, 61–75.

Bertelsen, A.; Harvald, B.; and Hague, M. 1977. A Danish study of manic-depressive disorders. *British Journal of Psychiatry*, 130, 330–351.

Bessman, S. P.; Williamson, M. L.; and Koch, R. 1978. Diet, genetics, and mental retardation interaction between phenylketonuric heterozygous mother and fetus to produce nonspecific diminution of IQ: Evidence in support of the justification hypothesis. *Proceedings of the National Academy of Sciences*, 78, 1562–1566.

Biederman, J.; Munir, K.; Knee, D.; Habelow, W.; Armentano, M.; Autor, S.; Hoge, S. K.; and Waternaux, C. 1986. A family study of patients with attention deficit disorder and normal controls. *Journal of Psychiatric Research*, 20, 263–274.

Bignami, G. 1965. Selection for high and low rates of conditioning in the rat. *Animal Behavior*, 13, 221–227.

Black, I. B.; Adler, J. E.; Dreyfus, C. F.; Friedman, W. F.; LaGamma, E. F.; and Roch, A. H. 1987. Biochemistry of information storage in the nervous system. *Science*, 236, 1263–1268.

Blixt, S. 1975. Why didn't Gregor Mendel find linkage? *Nature*, 256, 206.

Blizard, D. A. 1981. The Maudsley reactive and nonreactive strains: A North American perspective. *Behavior Genetics*, 11, 469–489.

Blizard, D. A., and Bailey, D. W. 1979. Genetic correlations between open-field activity and defecation: Analysis with the CxB recombinant-inbred strains. *Behavior Genetics*, 9, 349–357.

Blizard, D. A., and Fulker, D. W. 1981. *Rattus norvegicus* as a tool in mammalian behavior genetics: Introduction to a special issue of *Behavior Genetics*. *Behavior Genetics*, 11, 427–430.

Bodmer, W. F. 1986. Human genetics: The molecular challenge. *Cold Spring Harbor Symposia on Quantitative Biology*, 51, 1–14.

Bodmer, W. F., and Cavalli-Sforza, L. L. 1976. *Genetics, evolution, and man*. San Francisco: W. H. Freeman and Company.

Boerwinkle, E.; Chakraborty, R.; and Sing, C. F. 1986. The use of measured genotype information in the analysis of quantitative phenotypes in man. *Annals of Human Genetics*, 50, 181–194.

Bohman, M.; Cloninger, C. R.; Knorring, A.-L.; and Sigvardsson, S. 1984. An adoption study of somatoform disorders. III. Cross-fostering analysis and genetic relationship to alcoholism and criminality. *Archives of General Psychiatry*, 41, 872–878.

Bohman, M.; Cloninger, R.; Sigvardsson, S.; and von Knorring, A.-L. 1987. The genetics of alcoholism and related disorders. *Journal of Psychiatric Research*, 21, 447–452.

Bohman, M.; Sigvardsson, S.; and Cloninger, C. R. 1982. Maternal inheritance of alcohol abuse: Cross-fostering analysis of adopted women. *Archives of General Psychiatry*, 38, 965–969.

Böök, J. A. 1957. Genetical investigation in a north Swedish population: The offspring of first-cousin marriages. *Annals of Human Genetics,* 21, 191–221.

Boolootian, R. A. 1971. *Slide guide for human genetics.* New York: John Wiley.

Boomsma, D. I.; Martin, N. G.; and Neale, M. C. 1989. Structural modelling in the analysis of twin data. *Behavior Genetics,* 19, 5–8.

Borgaonkar, D. S. 1977. *Chromosomal variation in man.* 2d ed. New York: Alan R. Liss.

———. 1984. *Chromosomal variation in man: A catalog of variants and anomalies.* Baltimore: Alan R. Liss.

Borst, P., and Greaves, D. R. 1987. Programmed gene rearrangements altering gene expression. *Science,* 235, 658–667.

Bouchard, T. J. 1984. Twins reared together and apart: What they tell us about human diversity. In *Individuality and determinism,* ed. S. W. Fox, pp. 147–178. New York: Plenum.

Bouchard, T. J., and McGee, M. G. 1977. Sex differences in human spatial ability: Not an X-linked recessive gene effect. *Social Biology,* 24, 332–335.

Bouchard, T. J., Jr. 1987. Environmental determinants of IQ similarity in identical twins reared apart. Paper presented at the Seventeenth Annual Meeting of the Behavior Genetics Association, June 25, Minneapolis.

Bouchard, T. J., Jr., and McGue, M. 1981. Familial studies of intelligence: A review. *Science,* 212, 1055–1059.

Boué, J. G. 1977. Chromosomal studies in more than 900 spontaneous abortuses. Paper presented at the Teratology Society Meeting, 1974. Cited in D. W. Smith, 1977.

Bovet, D. 1977. Strain differences in learning in the mouse. In *Genetics, environment and intelligence,* ed. A. Oliverio, pp. 79–92. Amsterdam: North-Holland Publishing Company.

Bovet, D.; Bovet-Nitti, F.; and Oliverio, A. 1969. Genetic aspects of learning and memory in mice. *Science,* 163, 139–149.

Brill, H. 1975. Presidential address: Nature and nurture as political issues. In *Genetic research in psychiatry,* eds. R. R. Fieve, D. Rosenthal, and H. Brill, pp. 283–288. Baltimore: Johns Hopkins University Press.

Broadhurst, P. L. 1960. Experiments in psychogenetics. Applications of biometrical genetics to the inheritance of behaviour. In *Experiments in personality.* Psychogenetics and psychopharmacology, vol. 1, ed. H. J. Eysenck, pp. 1–102. London: Routledge & Kegan Paul.

———. 1975. The Maudsley reactive and nonreactive strains of rats: A survey. *Behavior Genetics.* 5, 299–319.

———. 1976. The Maudsley reactive and nonreactive strains of rats: A clarification. *Behavior Genetics,* 6, 363–365.

———. 1978. *Drugs and the inheritance of behavior.* New York: Plenum Press.

———. 1979. The experimental approach to behavioral evolution. In *Theoretical advances in behavior genetics,* ed. J. R. Royce and L. P. Mos, pp. 43–95. Alphen aan den Rijn, Netherlands: Sijthoff Noordhoff International.

Bronum-Nielsen, K.; Tommerup, N.; Poulsen, H.; Jacobson, P.; Beck, B.; and Mikkelsen, K. 1983. Carrier detection and X inactivation studies in the fragile (X) syndrome: Cytogenetic studies in 63 obligate and potential carriers of fragile (X). *Human Genetics,* 64, 240–245.

Brown, W. T.; Jenkins, E. C.; Cohen, I. L.; Fisch, G. S.; Wolf-Schein, E. G.; Gross, A.; Waterhouse, L.; Fein, D.; Mason-Brothers, A.; Ritvo, E.; Ruttenberg, B. A.;

Bentley, W.; and Castells, S. 1986. Fragile X and autism: A multicenter study. *American Journal of Medical Genetics*, 23, 341–352.

Brown, W. T.; Jenkins, E. C.; Krawczun, M. S.; Wisniewski, K.; Rudelli, R.; Cohen, I. L.; Fisch, G.; Wolf-Schein, E.; Miezejeski, C.; and Dobkins, C. 1986. The fragile X syndrome. *Annals of the New York Academy of Sciences*, 477, 129–150.

Bruell, J. H. 1964a. Heterotic inheritance of wheelrunning in mice. *Journal of Comparative and Physiological Psychology*, 58, 159–163.

———. 1964b. Inheritance of behavioral and physiological characters of mice and the problem of heterosis. *American Zoologist*, 4, 125–138.

———. 1967. Behavioral heterosis. In *Behavior-genetic analysis*, ed. J. Hirsch, pp. 270–274. New York: McGraw-Hill.

Brush, F. R.; Froehlich, J. C.; and Baron, S. 1979. Genetic selection for avoidance behavior in the rat. *Behavior Genetics*, 9, 309–316.

Bruun, K.; Markkanen, T.; and Partanen, J. 1966. *Inheritance of drinking behavior, a study of adult twins.* Helsinki: The Finnish Foundation for Alcohol Research.

Bulmer, M. G. 1970. *The biology of twinning in man.* Oxford: Clarendon Press.

Burt, C. 1955. The evidence for the concept of intelligence. *British Journal of Educational Psychology*, 25, 158–177.

Buss, A. H., and Plomin, R. 1984. *Temperament: Early developing personality traits.* Hillsdale, N.J.: Lawrence Erlbaum Associates.

Buss, D. M. 1984a. Marital assortment for personality dispositions: Assessment with three data sources. *Behavior Genetics*, 14, 111–123.

———. 1984b. Toward a psychology of person-environment (PE) correlation: The role of spouse selection. *Journal of Personality and Social Psychology*, 47, 361–377.

———. 1984c. Evolutionary biology and personality psychology: Towards a conception of human nature and individual differences. *American Psychologist*, 39, 1135–1147.

———. 1985. Human mate selection. *American Scientist*, 73, 47–51.

Caldwell, B. M., and Bradley, R. H. 1978. *Home Observation for Measurement of the Environment.* Little Rock: University of Arkansas.

Canter, S. 1973. Personality traits in twins. In *Personality differences and biological variations*, eds. G. Claridge, S. Canter, and W. I. Hume, pp. 21–51. New York: Pergamon Press.

Carter, H. D. 1932. Twin similarities in occupational interests. *Journal of Educational Psychology*, 23, 641–655.

Carter-Saltzman, L., and Scarr, S. 1975. Blood group, behavioral and morphological differences among dizygotic twins. *Social Biology*, 22, 373–374.

———. 1977. MZ or DZ? Only your blood grouping laboratory knows for sure. *Behavior Genetics*, 7, 273–280.

Caskey, C. T. 1987. Disease diagnosis by recombinant DNA methods. *Science*, 236, 1223–1229.

Caspersson, T.; Farber, S.; Foley, G. E.; Kudynowski, J.; Modest, E. J.; Simonsson, E.; Wagh, U.; and Zech, L. 1968. Chemical differentiation along metaphase chromosomes. *Experimental Cell Research*, 49, 219–222.

Caspersson, T., Lomakka, G., and Møller, A. 1971. Computerized chromosome identification by aid of the quinacrine mustard fluorescence technique. *Hereditas*, 67, 103–109.

Cassada, R.; Isnenghi, E.; Denich, K.; Radnia, K.; Schierenberg, E.; and Smith, K. 1980. Genetic dissection of embryogenesis in *Caenorhabditis elegans*. In *Devel-*

opmental biology using purified genes, ed. D. B. Brown, pp. 209–227. New York: Academic Press.

Castle, W. E. 1903. The law of heredity of Galton and Mendel and some laws governing race improvement by selection. *Proceedings of the American Academy of Sciences*, 39, 233–242.

Cattell, R. B. 1953. Research designs in psychological genetics with special reference to the multiple variance analysis method. *American Journal of Human Genetics*, 5, 76–93.

———. 1960. The multiple abstract variance analysis equations and solutions: For nature–nurture research on continuous variables. *Psychological Review*, 67, 353–372.

———. 1973. *Personality and mood by questionnaire*. San Francisco: Jossey-Bass.

Cech, T. R.; Zaug, A. J.; and Grabowski, P. J. 1981. In vitro splicing of the ribosomal RNA precursor of Tetrahymena: Involvement of a guanosine nucleotide in the excision of the intervening sequence. *Cell*, 27, 487–496.

Cederlöf, R.; Friberg, L.; Jonsson, E.; and Kaij, L. 1961. Studies on similarity diagnosis with the aid of mailed questionnaires. *Acta Genetica et Statistica Medica*, 11, 338–362.

Cederlöf, R., and Lorich, U. 1978. The Swedish twin registry. In *Twin research, Part B, Biology and Epidemiology*, ed. W. E. Nance, pp. 189–196. New York: Alan R. Liss.

Chambon, P. 1981. Split genes. *Scientific American*, 244, 60–71.

Changeux, J.-P. 1965. The control of biochemical reactions. *Scientific American*, 212, 36–45.

Chen, C.-S., and Fuller, J. L. 1976. Selection for spontaneous or priming-induced audiogenic seizure susceptibility in mice. *Journal of Comparative and Physiological Psychology*, 90, 765–772.

Chipuer, H. M.; Rovine, M.; and Plomin, R. 1989. LISREL Modelling: Genetic and environmental influences on IQ revisited. Manuscript submitted for publication.

Christiansen, K. O. 1977. A preliminary study of criminality among twins. In *Biosocial bases of criminal behavior*, eds. S. A. Mednick and K. O. Christiansen, pp., 89–108. New York: Gardner.

Cicchetti, D., and Serafica, F. C. 1981. Interplay among behavioral systems: Illustrations from the study of attachment, affiliation and wariness in young children with Down's syndrome. *Developmental Psychology*, 17, 36–49.

Clarke, B. 1975. The causes of biological diversity. *Scientific American*, 233, 50–60.

Cloninger, C. R.; Bohman, M.; and Sigvardsson, S. 1981. Inheritance of alcohol abuse: Cross-fostering analysis of adopted men. *Archives of General Psychiatry*, 38, 861–869.

Cloninger, C. R., and Gottesman, I. I. 1987. Genetic and environmental factors in antisocial behavior disorders. In *The causes of crime: New biological approaches*, eds. S. A. Mednick, T. E. Moffitt, and S. A. Stark, pp. 92–109. New York: Cambridge University Press.

Cloninger, C. R.; Martin, R. J.; Guze, S. B.; and Clayton, P. J. 1986. A prospective follow-up and family study of somatization in men and women. *American Journal of Psychiatry*, 143, 873–878.

Cohen, D. J.; Dibble, E.; Grawe, J. M.; and Pollin, W. 1973. Separating identical from fraternal twins. *Archives of General Psychiatry*, 29, 465–469.

————. 1975. Reliably separating identical from fraternal twins. *Archives of General Psychiatry*, 32, 1371–1375.

Cohen, R.; Bloch, N.; Flum, Y.; Kadar, M.; and Goldschmidt, E. 1963. School attainment in an immigrant village. In *The genetics of migrant and isolate populations*, ed. E. Goldschmidt, pp. 350–351. Baltimore: Williams & Wilkins.

Cohen, S.; Chang, A.; Boyer, H.; and Helling, R. 1973. Construction of biologically functional bacterial plasmids in vitro. *Proceedings of the National Academy of Sciences USA*, 70, 3240–3244.

Collins, R. L. 1970. A new genetic locus mapped from behavioral variation in mice: Audiogenic seizure prone (*asp*). *Behavior Genetics*, 1, 99–109.

Collins, R. L., and Fuller, J. L. 1968. Audiogenic seizure prone (*asp*): A gene affecting behavior in linkage group VIII of the mouse. *Science*, 162, 1137–1139.

Connell, J. P., and Tanaka, J. S. 1987. Special section: Structural equation modeling. *Child Development*, 58, 1–175.

Connolly, J. A. 1978. Intelligence levels of Down's syndrome children. *American Journal of Mental Deficiency*, 83, 193–196.

Connolly, K. 1968. Report on a behavioural selection experiment. *Psychological Reports*, 23, 625–627.

Cooper, R. M., and Zubek, J. P. 1958. Effects of enriched and restricted early environments on the learning ability of bright and dull rats. *Canadian Journal of Psychology*, 12, 159–164.

Corey, L. A., and Nance, W. E. 1978. The monozygotic half-sib model: A tool for epidemiologic research. In *Twin research, Part A, Psychology and methodology*, ed. W. E. Nance, pp. 201–210. New York: Alan R. Liss.

Corney, G. 1978. Twin placentation and some effects on twins of known zygosity. In *Twin research, Part B, Biology and epidemiology*, ed. W. E. Nance, pp. 9–16. New York: Alan R. Liss.

Cotton, N. S. 1979. The familial incidence of alcoholism: A review. *Journal of Studies in Alcohol*, 40, 89–116.

Court-Brown, W. M. 1969. Sex chromosomes aneuploidy in man and its frequency, with special reference to mental subnormality and criminal behavior. *International Review of Experimental Pathology*, 7, 31–97.

Crabbe, J. C.; Kosubud, A.; Young, E. R.; Tam, B. R.; and McSwigan, J. D. 1985. Bidirectional selection for susceptibility to ethanol withdrawal seizures in *Mus musculus*. *Behavior Genetics*, 15, 521–536.

Crick, F. 1979. Split genes and RNA splicing in evolution of eukaryotic cells. *Science*, 204, 264–271.

Crow, J. F. 1986. *Basic concepts in population, quantitative, and evolutionary genetics*. New York: W. H. Freeman and Company.

Crowe, R. R.; Noyes, R.; Pauls, D. L.; and Slymen, D. 1983. A family study of anxiety disorder. *Archives of General Psychiatry*, 40, 1065–1069.

Culliton, B. J. 1975a. Amniocentesis: HEW backs test for prenatal diagnosis of disease. *Science*, 190, 537–540.

————. 1975b. XYY: Harvard researcher under fire stops newborn screening. *Science*, 188, 1284–1285.

————. 1976. Genetic screening: States may be writing the wrong kinds of laws. *Science*, 191, 926–929.

Darwin, C. 1859. *On the origin of species by means of natural selection, or the preservation of favoured races in the struggle for life.* London: John Murray. (New York: Modern Library, 1967.)

———. 1868. *The variation of animals and plants under domestication.* New York: Orange Judd.

———. 1871. *The descent of man and selection in relation to sex.* London: John Murray. (New York: Modern Library, 1967.)

———. 1872. *The expression of the emotions in man and animals.* London: John Murray.

———. 1888. Letter from C. Darwin to C. Lyell, 1858. In *The life and letters of Charles Darwin*, vol. 2, ed. F. Darwin, p. 117. London: John Murray.

———. 1896. *Journal of researches into the natural history and geology of the countries visited during the voyage of H. M. S. Beagle round the world under the command of Capt. Fitz Roy, T. N.* New York: Appleton.

Davies, K. E.; Pearson, P. L.; Harper, P. S.; Murray, J. M.; O'Brien, T.; Sarfrazi, M.; and Williamson, R. 1983. Linkage analysis of two cloned DNA sequences flanking the Duchenne muscular dystrophy locus on the short arm of the human X chromosome. *Nucleic Acids Research*, 11, 2303–2305.

Dawkins, R. 1976. *The selfish gene.* New York: Oxford University Press.

Dawkins, R. 1982. *The blind watchmaker.* Essex, England: Longman.

Deckard, B. S.; Lieff, B.; Schlesinger, K.; and DeFries, J. C. 1976. Developmental patterns of seizure susceptibility in inbred strains of mice. *Developmental Psychobiology*, 9, 17–24.

Decker, S. N., and Vandenberg, S. G. 1985. Colorado twin study of reading disability. In *Biobehavioral measures of dyslexia*, ed. D. B. Gray and J. Kavanaugh, pp. 123–135. Baltimore: York.

DeFries, J. C. 1972. Quantitative aspects of genetics and environment in the determination of behavior. In *Genetics, environment, and behavior: Implications for educational policy.* eds. L. Ehrman, G. S. Omenn, and E. Caspari, pp. 6–16. New York: Academic Press.

DeFries, J. C.; Ashton, G. C.; Johnson, R. C.; Kuse, A. R.; McClearn, G. E.; Mi, M. P.; Rashad, M. N.; Vandenberg, S. G.; and Wilson, J. R. 1976. Parent–offspring resemblance for specific cognitive abilities in two ethnic groups. *Nature*, 261, 131–133.

DeFries, J. C.; Corley, R. P.; Johnson, R. C.; Vandenberg, S. G.; and Wilson, J. R. 1982. Sex-by-generation and ethnic group-by-generation interactions in the Hawaii Family Study of Cognition. *Behavior Genetics*, 12, 223–230.

DeFries, J. C., and Fulker, D. W. 1985. Multiple regression analysis of twin data. *Behavior Genetics*, 15, 467–473.

———. 1986. Multivariate behavioral genetics and development: An overview. *Behavior Genetics*, 16, 1–10.

———. (In press.) Multiple regression analysis of twin data: Etiology of deviant scores versus individual differences. *Acta Geneticae Medicae et Gemellologiae.*

DeFries, J. C.; Fulker, D. W.; and LaBuda, M. C. 1987. Evidence for a genetic aetiology in reading disability of twins. *Nature*, 329, 537–539.

DeFries, J. C.; Gervais, M. C.; and Thomas, E. A. 1978. Response to 30 generations of selection for open-field activity in laboratory mice. *Behavior Genetics*, 8, 3–13.

DeFries, J. C., and Hegmann, J. P. 1970. Genetic analysis of open-field behavior. In *Contributions to behavior-genetic analysis: The mouse as a prototype*, eds. G. Lindzey and D. C. Thiessen, pp. 23–56. New York: Appleton-Century-Crofts.

DeFries, J. C.; Hegmann, J. P.; and Weir, M. W. 1966. Open-field behavior in mice:

Evidence for a major gene effect mediated by the visual system. *Science*, 154, 1577–1579.

DeFries, J. C.; Johnson, R. C.; Kuse, A. R.; McClearn, G. C.; Polovina, J.; Vandenberg, S. G.; and Wilson, J. R. 1979. Familial resemblance for specific cognitive abilities. *Behavior Genetics*, 9, 23–43.

DeFries, J. C.; Kuse, A. R.; and Vandenberg, S. G. 1979. Genetic correlations, environmental correlations and behavior. In *Theoretical advances in behavior genetics*, ed. J. R. Royce and L. P. Mos, pp. 389–421. Alphen aan den Rijn, Netherlands: Sijthoff Noordhoff International.

DeFries, J. C.; LaBuda, M. C.; and Fulker, D. W. 1987. Genetic and environmental covariance structures among WISC-R subtests: A twin study. *Intelligence*, 11, 233–244.

DeFries, J. C.; Plomin, R.; and LaBuda, M. C. 1987. Genetic stability of cognitive development from childhood to adulthood. *Developmental Psychology*, 23, 4–12.

DeFries, J. C.; Vogler, G. P.; and LaBuda, M. C. 1986. Colorado Family Reading Study: An overview. In *Perspectives in behavior genetics*, eds. J. L. Fuller and E. C. Simmel, pp. 29–56. Hillsdale, N. J.: Lawrence Erlbaum Associates.

DeFries, J. C., and Plomin, R. 1978. Behavioral genetics. *Annual Review of Psychology*, 29, 473–515.

DeFries, J. C.; Thomas, E. A.; Hegmann, J. P.; and Weir, M. W. 1967. Open-field behavior in mice: Analysis of maternal effects by means of ovarian transplantation. *Psychonomic Science*, 8, 207–208.

DeFries, J. C.; Vandenberg, S. G.; and McClearn, G. E. 1976. The genetics of specific cognitive abilities. *Annual Review of Genetics*, 10, 179–207.

deGrouchy, J.; Turleau, C.; and Finaz, C. 1978. Chromosomal phylogeny of the primates. *Annual Review of Genetics*, 12, 289–328.

DeLisi, C. 1988. The human genome project. *American Scientist*, 76, 488–493.

DeLisi, L. E.; Mirsky, A. F.; Buchsbaum, M. S.; van Kammen, D. P.; and Berman, K. F. (1984). The Genain quadruplets 25 years later: A diagnostic and biochemical followup. *Psychiatric Research*, 13, 59–76.

Detera-Wadleigh, S. D.; Berrettini, W. H.; Goldin, L. R.; Boorman, D.; Anderson, S.; and Gershon, E. S. 1987. Close linkage of c-harvey-ras-1 and the insulin gene to affective disorder is ruled out in three North American pedigrees. *Nature*, 325, 806–808.

Detterman, D. 1987. Mental retardation, cognition, and achievement in twins. Research grant awarded by the National Institute of Child Health and Human Development.

Dobzhansky, T. 1964. *Heredity and the nature of man*. New York: Harcourt, Brace & World.

Donis-Keller, H.; Green, P.; Helms, C.; Cartinhour, S.; Weiffenbach, B.; Stephens, K.; Keither, T. P.; Bowden, D. W.; and Smith, D. R. 1987. A human gene map. *Cell*, 51, 319–337.

Driscoll, P., and Battig, K. 1982. Behavioral, emotional and neurochemical profiles of rats selected for extreme differences in active, two-way avoidance performance. In *Genetics of the brain*, ed. I. Lieblich, pp. 95–123. Amsterdam: Elsevier.

Dudai, Y. 1983. Mutations affect storage and use of memory differentially in *Drosophila*. *Proceedings of the National Academy of Sciences USA*, 80, 5445–5448.

Dudai, Y.; Jan, Y. N.; Byers, D.; Quinn, W. G.; and Benzer, S. 1976. Dunce, a mutant of *Drosophila* deficient in learning. *Proceedings of the National Academy of Sciences*, 73, 1684–1688.

Dudek, B. C., and Abbott, M. E. 1984. A biometrical genetic analysis of ethanol response in selectively bred long-sleep and short-sleep mice. *Behavior Genetics*, 14, 1–20.

Duerr, J. S., and Quinn, W. G. 1982. Three *Drosophila* mutations that block associative learning also affect habituation and sensitization. *Proceedings of the National Academy of Sciences USA*, 79, 3646–3650.

Dunn, J. F., and Plomin, R. 1986. Determinants of maternal behavior toward three-year-old siblings. *British Journal of Developmental Psychology*, 57, 348–356.

Dunner, D. L; Fleiss, J. L.; Addonizio, G.; and Fieve, R. R. 1976. Assortative mating in primary affective disorder. *Biological Psychiatry*, 32, 1134–1137.

Earls, F. 1987. On the familial transmission of child psychiatric disorder. *Journal of Child Psychology and Psychiatry*, 28, 791–802.

East, E. M., and Hayes, H. K. 1911. Inheritance in maize. *Bulletin of the Connecticut Agricultural Experiment Station*, 167, 1–142.

Eaves, L. J.; Kendler, K. S.; and Schulz, S. C. 1986. The familial sporadic classification: Its power for the resolution of genetic and environmental etiological factors. *Journal of Psychiatric Research*, 20, 115–130.

Eaves, L. J.; Last, K. A.; Young, P. A.; and Martin, N. G. 1978. Model-fitting approaches to the analysis of human behaviour. *Heredity*, 41, 249–320.

Eaves, L. J., and Young, P. A. 1981. Genetical theory and personality differences. In *Dimensions of personality*, ed. R. Lynn. Oxford: Pergamon Press.

Ebert, P. D. 1983. Selection for aggression in a natural population. *Aggressive behavior: Genetic and neural approaches*, ed. E. C. Simmel, M. E. Hahn, and J. K. Walters, pp. 103–127. Hillsdale, N.J.: Lawrence Erlbaum Associates.

Ebert, P. D., and Hyde J. S. 1976. Selection for agonistic behavior in wild female *Mus musculus*. *Behavior Genetics*, 6, 291–304.

Edwards, M. D.; Stuber, C. W.; and Wendel, J. F. 1987. Molecular-marker-facilitated investigations of quantitative-trait loci in maize. I. Numbers, genomic distribution and types of gene action. *Genetics*, 116, 113–125.

Egeland, J. A.; Gerhard, D. S.; Pauls, D. L.; Sussex, J. N.; and Kidd, K. K. 1987. Bipolar affective disorders linked to DNA markers on chromosome 11. *Nature*, 325, 783–787.

Ehrman, L., and Parsons, P. A. 1981. *The genetics of behavior*. Sunderland, Mass.: Sinauer Associates.

Ehrman, L., and Probber, J. 1978. Rare *Drosophila* males: The mysterious matter of choice. *American Scientist*, 66, 216–222.

Ehrman, L., and Seiger, M. B. 1987. Diversity in Hawaiian drosophilids: A tribute to Dr. Hampton L. Carson upon his retirement. *Behavior Genetics*, 17, 537–615.

Eiseley, L. 1959. Charles Darwin, Edward Blyth, and the theory of natural selection. *Proceedings of the American Philosophical Society*, 103, 94–158.

Elder, G. H., Jr. 1985. *Life course dynamics: Trajectories and transitions, 1968–1980*. Ithaca, N.Y.: Cornell University Press.

Eleftheriou, B. E., ed. 1975. *Psychopharmacogenetics*. New York: Plenum Press.

Eleftheriou, B. E. and Elias, P. K. 1975. Recombinant inbred strains: A novel genetic approach for psychopharmacogeneticists. In *Psychopharmacogenetics*, ed. B. E. Eleftheriou, pp. 43–71. New York: Plenum Press.

Elmer, G. I.; Meisch, R. A.; and George, F. R. 1987. Mouse strain differences in operant self-administration of ethanol. *Behavior Genetics*, 17, 439–452.

Elston, R. C., and Stewart, J. 1971. A general model for the genetic analysis of pedigree data. *Human Heredity*, 21, 523–542.

Emerson, R. A., and East, E. M. 1913. The inheritance of quantitative characters in maize. *University of Nebraska Research Bulletin*, 2, 5–120.

Epstein, C. J., and Golbus, M. S. 1977. Prenatal diagnosis of genetic diseases. *American Scientist*, 65, 703–711.

Erlenmeyer-Kimling, L., and Cornblatt, B. 1987. High-risk research in schizophrenia: A summary of what has been learned. *Journal of Psychiatric Research*, 21, 401–411.

Essen-Möller, E. 1941. Empirische Ahnlichkeitsdiagnose bei Zwillingen. *Hereditas*, 27, 1.

Estivill, X.; Farrall, M.; Scambler, P. J.; Bell, G. M.; Hawley, K. M. F.; Lench, N. J.; Bates, G. P.; and Kruyer, H. C. 1987. A candidate for the cystic fibrosis locus isolated by selection for methylation-free islands. *Nature*, 326, 840–845.

Eysenck, H. J., and Broadhurst, P. L. 1964. Experiments with animals: Introduction. In *Experiments in motivation*, ed. H. J. Eysenck, pp. 285–291. New York: Macmillan.

Falconer, D. S. 1960. *Introduction to quantitative genetics*. New York: Ronald Press.

———. 1965. The inheritance of liability to certain diseases estimated from the incidence among relatives. *Annals of Human Genetics*, 29, 51–76.

———. 1981. *Introduction to quantitative genetics*. London: Longman.

Fantino, E., and Logan, C. A. 1979. *The experimental analysis of behavior*. San Francisco: W. H. Freeman and Company.

Faraone, S. V., and Tsuang, M. T. 1985. Quantitative models of the genetic transmission of schizophrenia. *Psychological Bulletin*, 98, 41–66.

Farmer, A. E.; McGuffin, P.; and Gottesman, I. I. 1984. Searching for the split in schizophrenia: A twin study perspective. *Psychiatric Research*, 13, 109–118.

Fischer, M; Harvald, B.; and Hauge, M. 1969. A Danish twin study of schizophrenia. *British Journal of Psychiatry*, 115, 981–990.

Fisher, R. A. 1918. The correlation between relatives on the supposition of Mendelian inheritance. *Transactions of the Royal Society of Edinburgh*, 52, 399–433.

———. 1930. *The genetical theory of natural selection*. Oxford: Clarendon Press.

Floderus-Myrhed, B.; Pedersen, N.; and Rasmuson, I. 1980. Assessment of heritability for personality based on a short form of the Eysenck Personality Inventory: A study of 12,898 twin pairs. *Behavior Genetics*, 10, 153–162.

Følling, A.; Mohr, O. L.; and Ruud, L. 1945. *Oligophrenia phenylpyrouvica*, a recessive syndrome in man. *Norske Videnskaps/Akademi I Oslo, Matematisk-Naturvidenskapelig Klasse*, 13, 1–44.

Folstein, S., and Rutter, M. 1977. Infantile autism: A genetic study of 21 twin pairs. *Journal of Child Psychology and Psychiatry*, 18, 297–321.

———. 1988. Autism: Familial aggregation and genetic implications. *Journal of Autism and Developmental Disorders*, 18, 3–30.

Fowler, W. A. 1965. A study of process and method in three-year-old twins and triplets learning to read. *Genetic Psychology Monographs*, 72, 3–90.

Friedman, G. D., and Lewis, A. M. 1978. The Kaiser-Permanente twin registry. In *Twin research, Part B. Biology and epidemiology*, ed. W. E. Nance, pp. 173–178. New York: Alan R. Liss.

Friedmann, T. 1971. Prenatal diagnosis of genetic disease. *Scientific American*, 225, 34–42.

Fulker, D. W. 1979. Some implications of biometrical genetical analysis for psychological research. In *Theoretical advances in behavior genetics*, ed. J. R. Royce and L. P. Mos, pp. 337–380. Alphen aan den Rijn, Netherlands: Sijthoff Noordhoff International.

———. 1981. The genetic and environmental architecture of psychoticism, extraversion and neuroticism. In *A model for personality*, ed. H. J. Eysenck, pp. 88–122. New York: Springer.

Fulker, D. W.; Plomin, R.; Thompson, L. A.; Phillips, K.; Fagan, J. F. III; and Haith, M. M. 1988. Rapid screening of infant predictors of adult IQ: A study of twins and their parents. Manuscript submitted for publication.

Fuller, J. L. 1983*a*. Ethology and behavior genetics. In *Behavior genetics: Principles and applications*, eds. J. L. Fuller and E. C. Simmel, pp. 337–362. Hillsdale, N.J.: Lawrence Erlbaum Associates.

———. 1983*b*. Sociobiology and behavior genetics. In *Behavior genetics: Principles and applications*, eds. J. L. Fuller and E. C. Simmel, pp. 435–477. Hillsdale, N.J.: Lawrence Erlbaum Associates.

Fuller, J. L.; Easler, C.; and Smith, M. E. 1950. Inheritance of audiogenic seizure susceptibility in the mouse. *Genetics*, 35, 622–632.

Fuller, J. L., and Thompson, W. R. 1960. *Behavior genetics*. New York: John Wiley.

———. 1978. *Foundations of behavior genetics*. St. Louis: C. V. Mosby.

Galton, F. 1865. Hereditary talent and character. *Macmillan's Magazine*, 12, 157–166, 318–327.

———. 1869. *Hereditary genius: An inquiry into its laws and consequences*. London: Macmillan. (Cleveland: World Publishing Co., 1962.)

———. 1871. Gregariousness in cattle and in men. *Macmillan's Magazine*, 23, 353–357.

———. 1876. The history of twins as a criterion of the relative powers of nature and nurture. *Royal Anthropological Institute of Great Britain and Ireland Journal*, 6, 391–406.

———. 1883. *Inquiries into human faculty and its development*. London: Macmillan.

Garmezy, N., and Rutter, M. 1983. *Stress, coping, and development in children*. New York: McGraw-Hill.

Garrod, A. E. 1908. The Croonian lectures on inborn errors of metabolism, I, II, III, IV. *Lancet 2*, 1–7, 73–79, 142–148, 214–220.

Gehring, W. J. 1976. Developmental genetics of *Drosophila*. *Annual Review of Genetics*, 10, 209–252.

Gehring, W. J. 1987. Homeo boxes in the study of development. *Science*, 236, 1245–1252.

Gelderman, H. 1975. Investigations on inheritance of quantitative characters in animals by gene markers. I. Methods. *Theoretical and Applied Genetics*, 46, 319–330.

Gershon, E. S. (In press.) Single genes in the major psychiatric disorders. *Science*.

Gershon, E. S.; Schreiber, J. L.; Hamovit, J. R.; Dibble, E. D.; and Kaye, W. 1984. Clinical findings in patients with anorexia nervosa and affective illness in their relatives. *American Journal of Psychiatry*, 141, 1419–1422.

Gesell, A. L., and Thompson, H. 1929. Learning and growth in identical infant twins. *Genetic Psychology Monographs*, 6, 5–120.

Gholson, B., and Barker, B. 1985. Kuhn, Lakatos, and Laudan: Applications in the history of physics and psychology. *American Psychologist*, 40, 755–769.

Gilliam, T. C.; Bucan, M.; MacDonald, M. E.; Zimmer, M; Haines, J. L.; Cheng, S. V.; Pohl, T. M.; and Meyers, R. H. 1987. A DNA segment encoding two genes very tightly linked to Huntington's disease. *Science*, 238, 950–952.

Ginsberg, B. E., and Miller, D. S. 1963. Genetic factors in audiogenic seizures. *Colloques Nationaux du Centre National de la Recherche Scientifique, Paris,* 112, 217–225.

Glover, D. M. 1984. *Gene cloning: The mechanics of DNA manipulation.* London: Chapman and Hall.

Gluzman, Y., and Shenk, T. 1984. *Enhancers and eukaryotic gene expression.* Cold Spring Harbor: Cold Spring Harbor Laboratory.

Goldgaber, D.; Lerman, M. I.; McBride, O. W.; Saffiotti, U.; and Gajdusek, C. 1987. Characterization and chromosomal localization of a DNA encoding brain amyloid of Alzheimer's disease. *Science,* 235, 877–880.

Goldsmith, H. H., and Campos, J. J. 1986. Fundamental issues in the study of early temperament: The Denver Twin Temperament Study. In *Advances in Developmental Psychology,* eds. M. E. Lamb, A. L. Brown, and B. Rogoff, pp. 231–283. Hillsdale, N.J.: Lawrence Erlbaum Associates.

Gordon, J. W.; Scangos, G. A.; Plotkin, D. J.; Barbos, J. A.; and Ruddle, F. H. 1980. Genetic transformation of mouse embryos by microinjection of purified DNA. *Proceedings of the National Academy of Science USA,* 77, 7380–7384.

Gorlin, R. J. 1977. Classical chromosome disorders. In *New chromosomal syndromes,* ed. J. J. Yunis, pp. 59–118. New York: Academic Press.

Gottesman, I. I.; Carey, G.; and Hanson, D. R. 1983. Pearls and perils in epigenetic psychopathology. In *Childhood psychopathology and development,* eds. S. B. Guze, E. J. Earls, and J. E. Barrett, pp. 287–300. New York: Raven Press.

Gottesman, I. I., and Shields, J. 1976. A critical review of recent adoption, twin, and family studies of schizophrenia: Behavioral genetics perspectives. *Schizophrenia Bulletin,* 2, 360–401.

———. 1977. Twin studies and schizophrenia a decade later. In *Contributions to the psychopathology of schizophrenia,* ed. B. A. Maher. New York: Academic Press.

———. 1982. *Schizophrenia: The epigenetic puzzle.* Cambridge, England: Cambridge University Press.

Green, E. L. 1966. Breeding systems. In *Biology of the laboratory mouse,* 2d ed., ed. E. L. Green. New York: McGraw-Hill.

Griffiths, M. I., and Phillips, C. J. 1976. *Twin research (Birmingham, 1968–1972).* Institute of Child Health, University of Birmingham.

Grossfield, J., and Ringo, J. M. 1984. Courtship and learning in *Drosophila. Behavior Genetics,* 14, 381–557.

Gruneberg, H. 1952. *The genetics of the mouse.* The Hague: Martinus Nijhoff.

Guilford, J. P. 1967. *The nature of human intelligence.* New York: McGraw-Hill.

Gusella, J. F.; Tanzi, R. E.; Anderson, M. A.; Hobbs, W.; Gibbons, K.; Raschtchian, R.; Gilliam, T. C.; and Wallace, M. R. 1984. DNA markers for nervous system diseases. *Science,* 225, 1320–1326.

Gusella, J. F.; Wexler, N. S.; Conneally, P. M.; Naylor, S. L.; Anderson, M. A.; Tanzi, R. E.; Watkins, P. C.; and Ottina, K. 1983. Apolymorphic DNA marker genetically linked to Huntington's disease. *Nature,* 306, 234–238.

Guttman, R. 1974. Genetic analysis of analytical spatial ability: Raven's Progressive Matrices. *Behavior Genetics,* 4, 273–284.

Guze, S. B.; Cloninger, C. R.; Martin, R. L; and Clayton, P. J. 1986. A follow-up and family study of Briquet's syndrome. *British Journal of Psychiatry,* 149, 17–23.

Haggard, E. A. 1958. *Intraclass correlation and the analysis of variance.* New York: Holt, Rinehart & Winston.

Hahn, M. E., Hewitt, J. K., Adams, M., and Tully, T. 1987. Genetic influences on ultrasonic vocalizations in young mice. *Behavior Genetics*, 17, 155–166.

Haldane, J. B. S. 1936. A search for incomplete sex-linkage in man. *Annals of Eugenics* 7, 28–57.

Hall, C. S. 1934. Emotional behavior in the rat. I. Defecation and urination as measures of individual differences in emotionality. *Journal of Comparative Psychology*, 18, 385–403.

Hall, J. C. 1977a. Behavioral analysis in *Drosophila* mosaics. In *Genetic mosaics and cell differentiation*, ed. W. J. Gehring, pp. 1–78. New York: Springer.

———. 1977b. Portions of the central nervous system controlling reproductive behavior in *Drosophila melanogaster*. *Behavior Genetics*, 7, 291–312.

———. 1984. Genetic analysis of behavior in insects. In *Comprehensive Insect Physiology, Biochemistry and Pharmacology*, vol. 9, eds. G. A. Kerkut and L. I. Gilbert, pp. 1–87. Oxford: Pergamon.

Hamilton, W. D. 1964. The genetical theory of social behaviour (I and II). *Journal of Theoretical Biology*, 7, 1–52.

Hardy, G. H. 1908. Mendelian proportions in a mixed population. *Science*, 28, 49–50.

Harper, M. E.; Ulrich, A.; and Saunders, G. F. 1981. Localization of the human insulin gene to the distal end of the short art of chromosome 11. *Proceedings of the National Academy of Sciences*, 78, 4458–4460.

Hay, D. A. 1975. Strain differences in maze-learning ability of *Drosophila melanogaster*. *Nature*, 257, 44–46.

———. 1976. The behavioral phenotype and mating behavior of two inbred strains of *Drosophila melanogaster*. *Behavior Genetics*, 6, 161–170.

———. 1985. *Essentials of behaviour genetics*. Oxford: Blackwells.

Hay, D. A., and O'Brien, P. J. 1983. The La Trobe Twin Study: A genetic approach to the structure and development of cognition in twin children. *Child Development*, 54, 317–330.

Heath, A. C.; Kendler, K. S.; Eaves, L. J.; and Markell, D. 1985. The resolution of cultural and biological inheritance: Informativeness of different relationships. *Behavior Genetics*, 15, 439–466.

Hegman, J. P. 1975. The response to selection for altered conduction velocity in mice. *Behavioral Biology*, 13, 413–423.

Hegmann, J. P., and DeFries, J. C. 1970. Are genetic correlations and environmental correlations correlated? *Nature*, 226, 284–286.

Henderson, N. D. 1967. Prior treatment effects on open field behavior of mice—A genetic analysis. *Animal Behaviour*, 15, 365–376.

Henderson, N. D. 1970. Genetic influences on the behavior of mice can be obscured by laboratory rearing. *Journal of Comparative and Physiological Psychology*, 72, 505–511.

———. 1972. Relative effects of early rearing environment on discrimination learning in housemice. *Journal of Comparative and Physiological Psychology*, 79, 243–253.

———. 1982. Human behavior genetics. *Annual Review of Psychology*, 33, 403–440.

Henry, K. R. 1967. Audiogenic seizure susceptibility induced in C57BL/6J mice by prior auditory exposure. *Science*, 158, 938–940.

Henry, K. R., and Schlesinger, K. 1967. Effects of the albino and dilute loci on mouse behavior. *Journal of Comparative and Physiological Psychology*, 63, 320–323.

Herrnstein, R. J. 1973. *I.Q. in the meritocracy*. Boston: Little, Brown.

Herskowitz, I.; Blair, L.; Forbes, D.; Hicks, J.; Kassir, Y.; Kushner, P.; Rine, J.; Sprague, G.; and Strathern, J. 1980. Control of cell type of the yeast *Saccharomyces cerevisiae* and a hypothesis for development in higher eukaryotes. In *The molecular genetics of development*, eds. T. Leighton and W. F. Leighton, pp. 79–118. New York: Academic Press.

Heston, L. L. 1966. Psychiatric disorders in foster home reared children of schizophrenic mothers. *British Journal of Psychiatry*, 112, 819–825.

Heston, L. L., and Mastri, A. R. 1977. The genetics of Alzheimer's disease: Associations with hematologic malignancy and Down's syndrome. *Archives of General Psychiatry*, 34, 976–981.

Hewitt, J. K.; Fulker, D. W.; and Broadhurst, P. L. 1981. Genetics of escape-avoidance conditioning in laboratory and wild populations of rats: A biometrical approach. *Behavior Genetics*, 11, 533–544.

Hilgard, J. R. 1933. The effect of early and delayed practice on memory and motor performance studied by the method of cotwin control. *Genetic Psychology Monographs*, 14, 493–567.

Hirsch, J. 1963. Behavior genetics and individuality understood. *Science*, 142, 1436–1442.

Hirsch, J., and McCauley, L. A. 1977. Successful replication of, and selective breeding for classical conditioning in the blowfly *Phormia regina*. *Animal Behavior*, 25, 784–785.

Ho, H.-z.; Foch, T. T.; and Plomin, R. 1980. Developmental stability of the relative influence of genes and environment on specific cognitive abilities in childhood. *Developmental Psychology*, 16, 340–346.

Hockett, C. F. 1973. *Man's place in nature*. New York: McGraw-Hill.

Hodgkinson, S.; Sherrington, R.; Gurling, H.; Marchbanks, R.; and Reeders, S. 1987. Molecular genetic evidence for heterogeneity in manic depression. *Nature*, 325, 805–806.

Holland, A. J.; Hall, A.; Murray, R.; Russell, G. F. M.; and Crisp, A. H. 1984. Anorexia nervosa: A study of 34 twin pairs and one set of triplets. *British Journal of Psychiatry*, 145, 414–419.

Holmes, W. G., and Sherman, P. W. 1983. Kin recognition in animals. *American Scientist*, 71, 46–55.

Homyk, T., Jr. 1977. Behavioral mutants of *Drosophila melanogaster*. II. Behavioral analysis and focus mapping. *Genetics*, 87, 105–128.

Homyk, T., Jr., and Sheppard, D. E. 1977. Behavioral mutants of *Drosophila melanogaster*. I. Isolation and mapping of mutations which decrease flight ability. *Genetics*, 87, 95–104.

Honzik, M. P. 1957. Developmental studies of parent–child resemblance in intelligence. *Child Development*, 28, 215–228.

Hook, E. B. 1973. Behavioral implications of the human XYY genotype. *Science*, 179, 139–150.

———. 1982. Epidemiology of Down syndrome. In *Down syndrome: Advances in biomedicine and the behavioral sciences*, eds. S. M. Pueschel and J. E. Rynders, pp. 21–43. Cambridge, Mass.: Ware Press.

Horn, J. M.; Loehlin, J. C.; and Willerman, L. 1979. Intellectual resemblance among adoptive and biological relatives: The Texas adoption project. *Behavior Genetics*, 9, 177–208.

———. 1982. Aspects of the inheritance of intellectual abilities. *Behavior Genetics*, 12, 479–516.

Horowitz, G. P., and Dudek, B. C. 1983. Behavioral pharmacogenetics. In *Behavior genetics: Principles and applications*, eds. J. L. Fuller and E. C. Simmel, pp. 117–154. Hillsdale, N.J.: Erlbaum.

Hotta, Y., and Benzer, S. 1970. Genetic dissection of the *Drosophila* nervous system by means of mosaics. *Proceedings of the National Academy of Sciences*, 67, 1156–1163.

Hrubec, Z., and Neel, J. V. 1978. The National Academy of Sciences–National Research Council twin registry: Ten years of operation. In *Twin research, Part B. Biology and epidemiology*, ed. W. E. Nance, pp. 153–172. New York: Alan R. Liss.

Hrubec, Z., and Omenn, G. S. 1981. Evidence of genetic predisposition to alcohol cirrhosis and psychosis: Twin concordances for alcoholism and its biological end points by zygosity among male veterans. *Alcoholism: Clinical and Experimental Research*, 5, 207–215.

Hsia, D. Y.-Y. 1968. *Human developmental genetics*. Chicago: Year Book Medical Publishers.

———. 1970. Phenylketonuria and its variants. In *Progress in medical genetics*, vol. 7, eds. A. G. Steinberg and A. G. Bearn, pp. 29–68. New York: Grune & Stratton.

Hsia, D. Y.-Y.; Knox, W. E.; Quinn, K. V.; and Paine, R. S. 1958. A one-year controlled study of the effect of low-phenylalanine diet on phenylketonuria. *Pediatrics*, 21, 178–202.

Huang, W. M.; Ao, S.-Z.; Casjens, S.; Orlandi, R.; Zeikus, R.; Weiss, R.; Winge, D.; and Fang, M. 1988. A persistent untranslated sequence within bacteriophage T4 DNA topoisomerase gene 60. *Science*, 26, 1005–1012.

Huettel, M. D. 1986. *Evolutionary genetics of invertebrate behavior: Progress and prospects*. New York: Plenum Press.

Husén, T. 1959. *Psychological twin research*. Stockholm: Almqvist & Wiksell.

Hutchins, R. M., and Adler, M. J. 1968. The idea of equality. In *The great ideas today*, eds. R. M. Hutchins and M. J. Adler, pp. 302–350. Chicago: Encyclopaedia Britannica, 1968.

Huxley, J. 1949. *Heredity east and west: Lysenko and world science*. New York: Schuman.

Hyde, J. S. 1974. Inheritance of learning ability in mice: A diallel-environmental analysis. *Journal of Comparative and Physiological Psychology*, 86, 116–123.

Jacob, F., and Monod, J. 1961. On the regulation of gene activity. *Cold Spring Harbor Symposia on Quantitative Biology*, 26, 193–209.

Jacobs, P. A.; Brunton, M.; Melville, M. M.; Brittain, R. P.; and McClemont, W. F. 1965. Aggressive behaviour, mental sub-normality, and the XYY male. *Nature*, 208, 1351–1352.

Jary, M. L., and Stewart, M. A. 1985. Psychiatric disorder in the parents of adopted children with aggressive conduct disorder. *Neuropsychobiology*, 13, 7–11.

Jeffreys, A. J.; Wilson, V.; and Thein, S. L. 1985. Individual-specific "fingerprints" of human DNA. *Nature*, 316, 76–79.

Jensen, A. R. 1967. Estimation of the limits of heritability of traits by comparison of monozygotic and dizygotic twins. *Proceedings of the National Academy of Sciences*, 58, 149–156.

———. 1972. *Genetics and education*. New York: Harper & Row.

———. 1973. *Educability and group differences*. New York: Harper & Row.

———. 1974. The problem of genotype–environment correlation in the estimation of heritability from monozygotic and dizygotic twins. *Acta Geneticae Medicae et Gemellogiae*, 25, 86–99.

————. 1976. Test bias and construct validity. *Phi Delta Kappan*, December, 340–346.

————. 1978. Genetic and behavioral effects of nonrandom mating. In *Human variation: The biopsychology of age, race and sex*, eds. R. T. Osborne, C. E. Noble, and N. Weyl, pp. 51–105. New York: Academic Press.

————. 1980. *Bias in mental testing*. New York: Free Press.

Jinks, J. L., and Broadhurst, P. L. 1974. How to analyze the inheritance of behavior in animals—the biometrical approach. In *The genetics of behavior*, ed. J. H. F. vanAbeelen, pp. 1–41. Amsterdam: North-Holland.

Jinks, J. L., and Fulker, D. W. 1970. Comparison of the biometrical genetical, MAVA, and classical approaches to the analysis of human behavior. *Psychological Bulletin*, 75, 311–349.

Johnson, C. A.; Ahern, F. M.; and Johnson, R. C. 1976. Level of functioning of siblings and parents of probands of varying degrees of retardation. *Behavior Genetics*, 6, 473–477.

Johnson, R. C.; DeFries, J. C.; Wilson, J. R.; McClearn, G. E.; Vandenberg, S. G.; Ashton, G. C.; Mi, M. P.; and Rashad, M. N. 1976. Assortative marriage for specific cognitive abilities in two ethnic groups. *Human Biology*, 48, 343–352.

Johnson, R. C.; McClearn, G. E.; Yuen, S.; Nagoshi, C. T.; Ahern, F. M.; and Cole, R. E. 1985. Galton's data a century later. *American Psychologist*, 40, 875–892.

Johnson, T. E. 1986. Molecular and genetic analyses of a multivariate system specifying behavior and life span. *Behavior Genetics*, 16, 221–234.

Jöreskog, K. G., and Sörbom, D. 1984. *LISTREL VI: User's guide*. 3rd edition. Mooresville, Ind.: Scientific Software.

Kallmann, F. J. 1946. The genetic theory of schizophrenia: An analysis of 691 schizophrenic twin index families. *American Journal of Psychiatry*, 103, 309–322.

Kallman, F. J. 1955. Genetic aspects of mental disorders in later life. In *Mental disorders in later life*, ed. O. J. Kaplan, pp. 26–46. Stanford, Calif.: Stanford University Press.

Kan, Y. W., and Dozy, A. M. 1978. Polymorphisms of DNA sequence adjacent to human β-globin structural gene: Relationship to sickle mutation. *Proceedings of the National Academy of Sciences USA*, 75, 5631–5635.

Kaprio, J.; Sarna, S.; Koskenvuo, M.; and Rantasalo, I. 1978. The Finnish twin registry: Formation and compilation, questionnaire study, zygosity determination procedures, and research program. In *Twin research, Part B, Biology and epidemiology*, ed. W. F. Nance, pp. 179–184. New York: Alan R. Liss.

Karlsson, J. L. 1966. *The biologic basis of schizophrenia*. Springfield, Ill: Charles C. Thomas.

————. 1970. The rate of schizophrenia in foster-reared close relatives of schizophrenic index cases. *Biological Psychiatry*, 2, 285–290.

Keeler, C. 1968. Some oddities in the delayed appreciation of "Castle's law." *Journal of Heredity*, 59, 110–112.

Kekic, V., and Marinkovic, D. 1974. Multiple-choice selection for light preference in *Drosophila subobscura*. *Behavior Genetics*, 4, 285–300.

Kendler, K. S., and Gruenberg, A. M. 1984. An independent analysis of the Danish Adoption Study of Schizophrenia. *Archives of General Psychiatry*, 41, 555–564.

Kendler, K. S.; Gruenberg, A. M.; and Tsuang, M. T. 1985. A family study of the subtypes of schizophrenia. *American Journal of Psychiatry*, 145, 57–62.

Kendler, K. S., and Robinette, C. D. 1983. Schizophrenia in the National Academy of Sciences–National Research Council twin registry: A 16-year update. *American Journal of Psychiatry*, 140, 1551–1563.

Kennedy, J. L.; Giuffra, L. A.; Moises, H. W.; Cavalli-Sforza, L. L.; Pakstis, A. J.; Kidd, J. R.; Castiglione, C. M.; Sjogren, B.; Wetterberg, L.; and Kidd, K. K. 1988. Evidence against linkage of schizophrenia to markers on chromosome 5 in a northern Swedish pedigree. *Nature*, 336, 167–170.

Kent, J. 1985. *Genetic and environmental contributions to cognitive abilities as assessed by a telephone test battery.* Unpublished doctoral dissertation, University of Colorado, Boulder.

Kenyon, C. 1988. The nematode *Caenorhabditis elegans*. *Science*, 240, 1448–1453.

Kessler, S. 1975. Extra chromosomes and criminality. In *Genetic research in psychiatry*, eds. R. R. Fieve, D. Rosenthal, and H. Brill, pp. 65–73. Baltimore: Johns Hopkins University Press.

———. 1979. *Genetic counseling: Psychological dimensions.* New York: Academic Press.

Kessler, S., and Moos, R. H. 1970. The XYY karyotype and criminality: A review. *Journal of Psychiatric Research*, 7, 153–170.

Kety, S. S. 1987. The significance of genetic factors in the etiology of schizophrenia: Results from the national study of adoptees in Denmark. *Journal of Psychiatric Research*, 21, 423–430.

Kety, S. S.; Rosenthal, D.; Wender, P. H.; and Schulsinger, F. 1968. The types and prevalence of mental illness in the biological and adoptive families of adopted schizophrenics. *Journal of Psychiatric Research*, 6, 345–362.

———. 1971. Mental illness in the biological and adoptive families of adopted schizophrenics. *American Journal of Psychiatry*, 128, 302–306.

———. 1976. Studies based on a total sample of adopted individuals and their relatives: Why they were necessary, what they demonstrated and failed to demonstrate. *Schizophrenia Bulletin*, 2, 413–428.

Kety, S. S.; Rosenthal, D.; Wender, P. H.; Schulsinger, F.; and Jacobsen, B. 1975. Mental illness in the biological and adoptive families of adopted individuals who have become schizophrenic: A preliminary report based on psychiatric interviews. In *Genetic research in psychiatry*, eds. R. R. Fieve, D. Rosenthal, and H. Brill, pp. 147–166. Baltimore: Johns Hopkins University Press.

Kitcher, P. 1985. *Vaulting ambition: Sociobiology and the quest for human nature.* Cambridge, Mass.: MIT Press.

Klein, T. W.; DeFries, J. C.; and Finkbeiner, C. T. 1973. Heritability and genetic correlations: Standard errors of estimates and sample size. *Behavior Genetics*, 4, 355–364.

Kloepfer, H. W. 1946. An investigation of 171 possible linkage relationships in man. *Annals of Eugenics*, 13, 35–47.

Kopin, I. J. 1981. Neurotransmitters and the Lesch–Nyhan syndrome. *New England Journal of Medicine*, 305, 1148–1150.

Kringlen, E. 1966. Schizophrenia in twins, an epidemiological-clinical study. *Psychiatry*, 29, 172–184.

———. 1978. Norwegian twin registers. In *Twin research, Part B, Biology and epidemiology*, ed. W. F. Nance, pp. 185–188. New York: Alan R. Liss.

Kung, C.; Chang, S. Y.; Satow, Y.; Van Houten, J.; and Hansma, H. 1975. Genetic dissection of behavior in *Paramecium*. *Science*, 188, 898–904.

Kyriacou, C. P., and Hall, J. C. 1984. Learning and memory mutations impair acoustic priming of mating behaviour in Drosophila. *Nature*, 308, 62–65.

LaBuda, M., DeFries, J. C., and Fulker, D. W. 1987. Genetic and environmental covariance structures among WISC-R subtests: A twin study. *Intelligence*, 11, 233–244.

Lalouel, J. M.; Rao, D. C.; Morton, N. E.; and Elston, R. L. 1983. A unified model for complex segregation analysis. *American Journal of Human Genetics*, 35, 816–826.

Lande, R. 1979. Quantitative genetics analysis of multivariate evolution, applied to brain: body size allometry. *Evolution*, 33, 402–416.

Lander, E. S. 1988. Splitting schizophrenia. *Science*, 336, 105–106.

Lander, E. S., and Botstein, D. 1986. Mapping complex genetic traits in humans: New methods using a complete RFLP linkage map. *Cold Spring Harbor Symposia on Quantitative Biology*, 51, 49–62.

Langinvainio, H.; Koskenvuo, M.; Kaprio, J.; and Sistonen, P. 1984. Finnish twins reared apart: II. Validation of zygosity, environmental dissimilarity and weight and height. *Acta Geneticae Medicae et Gemellologiae*, 33, 251–258.

Lebo, R. V.; Gorin, F.; Fletterick, R. J.; Kao, F.-T., Cheung, M.-C.; Bruce, B. D.; and Kan, Y. W. 1984. High resolution chromosome sorting and DNA spot-blot analysis assign McArdles syndrome to chromosome 11. *Science*, 225, 57–59.

Lehtovaara, A. 1938. *Psychologische Zwillingsuntersuchungen*, Helsinki: Finnish Academy of Science.

LePape, G., and Lassalle, J. M. 1984. A developmental genetic analysis of locomotor activity in mice: Maternal effects in the BALB/c and C57BL/6 strains and heredity in F1 hybrids. *Behavior Genetics*, 14, 21–29.

Lerner, I. M., and Libby, W. J. 1976. *Heredity, evolution, and society*. San Francisco: W. H. Freeman and Company.

Lewandowski, R. C., and Yunis, J. J. 1977. Phenotypic mapping in man. In *New chromosomal syndromes*, ed. J. J. Yunis, pp. 369–394. New York: Academic Press.

Lewin, R. 1984. Why is development so illogical? *Science*, 224, 1327–1329.

Lewontin, R. C.; Rose, S.; and Kamin, L. J. 1984. *Not in our genes: Biology, ideology, and human nature*. New York: Pantheon.

Li, C. C. 1975. *Path analysis: A primer*. Pacific Grove, Calif.: Boxwood Press.

Littlefield, C. H., and Rushton, J. P. 1986. When a child dies: The sociobiology of bereavement. *Journal of Personality and Social Psychology*, 51, 797–802.

Loehlin, J. C. 1982. Are personality traits differentially heritable? *Behavior Genetics*, 12, 417–428.

Loehlin, J. C. 1987. *Latent variable models: An introduction to factor, path, and structural analysis*. Hillsdale, N.J.: Lawrence Erlbaum Associates.

Loehlin, J. C., and DeFries, J. C. 1987. Genotype–environment correlation and IQ. *Behavior Genetics*, 17, 263–278.

Loehlin, J. C.; Lindzey, G.; and Spuhler, J. N. 1975. *Race differences in intelligence*. San Francisco: W. H. Freeman and Company.

Loehlin, J. C., and Nichols, R. C. 1976. *Heredity, environment and personality*. Austin: University of Texas Press.

Loehlin, J. C.; Sharan, S.; and Jacoby, R. 1978. In pursuit of the "spatial gene": A family study. *Behavior Genetics*, 8, 227–241.

Loehlin, J. C.; Willerman, L.; and Horn, J. M. 1988. Human behavior genetics. *Annual Review of Psychology*, 39, 101–133.

Loehlin, J. C.; Willerman, L.; and Vandenberg, S. G. 1974. Blood group and behavioral differences among dizygotic twins: A failure to replicate. *Social Biology*, 21, 205–206.

Lubs, H. A. 1969. A marker X chromosome. *American Journal of Human Genetics*, 21, 231–237.

Lush, I. E. 1984. The genetics of testing in mice. III. *Genetic Research (Cambridge)*, 44, 151–160.

Lush, J. L. 1949. Heritability of quantitative characters in farm animals. *Hereditas*, suppl. vol., 356–375.

———. 1951. Genetics and animal breeding. In *Genetics in the twentieth century*, ed. L. C. Dunn, pp. 493–525. New York: Macmillan.

Lykken, D. T. 1982. Research with twins: The concept of emergenesis. *Psychophysiology*, 19, 361–373.

Lykken, D. T.; McGue, M.; and Tellegen, A. 1987. Recruitment bias in twin research: The rule of two-thirds reconsidered. *Behavior Genetics*, 17, 343–362.

Lynch, C. B. 1981. Genetic correlation between two types of nesting in *Mus musculus*: Direct and indirect selection. *Behavior Genetics*, 11, 267–272.

Lyon, M. F. 1958. Twirler: A mutant affecting the inner ear of the house mouse. *Journal of Embryology and Experimental Morphology*, part I, 105–116.

Lytton, H.; Martin, N. G.; and Eaves, L. 1977. Environmental and genetical causes of variation in ethological aspects of behavior in two-year-old boys. *Social Biology*, 24, 200–211.

McAskie, M., and Clarke, A. M. 1976. Parent–offspring resemblance in intelligence: Theories and evidence. *British Journal of Psychology*, 67, 243–273.

McClearn, G. E. 1960. Strain differences in activity of mice: Influence of illumination. *Journal of Comparative and Physiological Psychology*, 53, 142–143.

———. 1976. Experimental behavioural genetics. In *Aspects of genetics in paediatrics*, ed. D. Barltrop, pp. 31–39. London: Fellowship of Postdoctorate Medicine, 1976.

McClearn, G. E., and DeFries, J. C. 1973. *Introduction to behavioral genetics*. San Francisco: W. H. Freeman and Company.

McClearn, G. E.; Wilson, J. R.; and Meredith, W. 1970. The use of isogenic and heterogenic mouse stocks in behavioral research. In *Contributions to behavior-genetic analysis: The mouse as a prototype*, eds. G. Lindzey and D. D. Thiessen, pp. 3–22. New York: Appleton-Century-Crofts.

McClintock, B. 1957. Controlling elements and the gene. *Cold Spring Harbor Symposia on Quantitative Biology*, 21, 197–216.

McDougall, W. 1908. *An introduction to social psychology*. London: Methuen.

MacGillivray, L.; Nylander, P. P. S.; and Corney, G. 1975. *Human multiple reproduction*. London: Saunders.

McGlone, J. 1985. Can spatial deficits in Turner's syndrome be explained by focal CNS dysfunction or atypical speech lateralization? *Journal of Clinical and Experimental Neuropsychology*, 7, 375–394.

McGuffin, P., and Sturt, E. 1986. Genetic markers in schizophrenia. *Human Heredity*, 36, 65–88.

McGuire, T. R. 1984. Learning in three species of diptera: The blow fly Phormia regina, the fruit fly *Drosophila melanogaster*, and the house fly *Musca domestica*. *Behavior Genetics*, 14, 479–526.

McKusick, V. A. 1986a. *Mendelian inheritance in man*. 8th ed. Baltimore: Johns Hopkins University Press.

———. 1986b. The gene map of Homo sapiens: Status and prospectus. *Cold Spring Harbor Symposia on Quantitative Biology*, 51, 15–27.

McLaren, A. 1976. *Mammalian chimaeras*. Cambridge, Eng.: Cambridge University Press.

Magenis, R. W., and Chamberlin, J. 1981. Parental origin of nondisjunction. In

Trisomy 21 (Down Syndrome): Research perspectives, eds. F. F. de la Cruz and P. S. Gerald, pp. 79–93. Baltimore: University Park Press.

Magnus, P.; Berg, K.; and Nance, W. E. 1983. Predicting zygosity in Norwegian twin pairs born 1915–1960. *Clinical Genetics*, 24, 103–112.

Martin, J. B. 1987. Molecular genetics: Applications to the clinical neurosciences. *Science*, 238, 765–772.

Martin, N. G. 1975. The inheritance of scholastic abilities in a sample of twins. *Annals of Human Genetics*, 39, 219–229.

Martin, N. G., and Eaves, L. J. 1977. The genetical analysis of covariance structure. *Annals of Human Genetics*, 39, 219–229.

Martin, N. G.; Eaves, L. J.; Heath, A. C.; Jardine, R.; Feingold, L. M.; and Eysenck, H. J. 1986. Transmission of social attitudes. *Proceedings of the National Academy of Sciences USA*, 83, 4364–4368.

Martin, N. G., and Jardine, R. 1986. Eysenck's contributions to behaviour genetics. In *Hans Eysenck: Consensus and controversy*, eds. S. Modgil and C. Modgil, pp. 13–27. Philadelphia: Falmer.

Martin, N. G.; Jardine, R.; and Eaves, L. J. 1984. Is there only one set of genes for different abilities? A reanalysis of the National Merit Scholarship Qualifying Test (NMSQT) data. *Behavior Genetics*, 14, 355–370.

Matheny, A. P.; Wilson, R. S.; and Dolan, A. B. 1976. Relations between twins' similarity of appearance and behavioral similarity: Testing an assumption. *Behavior Genetics*, 6, 343–352.

Mather, K., and Jinks, J. K. 1982. *Biometrical genetics: The study of continuous variation*. New York: Chapman and Hall.

Maxson, S. C., and Cowen, J. S. 1976. Electroencephalographic correlates of the audiogenic seizure response of inbred mice. *Physiology and Behavior*, 16, 623–629.

Mayr, E. 1974. Behavior programs and evolutionary strategies. *American Scientist*, 62, 650–659.

———. 1982. *The growth of biological thought: Diversity, evolution, and inheritance*. Cambridge, Mass.: The Belknap Press.

Mednick, S. A.; Gabrielli, W. F., Jr.; and Hutchings, B. 1984. Genetic influences in criminal convictions: Evidence from an adoption cohort. *Science*, 224, 891–894.

Medvedev, Z. A. 1969. *The rise and fall of T. D. Lysenko*. New York: Columbia University Press.

Mendel, G. J. 1866. Versuche Ueber Pflanzenhybriden. *Verhandlungen des Naturforschunden Vereines in Bruenn*, 4, 3–47. (Translated by Royal Horticultural Society of London.) Available in Dodson, E. O. 1956. *Genetics*. Philadelphia: Saunders. Also available in Sinnott, E. W.; Dunn, L. C.; and Dobzhansky, T. 1950. *Principles of genetics*. New York: McGraw-Hill.

Meredith, W. 1973. A model for analyzing heritability in the presence of correlated genetic and environmental effects. *Behavior Genetics*, 3, 271–277.

Merriman, C. 1924. The intellectual resemblance of twins. *Psychological Monographs*, 33, 1–58.

Mirenva, A. N. 1935. Psychomotor education and the general development of preschool children. *Pedagogical Seminary and Journal of Genetic Psychology*, 46, 433–454.

Mittler, P. 1971. *The study of twins*. Harmondsworth, England: Penguin Books.

Mohr, J. 1954. *A study of linkage in man*. Copenhagen: Munksgaard.

Money, J. 1964. Two cytogenetic syndromes: Psychologic comparisons. I. Intelligence and specific-factor quotients. *Journal of Psychiatric Research*, 2, 223–231.

———. 1968. Cognitive deficits in Turner's syndrome. In *Progress in human behavior genetics*, ed. S. G. Vandenberg, pp. 27–30. Baltimore: Johns Hopkins University Press.

Morgan, T. H.; Sturtevant, A. H.; Muller, H. J.; and Bridges, C. B. 1915. *The mechanism of Mendelian heredity*. New York: Henry Holt.

Munsinger, H. 1975. The adopted child's IQ: A critical review. *Psychological Bulletin*, 82, 623–659.

———. 1977. The identical twin transfusion syndrome: A source of error in estimating IQ resemblance and heritability. *Annals of Human Genetics*, 40, 307–321.

Murphey, R. M. 1983. Phenylketonuria (PKU) and the single gene: An old story retold. *Behavior Genetics*, 13, 141–157.

Murray, A. M., and Szostak, J. W. 1987. Artificial chromosomes. *Scientific American*, 252, 62–68.

Murray, R. M.; Lewis, S.; and Reveley, A. M. 1985. Towards an aetiological classification of schizophrenia. *Lancet*, 1, 1023–1026.

Nance, W. E. 1976. Genetic studies of the offspring of identical twins. *Acta Genetica Medicae et Gemellologiae*, 25, 103–113.

Nance, W. E., and Corey, L. A. 1976. Genetic models for the analysis of data from families of identical twins. *Genetics*, 83, 811–826.

Nance, W. E.; Winter, P. M.; Segreti, W. O.; Corey, L. A.; Parisi-Prinzi, G.; and Parisi, P. 1978. A search for evidence of hereditary superfetation in man. In *Twin research, Part B, Biology and epidemiology*, ed. W. E. Nance, pp. 65–70. New York: Alan R. Liss.

Natans, J.; Piantanida, T. P.; Eddy, R. L.; Shows, T. B.; and Hogness, D. S. 1986. Molecular genetics of inherited variation in human color vision. *Science*, 232, 203–207.

Nee, L. E.; Eldride, R.; Sunderland, T.; Thomas, C. B.; and Katz, D. 1987. Dementia of the Alzheimer type: Clinical and family study of 22 twin pairs. *Neurology*, 37, 259–363.

Nei, M. 1987. *Molecular evolutionary genetics*. New York: Columbia University Press.

Neri, G.; Opitz, J. M.; Mikkelsen, M.; Jacobs, P. A.; Daviews, K., and Turner, G., eds. 1988. X-linked mental retardation 3. (Special issue.). *American Journal of Medical Genetics*, 30 (1, 2).

Nesbitt, M. N.; Guthrie, D.; Spence, M. A.; and Butler, K. 1981. Use of chimeric mice to study behavior. In *Genetic research strategies in psychobiology and psychiatry*, eds. E. S. Gershon, S. Matthysee, X. O. Breakefield, and R. D. Ciaranello, pp. 105–112. New York: Academic Press.

Nesbitt, M. N.; Spence, M. A.; and Butler, K. 1979. Behavior in chimeric mice combining differently behaving strains. *Behavior Genetics*, 9, 277–288.

Netley, C. 1986. Summary overview of behavioural developments in individuals with neonatally identified X and Y aneuploidy. *Birth defects: Original Article Series*, 22, 293–306.

Newman, J.; Freeman, F.; and Holzinger, K. 1937. *Twins: A study of heredity and environment*. Chicago: University of Chicago Press.

Nichols, P. L. 1984. Familial mental retardation. *Behavior Genetics*, 14, 161–170.

Nichols, P. L., and Broman, S. H. 1974. Familial resemblance in infant mental development. *Developmental Psychology*, 10, 442–446.

Nichols, R. C. 1965. The National Merit twin study. In *Methods and goals in human behavior genetics*, ed. S. G. Vandenberg, pp. 231–243. New York: Academic Press.

———. 1978. Twin studies of ability, personality, and interests. *Homo*, 29, 158–173.

Nichols, R. C., and Bilbro, W. C. 1966. The diagnosis of twin zygosity. *Acta Genetica*, 16, 265–275.

Nilsson-Ehle, H. 1908–1909. Einige Ergebnisse von Kruezungen bei Hafer und Weisen. *Botanische Notiser*, 257–294.

Noël, B.; Duport, J. P.; Revil, D.; Sussuyer, I.; and Quack, B. 1974. The XYY syndrome: Reality or myth? *Clinical Genetics*, 5, 387–394.

Nurnberger, J. I., and Gershon, E. S. 1981. Genetics of affective disorders. In *Depression and antidepressants: Implications for courses and treatment*, ed. E. Friedman, pp. 23–39. New York: Raven Press.

Nussbaum, R. L., and Ledbetter, D. H. 1986. Fragile X syndrome: A unique mutation in man. *Annual Review of Genetics*, 20, 109–145.

Nylander, P. P. S. 1978. Causes of high twinning frequencies in Nigeria. In *Twin research, Part B, Biology and epidemiology*, ed. W. E. Nance, pp. 35–44. New York: Alan R. Liss.

Oliverio, A.; Eleftheriou, B. E.; and Bailey, D. W. 1973. A gene influencing active avoidance performance in mice. *Physiology and Behavior*, 11, 497–501.

Omenn, G. S. 1978. Prenatal diagnosis of genetic disorders. *Science*, 200, 952–958.

Orvaschel, H. 1983. Maternal depression and child dysfunction: Children at risk. In *Advances in clinical child psychology*, eds. B. B. Lahey and A. E. Kazdin, pp. 169–197. New York: Plenum.

Packard, A. S. 1901. *Lamarck, the founder of evolution*. New York: Longmans, Green.

Padeh, B.; Wahlsten, D.; and DeFries, J. C. 1974. Operant discrimination learning and operant bar-pressing rates in inbred and heterogeneous laboratory mice. *Behavior Genetics*, 4, 383–393.

Paigen, K. 1980. Temporal genes and other developmental regulators in mammals. In *The molecular genetics of development*, eds. T. Leighton and W. F. Loomis, pp. 419–470. New York: Academic Press.

Palm, J. 1961. Transplantation of ovarian tissue. In *Transplantation of tissues and cells*, eds. R. E. Billingham and W. K. Silvers, pp. 49–56. Philadelphia: The Wistar Institute Press.

Parkinson, J. S. 1977. Behavioral genetics in bacteria. *Annual Review of Genetics*, 11, 397–414.

Pauls, D. L.; Towbin, K. E.; Leckman, J. F.; Zahner, G. E. P.; and Cohen, D. J. 1986. The inheritance of Gilles de la Tourette's syndrome and associated behaviors. *New England Journal of Medicine*, 315, 993–997.

Pearson, K. 1924. *The life, letters and labours of Francis Galton*, vol. 1, London: Cambridge University Press.

Pedersen, N. L.; Friberg, L.; Floderus-Myrhed, B.; McClearn, G. E.; and Plomin, R. 1984. Swedish early separated twins: Identification and characterization. *Acta Geneticae Medicae et Gemellologiae*, 33, 243–250.

Pedersen, N. L.; Plomin, R.; McClearn, G. E.; and Friberg, L. 1988. Neuroticism, extraversion and related traits in adult twins reared apart and reared together. *Journal of Personality and Social Psychology*, 55, 950–957.

Pembrey, M. A.; Winter, R. M.; and Davies, K. E. 1985. A premutation that generates a defect at crossing over explains the inheritance of fragile X mental retardation. *American Journal of Medical Genetics*, 21, 709–717.

Penrose, L. S., and Smith, G. F. 1966. *Down's anomaly*. London: J. & A. Churchill.

Petit, C. 1951. Le rôle de l'isolement sexuel dans l'évolution des populations de *Drosophila melanogaster*. *Bulletin Biologique de la France et de la Belgique*, 85, 392–418.

Plomin, R. 1981. Heredity and temperament: A comparison of twin data for self-report questionnaires, parental ratings, and objectively assessed behavior. In *Twin research 3: Intelligence, personality, and development*, eds. L. Gedda, P. Parisi, and W. E. Nance, pp. 269–278. New York: Alan R. Liss.

———. 1986a. *Development, genetics, and psychology*. Hillsdale, N.J.: Lawrence Erlbaum Associates.

———. 1986b. Multivariate analysis and developmental behavioral genetics: Developmental change as well as continuity. *Behavior Genetics*, 16, 25–44.

———. 1988. The nature and nurture of cognitive abilities. In *Advances in the psychology of human intelligence*, ed. R. J. Sternberg, pp. 1–33. Hillsdale, N.J.: Lawrence Erlbaum Associates.

Plomin, R., and Daniels, D. 1987. Why are children in the same family so different from each other? *Behavioral and Brain Sciences*, 10, 1–16.

Plomin, R., and DeFries, J. C. 1979. Multivariate behavioral genetic analysis of twin data on scholastic abilities. *Behavior Genetics*, 9, 505–517.

———. 1980. Genetics and intelligence: Recent data. *Intelligence*, 4, 15–24.

Plomin, R., and DeFries, J. C. 1985. *Origins of individual differences in infancy: The Colorado Adoption Project*. New York: Academic Press.

Plomin, R.; DeFries, J. C.; and Fulker, D. W. 1988. *Nature and nurture during infancy and early childhood*. New York: Cambridge University Press.

Plomin, R.; DeFries, J. C.; and Loehlin, J. C. 1977. Genotype–environment interaction and correlation in the analysis of human behavior. *Psychological Bulletin*, 84, 309–322.

Plomin, R.; DeFries, J. C.; and Roberts, M. K. 1977. Assortative mating by unwed biological parents of adopted children. *Science*, 196, 449–450.

Plomin, R.; Foch, T. T.; and Rowe, D. C. 1981. Bobo clown aggression in childhood: Environment, not genes. *Journal of Research in Personality*, 15, 331–342.

Plomin, R.; Loehlin, J. C.; and DeFries, J. C. 1985. Genetic and environmental components of "environmental" influences. *Developmental Psychology*, 21, 391–402.

Plomin, R.; McClearn, G. E.; Pedersen, N. L.; Nesselroade, J. R.; and Bergeman, C. S. 1988a. Genetic influence on childhood family environment perceived retrospectively from the last half of the life span. *Developmental Psychology*, 24, 738–745.

Plomin, R.; Pedersen, N. L.; McClearn, G. E.; Nesselroade, J. R.; and Bergeman, C. S. 1988b. EAS temperaments during the last half of the life span: Twins reared apart and twins reared together. *Psychology and Aging*, 3, 43–50.

Plomin, R., and Rowe, D. C. 1979. Genetic and environmental etiology of social behavior in infancy. *Developmental Psychology*, 15, 62–72.

Plomin, R., and Rowe, D. C. 1979. Genetic and environmental etiology of social behavior in infancy. *Developmental Psychology*, 15, 62–72.

Plomin, R., and Willerman, L. 1975. A cotwin control study and a twin study of reflection-impulsivity in children. *Journal of Educational Psychology*, 67, 537–543.

Plomin, R.; Willerman, L.; and Loehlin, J. C. 1976. Resemblance in appearance and the equal environments assumption in twin studies of personality traits. *Behavior Genetics*, 6, 43–52.

Polivanov, S. 1975. Response of *Drosophila persimilis* to phototactic and geotactic selection. *Behavior genetics*, 5, 255–267.

Porter, I. H. 1977. Evolution of genetic counseling in America. In *Genetic counseling: A monograph of the National Institute of Child Health and Human Development*, eds. H. A. Lubs and F. de la Cruz, pp. 17–34. New York: Raven Press.

Price, R. A.; Kidd, K. K.; Cohen, D. J.; Pauls, D. L.; and Leckman, J. F. 1985. A twin study of Tourette syndrome. *Archives of General Psychiatry*, 42, 815–820.

Price, W. H., and Jacobs, P. A. 1970. The 47, XYY male with special reference to behavior. *Seminars in Psychiatry*, 2, 30–39.

Pueschel, S. M. 1982. *A study of the young child with Down syndrome*. New York: Human Science Press.

Pueschel, S. M., and Rynders, J. E. 1982. *Down syndrome: Advances in biomedicine and the behavioral sciences*. Cambridge, Mass.: Ware Press.

Quinn, W. G., and Greenspan, R. J. 1984. Learning and courtship in *Drosophila*: Two stories with mutants. *Annual Review of Neuroscience*, 7, 67–92.

Quinn, W. G.; Harris, W. A.; and Benzer, S. 1974. Conditioned behavior in *Drosophila melanogaster*. *Proceedings of the National Academy of Sciences*, 71, 708–712.

Rao, T. V. 1978. Maternal age, parity, and twin pregnancies. In *Twin research, Part B, Biology and epidemiology*, ed. W. E. Nance, pp. 99–104. New York: Alan R. Liss.

Reading, A. J. 1966. Effect of maternal environment on the behavior of inbred mice. *Journal of Comparative and Physiological Psychology*, 62, 437–440.

Ready, D. F.; Hanson, T. E.; and Benzer, S. 1976. Development of the *Drosophila* retina, a neurocrystalline lattice. *Developmental Biology*, 53, 217–240.

Reed, E. W., and Reed, S. C. 1965. *Mental retardation: A family study*. Philadelphia: Saunders.

Reeder, S. T.; Breuning, M. H.; Davies, E.; Nicolls, R. D.; Jarman, A. P.; Higgs, D. R.; Pearson, P. L.; and Weatherall, D. J. 1985. A highly polymorphic DNA marker linked to adult polycystic kidney disease on chromosome 16. *Nature*, 317, 542–545.

Regier, D. A.; Goldberg, E. D.; and Taube, C. A. 1978. The de facto US mental health services system. *Archives of General Psychiatry*, 35, 685–693.

Reich, T.; Cloninger, C. R.; and Guze, S. B. 1975. The multifactorial model of disease transmission: I. Description of the model and its use in psychiatry. *British Journal of Psychiatry*, 127, 1–10.

Reich, T.; Van Eerdewegh, P.; Rice, J.; Mullaney, J.; Endicott, J.; and Klerman, G. L. 1987. The familial transmission of primary major depressive disorder. *Journal of Psychiatric Research*, 21, 613–624.

Rende, R., Plomin, R., & Vandenberg, S. G. In press. Who discovered the twin method? *Behavior Genetics*.

Renvoize, E. B.; Mindham, R. H. S.; Stewart, M.; McDonald, R.; and Wallace, D. R. D. 1986. Identical twins discordant for presenile dementia of the Alzheimer type. *British Journal of Psychiatry*, 149, 509–512.

Rice, J. P.; Reich, T.; Andreasen, N. C.; Endicott, J.; Van Eerdewegh, M.; Fishman, A.; Hirschfield, R. M. A.; and Klerman, G. L. 1987. The familial transmission of bipolar illness. *Archives of General Psychiatry*, 41, 441–447.

Rice, T.; Fulker, D. W.; and DeFries, J. C. 1986. Multivariate path analysis of specific cognitive abilities in the Colorado Adoption Project. *Behavior Genetics*, 16, 107–126.

Rice, T.; Fulker, D. W.; DeFries, J. C.; and Plomin, R. 1988. Path analysis of IQ during infancy and early childhood and an index of the home environment in the Colorado Adoption Project. *Intelligence*, 12, 27–45.

Ritchie-Calder, P. R. 1970. *Leonardo and the age of the eye.* New York: Simon & Schuster.

Ritvo, E. R.; Freeman, B. J.; Mason-Brothers, A.; Mo, A.; and Ritvo, A. M. 1985. Concordance for the syndrome of autism in 40 pairs of afflicted twins. *American Journal of Psychiatry*, 142, 74–77.

Roberts, C. A., and Johansson, C. B. 1974. The inheritance of cognitive interest styles among twins. *Journal of Vocational Behavior*, 4, 237–243.

Roberts, R. C. 1967. Some concepts and methods in quantitative genetics. In *Behavior-genetic analysis*, ed. J. Hirsch, pp. 214–257. New York: McGraw-Hill.

Roberts, R. J. 1983. Restriction and modification enzymes and their recognition sequences. *Nucleic Acids Research*, 11, 135–167.

Robins, L. N.; Helzer, J. E.; Weissmann, M. M.; Orvaschel, H.; Gruenberg, E.; Burke, J. D.; and Regier, D. A. 1984. Lifetime prevalence of specific psychiatric disorders in three sites. *Archives of General Psychiatry*, 41, 949–958.

Robinson, A.; Lubs, H.; and Bergsma, D. 1979. Sex chromosome aneuploidy: Prospective studies on children. *Birth defects: Original article series*, 15, 1–85.

Robinson, A., and Puck, T. 1967. Studies on chromosomal nondisjunction in man, II. *American Journal of Human Genetics*, 19, 112–129.

Roderick, T. H. 1980. Strain distributions of genetic polymorphisms in the mouse. In *Handbook on genetically standardized JAX mice*, eds. H.-J. Heiniger and J. J. Dorey, pp. 19–34. Bar Harbor, M.: Jackson Laboratory.

Rose, R. J.; Harris, E. L.; Christian, J. C.; and Nance, W. E. 1979a. Genetics variance in nonverbal intelligence: Data from the kinships of identical twins. *Science*, 205, 1153–1155.

Rose, R. J.; Miller, J. Z.; Dumont-Driscoll, M.; and Evans, M. M. 1979b. Twin-family studies of perceptual speed ability. *Behavior Genetics*, 9, 71–86.

Rose, R. J.; Miller, J. Z.; and Fulker, D. W. 1981. Twin-family studies of perceptual speed ability. II. Parameter estimation. *Behavior Genetics*, 11, 565–576.

Rosenthal, D. 1970. *Genetic theory and abnormal behavior.* New York: McGraw-Hill.

———. 1972. Three adoption studies of heredity in the schizophrenic disorders. *International Journal of Mental Health*, 1, 63–75.

Rosenthal, D. 1975a. Discussion: The concept of subschizophrenic disorders. In *Genetic research in psychiatry*, eds. R. R. Fieve, D. Rosenthal, and H. Brill, pp. 199–208. Baltimore: Johns Hopkins University Press.

———. 1975b. The spectrum concept in schizophrenic and manic-depressive disorders. In *Biology of the major psychoses*, ed. D. X. Freedman. New York: Raven Press.

Rosenthal, D.; Wender, P. H.; Kety, S. S.; Schulsinger, F.; Welner, J.; and Ostergaard, L. 1968. Schizophrenics' offspring reared in adoptive homes. *Journal of Psychiatric Research*, 6, 377–391.

Rosenthal, D.; Wender, P. H.; Kety, S. S.; Welner, J.; and Schulsinger, F. 1971. The adopted-away offspring of schizophrenics. *American Journal of Psychiatry*, 128, 307–311.

Roubertoux, P. L., and Carlier, M. 1988. Differences between CBA/H and NZB mice on intermale aggression. II. Maternal effects. *Behavior Genetics*, 18, 175–184.

Rovet, J., and Netley, C. 1982. Processing deficits in Turner's syndrome. *Developmental Psychology*, 18, 77–94.

Rowe, D. C. 1981. Environmental and genetic influences on dimensions of perceived parenting: A twin study. *Developmental Psychology*, 17, 203–208.

————. 1983*a*. Biometrical genetic models of self-reported delinquent behavior: twin study. *Behavior Genetics*, 13, 473–489.

————. 1983*b*. A biometrical analysis of perceptions of family environment: A study of twin and singleton sibling kinships. *Child Development*, 54, 416–423.

Ruddle, F. H., and Kucherlapati, R. S. 1974. Hybrid cells and human genes. *Scientific American*, 232, 37–44.

Russell, W. L., and Douglas, P. M. 1945. Offspring from unborn mothers. *Proceedings of the National Academy of Sciences*, 31, 402–405.

Sanchez, O., and Yunis, J. J. 1977. New chromosome techniques and their medical applications. In *New chromosomal syndromes*, ed. J. J. Yunis, pp. 1–54. New York: Academic Press.

Sarna, S.; Kaprio, J.; Sistonen, P.; and Koskenvuo, M. 1978. Diagnosis of twin zygosity by mailed questionnaire. *Human Heredity*, 28, 241–254.

Savic, S. 1980. *How twins learn to talk: A study of the speech development of twins from 1 to 3*. New York: Academic Press.

Sax, K. 1923. The association of size differences with seed coat pattern and pigmentation in *Phaseolus vulgaris*. *Genetics*, 8, 552–560.

Scarr, S. 1968. Environmental bias in twin studies. *Eugenics Quarterly*, 15, 34–40.

————. 1977. *Genetic effects on human behavior: Recent family studies*. Lecture, Annual Meeting of the American Psychological Association, San Francisco.

Scarr, S., and Carter-Saltzman, L. 1979. Twin method: Defense of a critical assumption. *Behavior Genetics*, 9, 527–542.

Scarr, S., and McCartney, K. 1983. How people make their own environments: A theory of genotype → environment effects. *Child Development*, 54, 424–435.

Scarr, S., and Weinberg, R. A. 1978*a*. The influence of "family background" on intellectual attainment. *American Sociological Review*, 43, 674–692.

————. 1978*b*. Attitudes, interests, and IQ. *Human Nature*, April, 29–36.

Schellenberg, G. D.; Bird, T. D.; Wijsman, E. M.; Moore, D. K.; Boehnke, M.; Bryant, E. M.; Lampe, T. H.; Nochlin, D.; Sumi, S. M.; Deeb, S. S.; Beyreuther, K.; and Martin, G. M. 1988. Absence of linkage of chromosome 21q21 markers to familial Alzheimer's disease. *Science*, 241, 1507–1510.

Schiff, M.; Duyme, M.; Dumaret, A.; Stewart, J.; Tomkiewicz, S.; and Feingold, J. 1978. Intellectual status of working-class children adopted early into upper-middle-class families. *Science*, 200, 1503–1504.

Schmidt, J. 1919. La valeur de l'individu à titre de generateur appreciée suivant la méthode du croisement diallele. *Comptes Rendus des Travaux du Laboratoire Carlsberg*, 14, 1–33.

Schoenfeldt, L. F. 1968. The hereditary components of the Project TALENT two-day test battery. *Measurement and Evaluation in Guidance*, 1, 130–140.

Schuckit, M. A.; Goodwin, D. W., and Winokur, G. 1972. A study of alcoholism in half siblings. *American Journal of Psychiatry*, 128, 1132–1135.

Schull, W. J., and Neel, J. V. 1965. *The effects of inbreeding on Japanese children*. New York: Harper & Row.

Scott, J. P. 1958. *Animal behavior*. Chicago: University of Chicago Press.

Scott, J. P., and Fuller, J. L. 1965. *Genetics and the social behavior of the dog*. Chicago: University of Chicago Press.

Scriber, C. R.; Kaufman, S.; and Woo, S.L.C. 1988. Mendelian hyperphenylalanemia. *Annual Review of Genetics*, 22, 301–321.

Searle, L. V. 1949. The organization of hereditary maze-brightness and maze-dullness. *Genetic Psychology Monographs*, 39, 279–325.

Sefton, A. E., and Siegel, P. B. 1975. Selection for mating ability in Japanese quail. *Poultry Science*, 54, 788–794.

Segreti, W. O.; Winter, P. M.; and Nance, W. E. 1978. Familial studies of monozygotic twinning. In *Twin research, Part B, Biology and epidemiology*, ed. W. E. Nance, pp. 55–60. New York: Alan R. Liss.

Shaffer, J. W. 1962. A specific cognitive deficit observed in gonadal aplasia (Turner's syndrome). *Journal of Clinical Psychology*, 18, 403–406.

Shah, S. A. 1970. *Report of the XYY chromosomal abnormality*. U. S. Public Health Service pub. no. 2103. Washington, D. C.: U.S. Government Printing Office.

Shapiro, J. 1983. *Mobile genetic elements*. New York: Academic Press.

Sherman, P. W. 1977. Nepotism and the evolution of alarm calls. *Science*, 197, 1246–1253.

Sherman, S. L.; Jacobs, P. A.; Morton, N. E.; Froster-Iskenius, U.; Howard-Peebles, P. N.; et al. 1985. Further segregation analysis of the fragile X syndrome with special reference to transmitting males. *Human Genetics*, 69, 289–299.

Sherrington, R.; Brynjolfsson, J.; Petursson, H.; Potter, M.; Dudleston, K.; Barraclough, B.; Wasmuth, J.; Dobbs, M.; and Gurling, H. 1988. Localization of a susceptibility locus for schizophrenia on chromosome 5. *Nature*, 336, 164–167.

Shields, J. 1962. *Monozygotic twins brought up apart and brought up together*. London: Oxford University Press.

———. 1978. MZA twins: Their use and abuse. In *Twin research, Part A, Psychology and methodology*, ed. W. E. Nance, pp. 79–93. New York: Alan R. Liss.

Shields, J.; Heston, L. L.; and Gottesman, I. I. 1975. Schizophrenia and the schizoid: The problem for genetic analysis. In *Genetic research in psychiatry*, eds. R. R. Fieve, D. Rosenthal, and H. Brill, pp. 167–197. Baltimore: Johns Hopkins University Press.

Siemens, H. W. 1927. The diagnosis of identity in twins. *Journal of Heredity*, 18, 201–209.

Sigvardsson, S.; Cloninger, C. R.; and Bohman, M. 1985. Prevention and treatment of alcohol abuse: Uses and limitations of the high risk paradigm. *Social Biology*, 32, 185–194.

Silverman, W.; Lubin, R.; Jenkins, E. C.; and Brown, W. T. 1983. Quantifying the strength of association between fra(X) chromosome marker presence and mental retardation. *Clinical Genetics*, 23, 436–440.

Skinner, B. F. 1938. *The behavior of organisms: An experimental analysis*. New York: Appleton-Century-Crofts.

Skodak, M., and Skeels, H. M. 1949. A final follow-up of one hundred adopted children. *Journal of Genetic Psychology*, 75, 85–125.

Smalley, S. L.; Asarnow, R. F.; and Spence, M. A. 1988. Autism and genetics. *Archives of General Psychiatry*, 45, 953–961.

Smith, C. 1974. Concordance in twins: Methods and interpretation. *American Journal of Human Genetics*, 26, 454–466.

Smith, D. W. 1977. Clinical diagnosis and nature of chromosomal abnormalities. In *New chromosomal syndromes*, ed. J. J. Yunis, pp. 55–117. New York: Academic Press.

Smith, J. W. 1978. The evolution of behavior. *Scientific American*, 239, 176–192.

Smith, R. T. 1965. A comparison of socio-environmental factors in monozygotic and dizygotic twins, testing an assumption. In *Methods and goals in human behavior genetics*, ed. S. G. Vandenberg, pp. 45–62. New York: Academic Press.

Smith, S. D., and Goldgar, D. E. 1986. Single-gene analyses and their application to learning disabilities. In *Genetics and learning disabilities*, ed. S. D. Smith, pp. 47–65. San Diego: College Hills Press.

Smith, S. M., and Penrose, L. S. 1955. Monozygotic and dizygotic twin diagnosis. *Annals of Human Genetics*, 19, 273–389.

Snyderman, M., and Rothman, S. 1987. Survey of expert opinion on intelligence and aptitude testing. *American Psychologist*, 42, 137–144.

Sokolowski, M. B.; Hansell, R. I. C.; and Rotin, D. 1983. *Drosophila* larval foraging behavior. II. Selection in the sibling species, *D. melanogaster* and *D. simulans*. *Behavior Genetics*, 13, 169–177.

Solomon, E., and Bodmer, W. F. 1979. Evolution of sickle variant gene. *Lancet*, 1, 923–925.

Southern, E. M. 1975. Detection of specific sequences among DNA fragments separated by gel electrophoresis. *Journal of Molecular Biology*, 98, 503–517.

Sprott, R. L., and Staats, J. 1975. Behavioral studies using genetically defined mice — A bibliography. *Behavior Genetics*, 5, 27–82.

———. 1978. Behavioral studies using genetically defined mice — A bibliography (July 1973–July 1976). *Behavior Genetics*, 8, 183–206.

———. 1979. Behavioral studies using genetically defined mice — A bibliography (July 1976–August 1978). *Behavior Genetics*, 9, 87–102.

———. 1980. Behavioral studies using genetically defined mice — a bibliography (August 1978–July 1979). *Behavior Genetics*, 10, 93–104.

———. 1981. Behavioral studies using genetically defined mice — a bibliography (August 1979–July 1980). *Behavior Genetics*, 11, 73–84.

Sprott, R. L., and Stavnes, K. 1975. Effects of situational variables on performance of inbred mice in active- and passive-avoidance situations. *Psychological Reports*, 37, 683–692.

Spuhler, J. N. 1968. Assortative mating with respect to physical characteristics. *Eugenics Quarterly*, 15, 128–140.

Stevenson, J.; Graham, P.; Fredman, G.; and McLoughlin, V. 1987. A twin study of genetic influences on reading and spelling ability and disability. *Journal of Child Psychology and Psychiatry*, 28, 229–247.

Stewart, D. A.; Bailey, J. D.; Netley, C. T.; Rovet, J.; and Park, E. 1986. Growth and development from early to midadolescence of children with X and Y chromosome aneuploidy. *Birth Defects: Original Articles Series*, 22, 119–182.

Stewart, D. W.; Netley, C. T.; and Park, E. 1982. Summary of clinical findings of children with 47,XXY, 47,XYY and 47,XXX karyotypes. *Birth Defects: Original Article Series*, 18, 1–5.

St. George-Hyslop, P. H.; Tanzi, R. E.; Polinsky, R. J.; Haines, J. L.; Nee, L.; Watkins, P. C.; Myer, R. H.; and Feldman, R. G. 1987. The genetic defect causing familial Alzheimer's disease maps on chromosome 21. *Science*, 235, 885–890.

Strayer, L. C. 1930. The relative efficacy of early and deferred vocabulary training studied by the method of cotwin control. *Genetic Psychology Monographs*, 8, 209–319.

Strober, M.; Morrell, W.; Burroughs, J.; Salkin, B.; and Jacobs, C. 1985. A controlled family study of anorexia nervosa. *Journal of Psychiatric Research*, 19, 239–246.

Sturtevant, A. H. 1915. Experiments on sex recognition and the problem of sexual selection in *Drosophila*. *Journal of Animal Behavior*, 5, 351–366.

Sumner, A. T.; Robinson, J. A.; and Evans, H. J. 1971. Distinguishing between X, Y, and YY-bearing human spermatozoa by fluorescence and DNA content. *Nature, New Biology*, 229, 231–233.

Sundet, J. M.; Tambs, K.; Magnus, P.; and Berg, K. 1988. On the question of secular trends in the heritability of intelligence test scores: A study of Norwegian twins. *Intelligence*, 12, 47–60.

Suomi, S. J. 1982. Sibling relationships in nonhuman primates. In *Sibling relationships*, eds. M. E. Lamb and B. Sutton-Smith, pp. 329–356. Hillsdale, N.J.: Lawrence Erlbaum Associates.

Tambs, K. 1987. No genetic effect on variation in field dependence: A study of rod-and-frame scores in families of monozygotic twins. *Behavior Genetics*, 17, 493–502.

Tambs, K.; Sundet, J. M.; and Magnus, P. 1984. Heritability analysis of the WAIS subtests: A study of twins. *Intelligence*, 8, 283–293.

———. 1986. Genetic and environmental contributions to the covariation between the Wechsler Adult Intelligence Scale (WAIS) subtests: A study of twins. *Behavior Genetics*, 16, 475–491.

Tanksley, S. D.; Medina-Filho, H.; and Rick, C. M. 1982. Use of naturally-occurring enzyme variation to detect and map genes controlling quantitative traits in an interspecific backcross of tomato. *Heredity*, 49, 11–25.

Taubman, P. 1976. The determinants of earnings: Genetics, family and other environments: A study of white male twins. *American Economic Review*, 66, 858–870.

Taylor, M. A., and Abrams, R. 1978. The prevalence of schizophrenia: A reassessment using modern diagnostic criteria. *American Journal of Psychiatry*, 135, 945–948.

Teasdale, T. W. 1979. Social class correlations among adoptees and their biological and adoptive parents. *Behavior Genetics* 9, 103–114.

Teasdale, T. W., and Owen, D. R. 1981. Social class correlations among separately adopted siblings and unrelated individuals adopted together. *Behavior Genetics*, 11, 577–588.

———. 1984. Heredity and familial environment in intelligence and educational level: A sibling study. *Nature*, 309, 620–622.

Tellegen, A.; Lykken, D. T.; Bouchard, T. J.; Wilcox, K.; Segal, N.; and Rich, S. 1988. Personality similarity in twins reared apart and together. *Journal of Social and Personality Psychology*, 54, 1031–1039.

Tennes, K.; Puck, M.; Bryant, K.; Frankenburg, W.; and Robinson, A. 1975. A developmental study of girls with trisomy X. *American Journal of Human Genetics*, 27, 71–86.

Theilgaard, A. 1981. The personalities of XYY and XXY men. In *Human behavior and genetics*, eds. W. Schmid and J. Nielsen, pp. 75–84. Amsterdam: Elsevier/North Holland Biomedica.

Theis, S. V. S. 1924. *How foster children turn out*. New York: State Charities Aid Association, Publication no. 165.

Thompson, J. C., and Thompson, M. W. 1986. *Genetics in medicine*. Philadelphia: W.B. Saunders.

Thompson, J. N., Jr., and Thoday, J. M., eds. 1979. *Quantitative genetic variation*. New York: Academic Press.

Thompson, W. R., and Bindra, D. 1952. Motivational and emotional characters of "bright" and "dull" rats. *Canadian Journal of Psychology*, 6, 116–122.

Thorndike, E. L. 1905. Measurement of twins. *Archives of Philosophy, Psychology, and Scientific Methods*, 1, 1–64.

Tienari, P. 1963. Psychiatric illnesses in identical twins. *Acta Psychiatrica Scandinavica*, 39, suppl. 171.

Tjio, J. H., and Levan, A. 1956. The chromosome number of man. *Hereditas*, 42, 1–6.

Tobach, E.; Bellin, J. S.; and Das, D. K. 1974. Differences in bitter taste perception in three strains of rats. *Behavior Genetics*, 4, 405–410.

Tolman, E. C. 1924. The inheritance of maze-learning ability in rats. *Journal of Comparative Psychology*, 4, 1–18.

Trivers, R. L. 1974. Parent–offspring conflict. *American Zoologist*, 14, 249–264.

Tsuang, M. T.; Winokur, G.; and Crowe, R. R. 1980. Morbidity risks of schizophrenia and affective disorders among first-degree relatives of patients with schizophrenia, mania, depression and surgical conditions. *British Journal of Psychiatry*, 137, 497–504.

Tully, T. 1984. *Drosophila* learning: Behavior and biochemistry. *Behavior Genetics*, 14, 525–557.

Tyler, P. A. 1969. A quantitative genetic analysis of runway learning in mice. Ph.D. dissertation, University of Colorado.

van Abeelen, J. H. F., and van der Kroon, P. H. W. 1967. *Nijmegen waltzer*—a new neurological mutant in the mouse. *Genetical Research*, 10, 117–118.

van Abeelen, J. H. F., and van Nies, J. H. M. 1983. Effects of intrahippocampally-injected naloxone and morphine upon behavioral responses to novelty in mice from two selectively-bred lines. *Psychopharmacology*, 81, 232–235.

Vandenberg, S. G. 1968. The contribution of twin research to psychology. *Psychological Bulletin*, 66, 327–352.

———. 1971. What do we know today about the inheritance of intelligence and how do we know it? In *Intelligence: Genetic and environmental influences*, ed. R. Cancro, pp. 182–218. New York: Grune & Stratton.

———. 1972. Assortative mating, or who marries whom? *Behavior Genetics*, 2, 127–157.

———. 1976. Twin studies. In *Human behavior genetics*, ed. A. R. Kaplan, pp. 90–150. Springfield, Ill.: Thomas.

Vandenberg, S. G.; Singer, S. M.; and Pauls, D. L. 1986. *The heredity of behavioral disorders in adults and children*. New York: Plenum.

Vandenberg, S. G.; Stafford, R. E.; and Brown, A. M. 1968. The Louisville twin study. In *Progress in human behavior genetics*, ed. S. G. Vandenberg, pp. 153–204. Baltimore: Johns Hopkins University Press.

Varmus, H. 1988. Retroviruses. *Science*, 240, 1427–1435.

Vernon, P. E. 1979. *Intelligence: Heredity and environment*. San Francisco: W. H. Freeman and Company.

von Linné, Karl. 1735. *Systema Naturea*, Regnum Animale, L. Salvii, Holminae.

Wade, N. 1976. Sociobiology: Troubled birth for new discipline. *Science*, 191, 1151–1155.

Wallace, B. 1968. *Topics in population genetics*. New York: W. W. Norton.

Ward, R. 1985. Genetic polymorphisms and additive genetic models. *Behavior Genetics*, 15, 537–548.

Ward, S. 1977. Invertebrate neurogenetics. *Annual Review of Genetics*, 11, 415–450.

Wareham, K. A.; Lyon, M. F.; Glenister, P. H.; and Williams, E. D. 1987. Age related reactivation of an X-linked gene. *Nature*, 327, 725–727.

Warren, S. T.; Zhang, F.; Licameli, G. R.; and Peters, J. F. 1987. The fragile X site in somatic cell hybrids: An approach for molecular cloning of fragile sites. *Science*, 237, 420–423.

Watanabe, T. K., and Anderson, W. W. 1976. Selection for geotaxis in *Drosophila melanogaster.* Heritability, degree of dominance, and correlated responses to selection. *Behavior genetics,* 6, 71–86.

Watson, J. B. 1930. *Behaviorism.* New York: W. W. Norton.

Watson, J. D., and Crick, F. H. C. 1953a. Genetical implications of the structure of deoxyribonucleic acid. *Nature,* 171, 964–967.

———. 1953b. Molecular structure of nucleic acids. A structure for deoxyribose nucleic acids. *Nature,* 171, 737–738.

Watt, N. F.; Anthony, E. J.; Wynne, L. C.; and Rolf, J. E. 1984. *Children at risk for schizophrenia: A longitudinal perspective.* Cambridge, England: Cambridge University Press.

Weiss, M. C., and Green, H. 1967. Human–mouse hybrid cell lines containing partial complements of human chromosomes and functioning human genes. *Proceedings of the National Academy of Sciences,* 58, 1104–1111.

Weissman, M. M. 1988. The epidemiology of anxiety disorders: Rates, risks and familial patterns. *Journal of Psychiatric Research,* 22, 99–114.

Weissman, M. M.; Kidd, K. K.; and Prusoff, B. P. 1987. Variability in rates of affective disorders in relatives of depressed and normal probands. *Archives of General Psychiatry,* 39, 1397–1403.

Wender, P. H.; Kety, S. S.; Rosenthal, D.; Schulsinger, F.; Ortmann, J.; and Lunde, I. 1986. Psychiatric disorders in the biological and adoptive families of adopted individuals with affective disorders. *Archives of General Psychiatry,* 43, 923–939.

Wender, P. H.; Rosenthal, D.; and Kety, S. 1968. A psychiatric assessment of the adoptive parents of schizophrenics. In *The transmission of schizophrenia,* eds. D. Rosenthal and S. Kety, pp. 235–250. Oxford, England: Pergamon Press.

Wender, P. H.; Rosenthal, D.; Kety, S. S.; Schulsinger, F.; and Welner, J. 1974. Crossfostering: A research strategy for clarifying the role of genetic and experimental factors in the etiology of schizophrenia. *Archives of General Psychiatry,* 30, 121–128.

Wender, P. H.; Rosenthal, D.; Rainer, J. D.; Greenhill, L.; and Sarlin, M. B. 1977. Schizophrenics' adopting parents: Psychiatric status. *Archives of General Psychiatry,* 34, 777–784.

West-Eberhard, M. J. 1975. The evolution of social behavior by kin selection. *Quarterly Review of Biology,* 50, 1–33.

White, R., and Caskey, C. T. 1988. The human as an experimental system in molecular genetics. *Science,* 240, 1483–1488.

White, R.; Leppert, D.; Bishop, T.; Barker, D.; Berkowitz, J.; Brown, C.; Callahan, P.; and Holm, T. 1985. Construction of linkage maps with DNA markers for human chromosomes. *Nature,* 313, 101–105.

Whitney, G., and Harder, D. B. 1986. Single-locus control of sucrose octaacetate tasting among mice. *Behavior Genetics,* 16, 559–574.

Whitney, G.; McClearn, G. E.; and DeFries, J. C. 1970. Heritability of alcohol preference in laboratory mice and rats. *Journal of Heredity,* 61, 165–169.

Wilder, H. H. 1904. Duplicate twins and double monsters. *American Journal of Anatomy,* 3, 387–472.

Willerman, L. 1979a. Effects of families on intellectual development. *American Psychologist,* 34, 923–929.

———. 1979b. *The psychology of individual and group differences.* San Francisco: W. H. Freeman and Company.

Willerman, L. 1987. *Where are the shared environmental influences on intelligence and personality?* Paper presented at the Society for Research on Child Development, Baltimore, Maryland, April.

Wilson, E. O. 1975. *Sociobiology, the new synthesis*. Cambridge, Mass.: Harvard University Press.

———. 1978. *On human nature*. Cambridge, Mass.: Harvard University Press.

Wilson, J. Q., and Herrnstein, R. J. 1985. *Crime and human nature*. New York: Simon & Schuster.

Wilson, P. T. 1934. A study of twins with special reference to heredity as a factor determining differences in environment. *Human Biology*, 6, 324–354.

Wilson, R. S. 1983. The Louisville Twin study: Developmental synchronies in behavior. *Child Development*, 54, 298–316.

———. 1986. Continuity and change in cognitive ability profile. *Behavior Genetics*, 16, 45–60.

Wilson, R. S., and Matheny, A. P., Jr. 1986. Behavior-genetics research in infant temperament: The Louisville Twin Study. In *The study of temperament: Changes, continuities and challenges*, eds. R. Plomin and J. F. Dunn, pp. 81–97. Hillsdale, N.J.: Lawrence Erlbaum Associates.

Witkin, H. A.; Mednick, S. A.; Schulsinger, F.; Bakkestrom, E.; Christiansen, K. O.; Goodenough, D. R.; Hirschorn, K.; Lundsteen, C.; Owen, D. R.; Philip, J.; Rubin, D. B.; and Stocking, M. 1976. Criminality in XYY and XXY men. *Science*, 193, 547–555.

Witt, G., and Hall, C. S. 1949. The genetics of audiogenic seizures in the house mouse. *Journal of Comparative and Physiological Psychology*, 42, 58–63.

Woodworth, R. S. 1948. *Contemporary schools of psychology*. Rev. ed. New York: Ronald Press.

Wright, S. 1921. Systems of mating. *Genetics*, 6, 111–178.

———. 1977. *Evolution and the genetics of populations*. Experimental results and evolutionary deductions, vol. 3. Chicago: University of Chicago Press.

Wyman, A. R., and White, R. L. 1980. A highly polymorphic locus in human DNA. *Proceedings of the National Academy of Sciences*, 77, 6754–6758.

Wynne-Edwards, V. C. 1962. *Animal dispersion in relation to social behaviour*. New York: Hafner.

Wyshak, G. 1978. Statistical findings on the effects of fertility drugs on plural births. In *Twin research, Part B, Biology and epidemiology*, ed. W. E. Nance, pp. 17–34. New York: Alan R. Liss.

Wysocki, C. J.; Whitney, G.; and Tucker, D. 1977. Specific anosmia in the laboratory mouse. *Behavior Genetics*, 7, 171–188.

Zazzo, R. 1960. *Les jumeaux, le couple et la personne*. Paris: Presses Universitaires de France.

Zielenicwski, A. M.; Fulker, D. W.; DeFries, J. C.; and LaBuda, M. C. 1987. Multiple regression analysis of twin and sibling data. *Personality and Individual Differences*, 8, 787–791.

Index of Names

Index of
Topics